BETWEEN GLOBAL AND LOCAL

For Veronika, Jakob and Albert
with thanks and love

Between Global and Local

Marginality and Marginal Regions in the Context of Globalization and Deregulation

WALTER LEIMGRUBER
University of Fribourg, Switzerland

LONDON AND NEW YORK

First published 2004 by Ashgate Publishing

Reissued 2018 by Routledge
2 Park Square, Milton Park, Abingdon, Oxon OX14 4RN
711 Third Avenue, New York, NY 10017, USA

Routledge is an imprint of the Taylor & Francis Group, an informa business

First issued in paperback 2018

A Library of Congress record exists under LC control number: 2003056927

ISBN 13: 978-0-815-38772-5 (hbk)
ISBN 13: 978-1-138-61893-0 (pbk)
ISBN 13: 978-1-351-16272-2 (ebk)

Contents

List of Figures

List of Tables

List of Abbreviations

AI	Amnesty International
APEC	Asia-Pacific Economic Cooperation
ASEAN	Association of Southeast Asian Nations
AU	African Union
CEO	Chief executive officer
CIS	Commonwealth of Independent States
ECOWAS	Economic Community of West African States
EU	European Union
FDI	Foreign direct investment
FRG	Federal Republic of Germany
GATT	General Agreement on Tariffs and Trade
GDP	Gross domestic product
GDR	German Democratic Republic
HDI	Human development index
ICRC	International Committee of the Red Cross
IGU	International Geographical Union
ILO	International Labour Organization
IMF	International Monetary Fund
IRRI	International Rice Research Institute
LIM	Law on Investment in Mountain regions
MAB	Man and Biosphere
MSF	*Médecins Sans Frontières* (Doctors without borders)
NAFTA	North American Free Trade Agreement
NATO	North Atlantic Treaty Organization
NEPAD	New Partnership for Africa's Development
NGO	Non-governmental organization
NIDL	New industrial division of labour
NPM	New public management
OAU	Organization of African Unity
OECD	Organization for Economic Co-operation and Development
OIDL	Old industrial division of labour
PMSC	Punjab Health Systems Organization
R&D	Research and development
TGV	*Train à grande vitesse* (high-speed train)
TICAD	Tokyo International Conference on African Development
UK	United Kingdom

UN	United Nations
UNDP	United Nations Development Program
UNEP	United Nations Environmental Programme
UNESCO	United Nations Educational, Scientific and Cultural Organization
UNHCR	United Nations High Commissioner for Refugees
US	United States
USSR	Union of Socialist Soviet Republics
WTO	World Trade Organization
WWF	World Wide Fund for Nature

Preface

This book has been written at a time when the world was witnessing increasing threats to global security. Discourses on the war against terrorism and the substitution of the terrible regime in Iraq dominated the deadlines, while the similarly critical situation on the Korean peninsula had flared up temporarily but had then been shelved – two major conflicts on similar issues are too much even for a superpower.

Writing a book on marginalization in 2002 and 2003 cannot be done without reference to the events of September 11, 2001. The image of the Twin Towers collapsing amidst a cloud of smoke and dust reminded one of the precariousness of the existence of humans and their artefacts. It recalled the fact that the history of mankind has been dominated by appropriation and exclusion, that the theme of marginalization is a red thread through it. Terrorism is not simply an act out of nothing but an expression of despair, a feeling of helplessness, of marginalization by forces impossible to control and to stop.

It is not our intention to concentrate on September 11 but rather to try to understand what lies behind this and similar acts of violence. The war against Afghanistan can be seen as a reaction out of the same feeling that motivated the terror attacks on the US: helplessness and anger. More precisely, it is helplessness against covert and highly volatile actors, and anger at the destruction of the symbols of power and hegemony (World Trade Centre, Pentagon). The violent and excessive reaction arose out of the shock the entire nation had suffered. The centre had been attacked – not from the periphery but unexpectedly from the margin. The answer was one of condescendence but this is not the right way to deal with disparities and inequality in the world. Humanity is too fragile in its existence to allow for further experiments in exploitation and exclusion – this may be the lesson from 2001, the *annus horribilis,* paid at a high price, but maybe not too high for future generations.

The background of this book is the author's activity within a group of researchers (geographers and others) inside the IGU (International Geographical Union). The root, however, lies his interest in the problems of borders. Marginality research is strongly related to boundary research, and they both concern domains in politics, social life, economics, culture, and environment. It is these strong ties between the two domains that have acted as incentives to write this book.

When writing on marginality and marginal regions, one is confronted with the problem of discussing a vague and ill-defined (and definable) topic that exists in many fields and at many scales. In a first tentative definition, marginal regions could be characterized as regions lying outside essential (social, economic, cultural) processes, in a sort of vacuum. Contrary to a periphery, marginal regions are seen to face a bleak future, and it would require a particular effort to get out of that vacuum. However, this pessimistic view is often proved wrong: marginality need not be dead end, there are ways out of it.

The book adopts a humanistic perspective, i.e. it considers marginality not an objective state of things but as depending on people's perceptions and evaluations. This idea may run against prevalent opinions that want to define marginality and associate notions through in a clear-cut (possibly quantitative) way. After having extensively consulted many studies and books on the issues, the author's attitude has become more prudent (less categorical), allowing ample space for the variations life holds in stock. It is a personal view, and I have to bear the responsibility for it.

Acknowledgements

A first version of this text has been delivered as a lecture for advanced students in Human Geography at the University of Fribourg. Both the discussions on the topic in the course of the lecture and the ideas gathered during a number of research seminars on 'Globalization, Deregulation and Marginalization', organised since 1993, have furnished important inputs into the final text. I am very grateful to all of my students for this support. In these thanks I include my colleagues at the Geography Unit of the Department of Geosciences in the University of Fribourg (Switzerland) who provided a stimulating intellectual atmosphere. In particular, I recall the discussions with Jean-Luc Piveteau (head of the Geography Institute until 1996) who has had a considerable influence on my way of thinking.

I am also grateful to my friends and colleagues in the IGU Commission on Evolving Issues of Geographical Marginality in the Early 21st Century World (and the preceding Commission and Study Groups) whose presence and discussions in the numerous meetings around the globe have encouraged me in the presented endeavour. In particular, I should like to thank Professor Lennart Andersson (Karlstad) who has read the manuscript and commented on it. I owe him new insights and moral support for this undertaking. Thanks are also due to my wife Veronika, who has read and commented on part of the manuscript, and my two sons Jakob and Albert for their patience and support during the writing of this book.

Last but not least I want to thank the team at Ashgate Publishers for their patience and their support. In particular Mrs Valerie Rose, Commissioning editor for Geography, has put up with many questions and problems, but also the other members of the team have done excellent work. Their constant encouragement has helped to carry me through this work.

Introduction

LENNART ANDERSSON

This introduction to the book before us is based on research that I have been involved in and thoughts and contexts which have been updated on that basis.

My approach is founded on the geographic perspective. In geography the earth as a whole is interesting but even more interesting, for some issues, are the similarities between different spaces on earth and, for other issues, the disparities between these spaces. If we want to gain a greater understanding of the world from a geographic perspective, there are many roads and angles of incidence we may choose. With regard to the issue raised in the present context I will start by indicating two basic relationships. The one is the relation between time and space and the other that between functions and territories.

The history of the world tells the story of the relationships between time and space. The situation of a specific geographic region changes as time passes. Its relations to other regions also change. Sometimes the changes can be classified as development. They can be traced to many causes, some being primarily the result of unexpected and abrupt physical changes: earthquakes, flooding etc. However, in managing even such changes the relations between functions and territories play a role. The importance for the situation in space of the interplay between functions and territories has been obvious to geographers for long time. For me, the concept of territory stands for the basic fact that people, organizations etc. can gain some kind of ownership over a part of the earth's surface, an ownership that can be of various kinds, ranging from legal ownership to different kinds of influence. The ownership is valid for a more or less clearly delimited space and, sometimes, also for a limited time. Both horizontal and functional – sectoral – forces influence the state of the territory. These two types of forces represent different dimensions which often have incompatible assumptions and objects. Different groups and organizations at various decision-levels have their own goals and perspectives. Some trends are long-term, some are short-term, some concern the neighbourhood, and some are global. But the aspect of power is important to them all. By studying and analysing the struggle for power over the admission to space and time of individuals and occurrences caused by conflicting horizontal and functional forces on different levels, it is possible to discover anomalies and find solutions to unsatisfactory situations.

Specific individuals and organizations look at things from their specific perspectives. Even researchers look at things from different perspectives. The meaning of the concept of truth is relative, as is the meaning of other concepts. As a consequence, the formulation of the state of a territory or group of individuals is always connected to the perspective of the evaluators. But there is, in some sense, a common denominator in all perspectives expressed by decision-makers or people who have influence: they believe that their own solution is the best for the issue and space in question.

The normative aspect is thus fundamental. Often special concepts are used to characterize a particular set of values. One such concept used to evaluate people, places and territories is the concept of marginality. The usual concepts for expressing disparities may sound too neutral to express just those values you want to stress. The use of this strong concept to characterize a spatial unit, a group of people or a designated societal class functions as an alarm clock.

There is room for a warning here. In research it is not unusual to practise the methodology of basing an analysis on contrasting relationships. There is often a need, sometimes a wish, to strengthen the disparities by putting people and spaces in one of two scales with opposing qualities. Arguments of the kind 'we – the others' are often used. There is a risk that the conflict perspective can overshadow the perspectives of development and improvement. One such pair of contrasting concepts, often used in geographic research, are the concepts of city and countryside. However, the basic problems facing people in life are the same for individuals living in both cities and countryside but the realities are different in the two environments. Moreover, cities are not homogenous, nor is the countryside. Spatial relationships can also be characterized by weaker concepts where feelings of 'both – and' are included. The choice of concept and language is really important. Sometimes the concept of marginality is used in order to stress the conflict perspective but sometimes the concept in itself may cause conflict. It could act as a stamp placed on an area or a group of people. It is, however, a very useful concept, especially for exploring uneven conditions.

The subjective character of what marginality stands for may result in totally different standpoints. During the period 1989-1999 I participated in the research network named PIMA – The Study of Perceived Planning Issues in Marginal Areas – involving researchers mostly from USA, Ireland and Sweden. The research papers published by the group noted that many people choose to live in what in terms of so-called objective criteria could be termed marginal areas. This highlights the fact that when you talk about marginal areas, marginal regions etc. you generally and in some ways automatically link the concept with economic conditions. But statements from inhabitants and other people involved indicate that there are other factors than economic considerations which influence the perceived quality of life. There is an inherent conflict in equating economic growth with development in general. True development must involve human development and the empowerment of local communities.

The reasoning above also emphasizes the geographical position that maintains that every territory or community, seen as a whole, is unique with unique characteristics. This uniqueness generally has a cultural background and stems from specific relations with the environment, space and place. It is obvious that every territory is unique even in a world which is dominated by the trend towards decreasing the friction of distance and increasing opportunities for communication of various kinds. In my opinion we have to show confidence in the uniqueness. The uniqueness of a local community may be based on various criteria such as specific bodies of local traditions and knowledge, and specific life styles and relations with the environment.

Consequently, it is necessary to deal with the concept of marginality carefully and really make clear what the points of departure and the perspectives of those concerned are.

As marginality includes an aspect of lower position, an important factor in research on marginality is to find ways of achieving a better situation and of eliminating the causes behind the classification. The issue of how to succeed in development work to eliminate marginality has been an important research field. Both the research networks I have been involved in, besides the above-mentioned PIMA network the International Geographical Union's Commission on marginality problems, have stressed research on that issue. The results may be looked upon as confusing. The interplay between organizations, decision-makers, 'usual' people and territories at different societal levels is really not easy to understand fully. A key issue here however, is what happens in the encounter between those who represent the functional decisions, which are intended to promote vertical effectiveness, and those decision-makers and individuals who represent the simultaneous efforts to achieve regional and local welfare, which is territorial in character. On many occasions there is no encounter; the intentions for functional reform never encounter their target groups. Other groups who have the necessary ability take advantage of the situation. Sometimes the intended target groups ignore the intentions, and sometimes collisions occur. As an example, I will refer to what has happened in Sweden with the strenuous and long-term efforts to diffuse higher education to groups of people with a social background which has not favoured transition to higher education or who are living in peripheral areas. The transition to higher education has increased only slowly among these groups. Instead the transition rate has increased in areas which already had a favoured situation.

There are many complicating factors behind the interplay between functions and territories on marginality issues. Intentions and goals of varying strengths are attached to many actors at different levels of both the functional and the territorial dimensions. The factors of power and politics always influence the development process. Politicians deal with specific political issues when these are interesting for them and not when and if they fit into the development process. All development in a local territory must be culturally normalised if it is to be truly sustainable. Different actors have different time scales. Besides, the functional perspectives are often expressed by a mixture of heterogeneous organizations and are hard to grasp. Experiences from many cases in different parts of the world tell us that traditional regional policy measures usually do not wholly produce the intended results. What occurs may often be described as unintended and unacceptable implications and deformations of good intentions. It is crucial to find successful measures to manage these complications in every individual case.

Independent of the existence of the imperfect mechanism of human communication and the lack of effectiveness of different strategies, it is clear that both societal organizations and private enterprises are important factors behind marginalization. Society changes with time. The division of strength and power also changes both geographically and functionally. It is natural that one determining factor is the relation between the public and the private sectors. These relations differ between different eras, parts of the world and political positions. In the case of Sweden the public sector has been strong and indisputable for decades but now there is a tendency to open some functions, previously only handled publicly, to competition with private enterprises: the railway sector, the health services and the school sector. Disadvantaged regions may find themselves with less support. The rules governing the relations between the state, the dominant organizations and individuals are changing continuously. Apart from these changes at the national level, there is the process of increased foreign ownership, linked with the expansion of large multinational enterprises and organizations. Political mergers and unions, e.g. the European Union in Europe and NEPAD in Africa, play an important role in the process of change. Even if the results are not unambiguous, the changes occurring today in many countries will logically lead to increasing disparities that is increasing marginalization. This conclusion, even if it is not generally valid, is really alarming.

Many studies in the PIMA network presented strategies for reducing or eliminating marginality in local communities: bottom-up strategies, top-down strategies and integrated strategies. Bottom-up strategies were generally favoured. A reasonable conclusion was that in issues of local and regional societal planning, which were the main sphere of interest of PIMA, the bottom-up approach was something of a winner during the 1990s. How to organize partnerships in order to avoid marginality is now a frequent topic for discussion. The PIMA preference for development from below had its background in its trust in people living in marginal areas and communities. This trust gave the network a special quality. The territorial aspects were basic. Planning from the bottom became a commonly accepted assumption for effectiveness in the work of reducing marginality.

But at the same time as these tendencies were identified, another tendency was obvious during the ten years of PIMA research on development issues in European countries like Ireland and Sweden. The efforts of the European Union to improve the situation in its geographically marginal and peripheral parts have resulted in the strengthening of the supranational level. The EU measures are concentrated on projects of various kinds, trying to combine both functional and territorial aspects. Projects are by definition delimited in both time and space. Part of the crucial holistic perspective that was inherent in the development work during the 1980s has been lost. There is also a need for experts and application services, which further contributes to the appearance of a centralization and top-down tendency, which is contrary to the previously prioritized place and territorial aspects. The slogan of 'The Europe of Regions' also encourages the forming of new regions, which are politically initiated from above and which successively have to seek popular acceptance. The region on both sides of the Sound between southern

Sweden with the city of Malmö and the Copenhagen area in Denmark is one example.

We may note then that even periods of 10 to 15 years can cause major changes in the situation for areas identified as marginal. The following summary of PIMA research is illustrative. It may be seen as a journey from research in the late 1980s on depressed individual communities in extensive marginal areas to an emphasis on the new possibilities of the globalized world of the new millennium. In that new world there are opportunities to become a winner by being successful in market competition and in the game where one can capture economic support from the funds of the European Union and their equivalents. In reality there are, surely, more losers than winners among the marginal communities. The picture reveals the increased dynamics in the changes to and from marginality and the accelerated importance of the forces of globalization. This latter concept is, however, not something that was invented at the turn of the last millennium but has always been in progress but with less strength and less attention.

World history has clearly demonstrated that marginality and marginalization have moved from one part of the world to another in long and short-term waves. As far as I am aware, historical research has not used the concept of marginality but many spatial changes can easily be identified as just waves of marginalization. The result is that countries, which were formerly marginal areas, are now flourishing countries and placed on the other side of the balance and vice verse. Many waves could easily be identified over the span of human history. This identification and analysis of former marginality and its causes is however, I think, a research field that still awaits interested researchers.

Globalization may logically imply that the friction of distance has decreased. It paves the way for the erasure of space, thereby facilitating communication and exchange of goods and ideas. An interesting observation in connection with the increased tendency towards globalization is that, at the same time, there is a clearly growing interest in the local level, the native place. This interest in local history, culture, traditions etc., involves more people today than ever. It may be looked upon as an application of the well-known slogan 'Think global, act local'. This is some kind of unintentional implication in a world characterized by the more and more aggressive attacks of globalization.

Globalization focuses on the basic issues and problems common to all people. But still research in development geography is divided into research dealing with issues of the North and that dealing with the South.

With the object of penetrating the role of modernization in marginal regions I have been involved in studies of two local communities in different parts of the world, one in Sweden and one in India. It is seldom that development issues in such different local environments, situated in very different parts of the world, are discussed in one context. As a starting-point for the case studies we put forward two different development strategies, illustrating strategies of diverse range: global modernization and local self-reliance. The global modernization strategy assumes that progress and prosperity are best brought to the world and its peoples and places through continued modernization and an industrialized type of production rationality on a global scale. Central to this strategy is presumably the reliance on large-scale technological and organizational solutions to development problems. Conversely, the local self-reliance strategy focuses on local needs and local resources. The two perspectives and strategies were considered as opposing extremes on a scale.

In the case studies of the two local communities in Sweden and in India respectively, we found that the general development problems were the same but that the two communities had different positions on the 'modernization curve'. The Swedish case suggests that, seen from the local standpoint, the generally positive implications of modernization have come to an end. We identified development work with its foundation in groups with a strong local perspective and a wish to change the situation by means of a strategy based on self-reliance. In the Indian illustration the modernization period was still in its introductory phase. The implications were mostly seen as positive even when they were considered from the local view. Work based on the self-reliance strategy was weak and probably more of an instinctive character.

The study, described above, supports the idea of carrying out more comparative studies, including environments in different parts of the world. Such studies can lead to mutual development cooperation at the local and regional levels.

Lastly I will highlight one factor which I think is still underestimated in the efforts to avoid or reduce marginalization, and that is the work to achieve an increased and equal level of knowledge. Increased knowledge is an effective measure to overcome poverty and create equality between regions, classes and sexes. A positive sign here is the fact that the new model for local development emphasizes renewal and growth as the goals of regional policy. In the new development paradigm education, knowledge and competence are key words. Knowledge is generated wherever two people meet. In that sense all knowledge is local. To a considerable extent, the production and use of knowledge may be dealt with both in ways that correspond to those used for other development issues but also in ways that are special to the complicated concept of knowledge. We need a geographically inspired way of approaching the issue of knowledge. The issue may be analysed from the starting-point of the classical relation between function and territory. It is a matter of the proper balance between territorial everyday knowledge based on the proximity factor and the local organization, and functional knowledge produced by the school system and the scientists in the sectoral research institutes. It is important to find the optimal local educational 'climate'.

In this introduction the geographical aspect has been highlighted. Many disciplines have generated knowledge relevant to an understanding of the issues which I have discussed. Discovering and analysing marginality not only from a geographer's point of view but also from all relevant perspectives and on different scales may act as an alarm clock for the world as well as for the local territory, depending on the context involved. It will, in its turn, provide the signal to start development work with the object of creating a world more worth living in for more people. These factors are reason enough to write a book on marginality, highlighting the current relations between globalization, deregulation and marginalization.

Prologue

No matter what he does, every person on earth plays a central role in the history of the world. And normally he doesn't know it.
(Paolo Coelho, The Alchemist)

How can Coelho's alchemist pronounce these final words, before he bids the shepherd boy good-bye? Is not the shepherd one of the many marginalized individuals? Why is he at the centre of history, just as everybody else? How can a beggar in New York or Paris, a street vendor in Buenos Aires, Calcutta or Dakar, an inhabitant of the Aborigine ghetto of Redfern in Sydney (Pilger, 2002, p.160) be as central to history as the President of a Superpower, the CEO of a transnational company, or a Pope? The Brazilian author leaves the reader to look for the answer himself. Maybe at the centre of history means also at the centre of the Universe – and this could also be true, if we looked at reality not only from a materialistic point of view but took a more detached stance.

This book intends to take the reader not only into the concrete world and discuss marginality, marginalization and marginal regions from a purely overt, visible, and statistical angle but to propose also a look at the covert, invisible background that guides human thinking and decisions. As will be pointed out in Chapter 3, this side is essential to the understanding of processes and phenomena (see also Leimgruber 2003b, forthcoming). It may not be an object of scientific investigation, but it stands behind what we can study with scientific methods – it even guides the way we as scientists work.

In order to familiarize the reader with the topic of this book, we open our arguments with two accounts of events that occurred after 1980 and mirror to some extent the broad field of marginality studies. Their subjects differ considerably, but the choice is intentional because the two examples illustrate how strongly regional problems are related to processes that occur both nationally and internationally.

The first tale is located in the country that was the cradle of the Industrial Revolution. Between 1965 and 1985, the industrial region of Teesside passed from an industrial growth area to an unemployment black spot (Foord et al., 1985, p.6). What had looked like a promising take-off after the implementation of the Hailsham plan (1963) turned into a disaster in the course of the 1970s and 1980s. And Teesside was only one region among others. During this period, northern England watched its economy decline. The Consett steelworks, for example, which had been completely renewed between 1957 and 1963, closed down in 1980 (see below, p.41), the National Coal Board was privatized, and the Northeast entered into a period of economic depression, which at certain times left more than 50 per cent of the workforce unemployed (Manners et al., 1980, p.24) and reduced the region to a net-receiver of social assistance from London. A region of formerly economic importance had been marginalized through the combined processes of deindustrialization and growing international competition on the steel market.

The second story refers to political processes in Switzerland that are related to environmental protection. In order to appreciate them, I have to explain a particularity of the Swiss legal system. The people have the right to demand an amendment to the Constitution (the modification of an article or the introduction of a new article) by formulating a text and supporting it with a minimum of 100,000 signatures of people with the right to vote. Such an amendment is first discussed in Parliament and then presented to the population to vote in a referendum. In order to take effect, the double majority of the people and of the cantons is required. Usually it is a political party or a party coalition who takes the initiative, but it is possible for an individual or an interest group to do the same. In 1987, 47 percent of the Swiss voters participated in a referendum on the protection of bogs. With 58 per cent affirmative votes (1,153,448) and a large majority of cantons (20 of 23) this ecological issue was accepted. Bogs and moors are now protected through the Constitution. In 1994, 41 per cent of the Swiss voters participated in a referendum and accepted an amendment to the Constitution aiming at protecting the Alps from excessive transit traffic. The result was 52 per cent affirmative votes (954,491 votes) and 16 of the 23 cantons in favour of the new article in the Constitution. Within seven years, the Swiss had reacted twice to the increasing marginalization of both the environment and the beauty of the landscape by the new economy and the international division of labour and production processes. This is all the more remarkable as both referenda originated from a bottom-up political process, i.e. people and associations who are standing in for nature and environment. Supporters in both cases came not only from Pro Natura (Swiss Nature Protection Association) but also from WWF Switzerland. Unusually, constitutional initiatives stand limited chances of success (in both cases, Parliament was opposed to the requests), but in the 1980s, the tide has begun to turn.

These two examples may appear to lie far apart, but they show how intimately the notions in the subtitle are interrelated. Although this book is devoted to marginality and marginal regions, we have to put them into the general perspective of modernity with its emphasis on unbounded mobility, global interconnectivity, and the redefinition of the role of the state (the public sector). This book attempts to furnish a contribution to the ongoing debate that has been taking place in the IGU Commission on Evolving Issues in Geographical Marginality in the early 21st Century World and its forerunners (Leimgruber, 1998a), where the author obtained many of the ideas presented in the present context. It also draws on many discussions with geography students in Fribourg university (CH) who share similar concerns.

PART I
GENERAL ASPECTS

Chapter 1

Setting the Frame

Marginality as a permanent issue

In a paper, published just before the major turnaround of the European political map, Brücher (1989) asks the question if the Saar-Lor-Lux transborder region covering the Saar Bundesland (Germany), the Lorraine Region (France) and the small state of Luxemburg was a border region, a periphery or the core area of the European Community. The title of his paper points to one of the central problems in discussing marginal regions: the issue of *(spatial) scale*. It will accompany us throughout this book, it is always present and will be illustrated in the case studies presented. Brücher himself cannot answer the question he asks. Looking at its location within the European Union, the Saar-Lor-Lux region is doubtlessly central. From an economic and traditional resource perspective (coal and iron ore, steelmaking), on the other hand, it is rather marginal. Similarly, the politico-administrative heterogeneity of the three countries involved constitutes a significant drawback. However, the existence of peaceful and intensive transborder contacts since World War II speaks in favour of centrality.

A second major point in this discussion, the role of time, is illustrated by Reitel (1989) in his paper on the transformation of mining (iron ore and coal) and the steel industry in the same region: once the most important economic factor, they have become minor activities within the period of half a century and have by now almost completely disappeared. His paper emphasizes the dynamics of regions that depend on internal and external factors and reminds us of the question of *(temporal) scale*: processes are both short-term oscillations and long-term cycles.

Marginality in space and time – the two essential dimensions for geographic research – is the topic of this study. The fundamental concept will be discussed both theoretically and in connection with specific domains in Chapter 2 and, more in depth, in Part II. Most frequently, geographers have investigated *economic* marginality, looking at regional disparities and disadvantaged regions. More recently, *social* and *cultural* aspects have been studied; in this field, geographers cooperate with sociologists, social anthropologists and ethnologists in a multidisciplinary way. The *political* realm has so far been treated very little, probably because politics is an issue that dominates both economy and social life to a considerable extent. To these traditional domains in the social sciences we shall add the *environment* (our physical living space). The perspective in this case will be somewhat different. The environment is central to human existence, but all too often has it been marginalized in our decisions to allow for greater economic profit. Marginality in this case emphasizes the way humanity looks at it, perceives it, and has been dealing with it over time. It would be wrong to simply discount it; any geographic approach has to take both man *and* nature into account (see Chapters 4 and 7). Humans have to live with the environment, they modify it according to their ideas, but they are also part of it, they also have their own nature. Man has created a considerable range of answers to the challenges of the environment, and this variety persists until today: 'So there is a physical or ecological envelope, but within this, human technology and knowledge allow a variety of adjustments to the resources of the planet.' (Simmons, 1989, p.6). The significance of culture as the background to our actions becomes apparent.

The latter statement takes us to a third important point: the study of marginal regions and marginality is not simply 'objective' research but is related to human *perception*. The phrase 'marginality is a state of mind' may sound simplistic (after all, there are 'objective' criteria for measuring marginality and delimiting marginal regions), but it does hold some truth. A newspaper article published a few years ago in a regional Swiss paper furnishes a telling example. Its title 'Innovations: Switzerland runs the risk of being marginalized' expressed the fear that in the field of technological innovation, Switzerland risked to lose its position at the forefront (La Liberté 09.09.1999). To the readers of the paper, this article sounded the alarm bell (it did so for this author), but when reading the text carefully, they understood that reference was made to innovation within Switzerland only, not to innovation achieved by Swiss firms abroad. This puts the article into a different perspective: we may safely assume that information on innovations will be circulated inside the firms and in this way flow back to headquarters and plants in Switzerland. Marginality, in this case, was clearly an invention of the journalist who gave his own interpretation of a report of the Swiss Council for Science. While pointing to potential problems in Research and Development within the country, it also highlighted positive elements for Switzerland: the general framework for research and development (R&D), the massive investment of 9.9 billion Swiss francs in R&D (67 per cent by private enterprises, 27 per cent by public funds), to say nothing of the favourable political and human resources situation were quoted as important trump-cards for the country.

The title of the newspaper article conveyed a pessimistic message only, missing out on the positive aspects. The question remains open if the journalist intended to set a political signal towards the promotion of research in general and universities in particular, or whether his was a purely local perspective given the need for funds in scientific research.

The examples mentioned above may look insufficient to our cause. Let us therefore add another spatial dimension. The late 20th and early 21st centuries have witnessed an increasing rift between a 'central' and a 'marginal' world, between the rich and powerful North and the poor and weak South, a new *limes* between the 'Empire' and the 'New Barbarians' (Rufin, 1991, see Figure 1.1). The marginal world is that part of the globe that experiences most political unrest and civil wars, and that as a consequence receives most refugees in innumerable temporary camps, the 'archipelago of misery' as Rufin calls them (ibid., pp.64 ff.; see below, p.166 f. and Figure 6.2), that risk to become the home of many for an unknown length of time. These camps (or *gulags*) are marginal regions within marginalized countries that belong to the marginal world – a perfect illustration of the *babushka* principle (the Russian puppets where a smaller puppet is inserted in a bigger one).

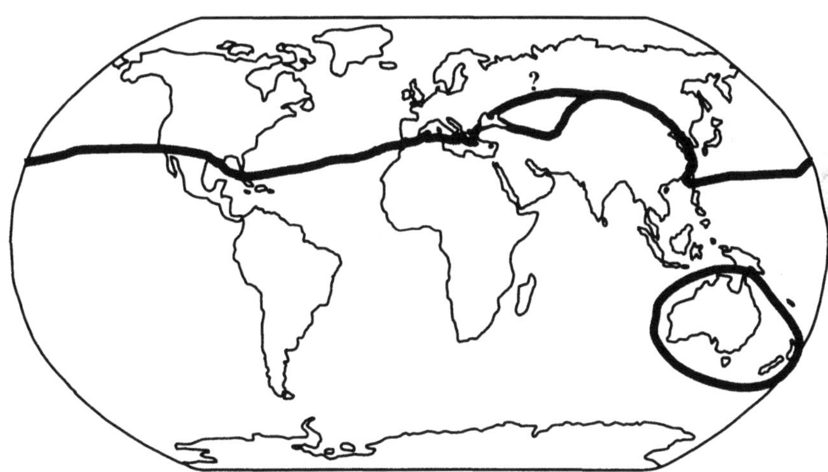

Figure 1.1 The global North-South divide

Source: Rufin, 1991, p.147 (modified)

It is cynic to pretend that marginality has always been part of history and will have to continue like that into the future. To think in this way means to deprive people of all hope to step out of a hopeless situation. It is imperative to fight marginality even if we know that we cannot eradicate it. We must do our best to eliminate its negative impact on the people and the environment.

To do justice to the topic, the temporal dimension has to be recalled as well. The political situation in the Middle East offers a good example. Depending on circumstances and power perspectives, centrality and marginality are interchangeable at will. In their regional geopolitical calculations, the US had attributed to Iraq a strategic position and supported the coup of 1963 that eventually brought Saddam Hussein into power (1979). From the 1960s onwards, they supplied the country with money, arms, and technology. 'So enduring was America's ardour, or rather its gratitude to Iraq for protecting its client Arab states from Iran's revolutionary virus, that Saddam was given everything he wanted, almost up to the day he invaded Kuwait in August 1990.' (Pilger, 2002, p.66). Suddenly, after the invasion, the image of this same person was reversed, and Saddam Hussein became a sort of public enemy number one, who had to be ousted from Kuweit. His fault was that he had tried to become independent of the great benefactor, but the Americans 'want another Saddam Hussein, rather like the one they had before 1991, who did as he was told.' (ibid., p.81). Central to the entire issue is not the country or who runs it but the chief energy source, oil. The regime as such does not matter.

These introductory remarks demonstrate that the theme of marginality is always present and will remain on the agenda in the future. It will persist on all possible spatial scales, and it may in certain cases be temporary, in others almost permanent. In our time, however, it has to be seen before the background of two important processes underway, globalization and deregulation. Indeed, marginalization can be seen as a consequence (maybe the major one) of both of them. For this reason, it is necessary to briefly discuss them and look at the foundation they lay for marginalization. Let it be made clear, however, that *marginality is not the result of globalization and deregulation*; the two processes have simply reinforced what is as old as human history.

Aspects of globalization

Globalization has become a fashion word in our time, used indiscriminately by everyone to describe almost anything that looks negative or seems to have a negative impact on our life – it is often pronounced almost as a swearword. The popular interpretation of the word, however, does not take the complex reality into account that merits serious reflection. Being vague, the term leaves a number of issues open, but is certainly not simply negative.

Looking at it from a semantic perspective, the adjective 'global' has two meanings: on the one hand it refers to the entire world (the globe, meaning worldwide), on the other hand it is a synonym to inclusive or all-encompassing, both in the concrete and in the abstract sense. By its very essence, it is a static term. The verb 'to globalize' and the noun 'globalization', on the other hand, suggest movement, dynamics, not a steady state. The word-family is value-free. It is therefore astonishing to observe how globalization has been used in a very diffuse manner. On the one hand it is presented to the public as inescapable destiny, on the other it has been loaded with strongly negative connotations. This may lead to contradictions and confusion, as is the case of the anti-globalization movement, which is itself organized globally. The choice of name masks the real target of the opposition, which is in fact the unbounded market liberalism.

A closer look at the first meaning of 'global' imposes itself. The World is commonly understood as the Planet Earth, global therefore refers to just this one scale. This, however, is our modern understanding of the term, based on our daily practice of shopping in the supermarket and watching TV news. Through modern information, communication and transportation technology, the globe has shrunk and physical (as well as time) distance has lost much of its former significance (Baumann, 1998, p.14 f.). However, the World is not an objective reality; it is perceived by humans from different perspectives and can be associated with the living or activity space of a particular group or society. 'Global' then comes to mean the totality of the known part of the planet at a moment in time (history), and it has held this meaning with the Romans, the Chinese, and the Arabs who traded and conquered lands within a certain range of their centres of power. The colonial empires eventually changed the scale of operation from 'regionally global' to 'truly global'. From this perspective, globalization is a very relative term that must not be reserved for the 20th and 21st centuries.

The history of globalization can be traced back to the early periods of human history, provided we accept to look at 'global' in relative terms and to take the slow speed of transportation and diffusion into account. The dissemination of cultural elements, the mixing of cultures, and the ensuing creation of new cultures or cultural elements are part of our heritage. No group or society could evolve in isolation, as a closed system; outside influences have always penetrated and exercised their influence. Elements that play a particularly strong role in a people's identity will even be appropriated to serve as the main feature of a people. 'How many Italians, for example, realize that pasta originates from China?' (Holton, 1998, p.28). Even this book is a witness to the complex historic globalization process, if we look at the language (English, as a mixture of Germanic and French elements), the writing (the alphabet developed in former Phoenicia), the numbers based on the Arabs 'who in turn learnt them from Indians, who had earlier invented positional notation.' (ibid.). Printing, finally, although now with a new technology, was invented in China (ibid.).

It would be unwise to condemn globalization as the evil of the 20th and 21st centuries. It is true that there are negative effects that have to be criticized and eliminated or corrected, but there are also positive ones, and they have to be cultivated. The domination of the entire world by one economic philosophy and the political interests behind is doubtlessly harmful to the cultural and natural diversity on earth. The fact that Human Rights' violations can be denounced on a worldwide scale within almost real time, that we have become aware of global problems that transcend our daily ones, or even the possibility to enjoy theatre, music and sports events wherever they occur is a positive side of the globalization process – this is a means to strengthen solidarity among the inhabitants of the globe.

In order to clarify what we are talking about, it is important to approach the globalization debate with a solid conceptual basis. The German sociologist Ulrich Beck (2000) distinguishes between the dynamic and the static semantic fields and has coined respective terms with precise contents: 'By *globalism* I mean the view that the world market eliminates or supplants political action – that is, the ideology of rule by the world market, the ideology of neoliberalism.' (Beck 2000, p.9). Anti-globalists therefore fight globalism in the shape of the neoliberalist credo. '*Globality* means that *we have been living for a long time in a world society*, in the sense that the notion of closed spaces has become illusory. No country or group can shut itself off from others.' (p.11; emphasis in the original). This is not new, as Adamo's statement about the Great Depression of the 1930s illustrates: 'The crisis of 1929 has demonstrated a) that the earth is no longer just an ecosystem but is to a large part contained in one single socio-system; b) that the dominant economy in the system earth is no longer Great Britain but the United States of America.' (Adamo, 2001, p.596; transl. WL). Nowadays, everybody on the entire planet takes part in this reality as soon as he/she watches television, listens to the radio, surfs on the Internet – or is affected by the war on terrorism or some sort of environmental pollution. Participation varies in space and time. '*Globalization*, on the other hand, denotes the *processes* through which sovereign national states are criss-crossed and undermined by transnational actors with varying prospects of power, orientations, identities and networks.' (p.11). Such transborder processes have been part of human history, intensified, however, from the 16th century onwards, when the European colonial powers sailed the oceans and annexed lands around the globe. From our European perspective, the great voyages of discovery constituted a new step in the ongoing process of globalization, a process that before had been limited to overland trade between Europe and Asia (the Silk Route). The discoveries, the colonization (appropriation), and the exploitation of foreign lands overseas constituted a fundamental break in globalization: the size and speed of the enterprise grew considerably, and the European influence went beyond exploiting natural and human resources and touched the social systems. This is where proto-globalization ended and the old globality system (the colonial system) came into existence.

What we term 'old globality system' has been characterized by a functional construction (Adamo, 2001, p.594), based on the linear interdependence between motherland and colonies (Figure 1.2). Every colony exported its specific resources to and imported manufactured products from the motherland. This system was very static and conceived on a long-term basis, and the flows of goods were always the same. This system was not contested, because it brought wealth and power to Europe. The gradual technological improvements following the Industrial Revolution increased the speed at which it worked (the construction of steamboats) and facilitated the penetration of the colonies (the building of railways) and the communication within the system (telegraph and telephone).

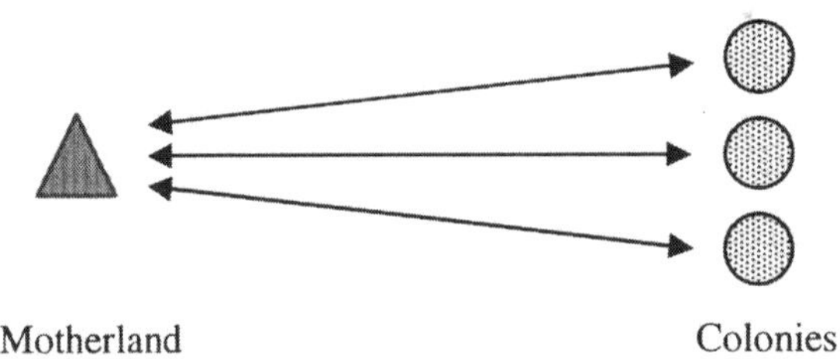

Figure 1.2 The old globality system

The old globality system broke down definitely with decolonization but signs of a change had been visible before. The globalization process, on the other hand, simply took a new turn. When the United States (a former colony) entered the world market with wheat in the 1880s, they triggered off a major crisis in European agriculture. Their advent as a global player demonstrated that the old, bilaterally constructed globality system was defunct. They were no longer part of it, and they clearly intended to contribute to its downfall.

In the 20th century, a new globality system established itself. Based on further technological advances (airplane, computer, satellite communication), it soon conquered the world, and the globalization process swept across Latin America, for example, 'like a hurricane' (Hinkelammert, 1998, p.92). Its foundation was the law of the market, i.e. the neoliberal credo that things will settle down easily if only every actor had complete freedom to act. This simplicity, however, is based on a tautology: the market price is based on free competition which exists when the price is a market price. The salaries obviously follow the same logic, but nobody asks if one can survive with a market salary (ibid., p.94). This statement obviously holds good for African and Asian countries as well.

The modern system (Figure 1.3) is very flexible and characterized by a hierarchy of firms that are linked in a subcontracting relationship (Dicken, 1998, pp.230 ff.). Around a pivotal centre (e.g. the headquarters of a transnational company) we witness a varying number of 'satellites', interconnected in a more or less complicated network (Figure 1.3 A). The centre furnishes investment capital (and cashes the profits), defines the outlines of production (the types, quantity, and quality of the products), engages in research and development, and exercises control over the entire exercise. Depending on the products, the 'satellites' may themselves be surrounded by 'sub-satellites' (subcontractors) that produce parts or are engaged in specific sections of the production process. Decisions are taken by the centre and may affect the 'satellites' to a varying degree, but they are usually taken on a short-term basis; the network is dynamic and is bound to change from time to time. Large enterprises buy and sell other firms, expand their activity and return to their core activities according to the world economic situation. Thus, while the centre remains stable, the entire system may look different after a certain period (Figure 1.3 B). In both the 'old' and the 'new' system manpower was and is exploited, but exploitation was much more stable over time and in a way also more predictable in the colonial system than it is in the globality system. Plants are nowadays opened and closed down following profitability decisions in the centre, irrespective of local and regional conditions. Both systems are therefore also examples for marginalization.

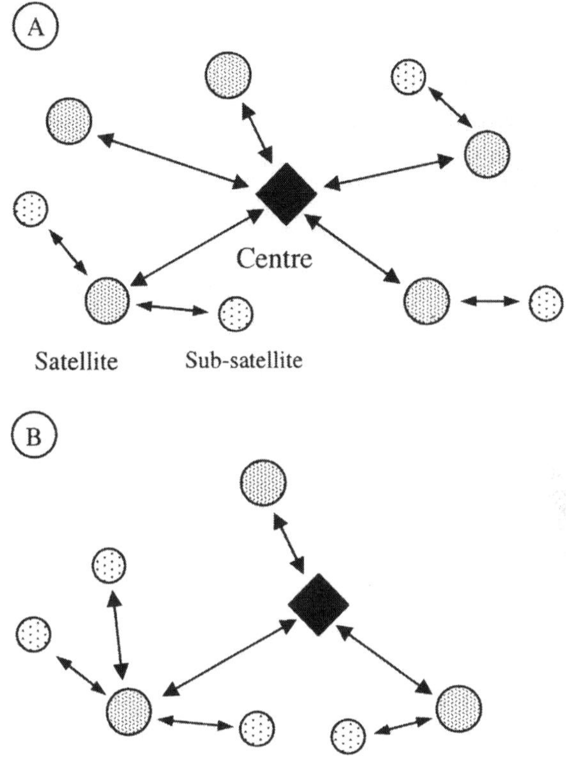

Figure 1.3 The new globality system and its dynamics

During the last decade, globalization has been contested to an increasing degree, and the cries of 'down with globalization' have become standard expressions in the media. However, the critics (who are themselves organized globally) are beyond the point. Their aim is not globalization (otherwise they would have to abandon their worldwide organizations) but the neoliberal philosophy and the lack of responsibility by many of its representatives. The numerous protests since Seattle (Smith, 2000; Verhofstadt, 2001) cannot stop the globalization process as such. They may exercise some influence on it, but it may sooner or later take turns we cannot foresee at the moment, the process being not linear. Its modern variety (which is under fire from the critics) is best illustrated by the President of Time Inc., James Linen: 'We are trying to create a new climate, in which private enterprise and developing countries work together ... for the greater profit of the free world. This world of international enterprise is more than governments ... It is the seamless web of enterprise, which has been shaping the global environment at revolutionary speed.' (quoted in Pilger, 2002, pp.38 f.). No comment ...

The misunderstandings about and the widespread resistance to globalism stem from the fetish-like image economic thinking (i.e. thinking in short-term profit and maximum productivity) holds in our time (Galtung 1998, p.176). The liberal market economy, based on private property and a global activity space has eclipsed other competing systems such as national market economies, planning economies, or local economies (ibid.). It is a model based on the quasi-monopoly the western (in particular the US) economy holds in a worldwide context and which combines economic, political and military power. If this monopoly should be broken, history might well take another course. 'An East Asian common market uniting the countries of Mahayana-Buddhism and Confucianism (China, Japan, Vietnam and Korea), who can produce goods in Japanese quality (Q) at Chinese prices (P) will have a much larger effect – such as the end of "globalization".' (ibid., p.178). This statement may be a bit too optimistic, but it shows that the current process of economic globalization is not necessarily the only possible way, and that it can take another course. By diffusing its knowledge and technology, the West has in fact created the basis for a possible reversal of the situation. What may first apply to the economic aspects of globality might later extend to the military and the political domains.

We cannot stop the process of globalization as such; it is part of a long wave of development and currently seems to be still on the upsurge. It is irreversible in the sense that we cannot turn back the wheel of history, but it can slow down and loose intensity. Globality is a reality we have created centuries ago and propelled through the development of new transportation and communication infrastructures, through new consumer behaviour and new demands such as travelling and holidays. It is not a fatality but simply the result of a process that has taken off with the Age of Enlightenment and has been carried on and intensified since. Criticism of and opposition to globality must, however, consider a vital point: western welfare is the result of the opening up of the markets, a process that benefited essentially the West. And the West defends his monopolies as much as it can. 'For the rich West, free trade is naturally something that should be embraced wholeheartedly ... as long as it is not in products that can harm Western economies.' (Verhofstadt 2001). True free trade with equal conditions for all partners would probably benefit humanity as a whole, but it is the industrialized world that dictates the rules. This problem continues to haunt us (Verhofstadt, 2002).

The globality debate is chiefly led by people north of the new *limes* (Figure 1.1), in the (geographical) North, in particular in the Triad. Looked at from the southern component of the North (i.e. Australia and New Zealand, an 'island' with a new *limes* of its own), the picture looks different. Here, 'the enormity of the globe is an ever present and costly factor.' (Le Heron and Roche, 1995, p.25). Both New Zealand and Australia are far away from the markets, and the perishable export goods either require special treatment, or have to be processed or preserved in one way or another, or they depend on fast (air) transport from the producer to the consumer. Whatever way is chosen, it is an additional cost factor unknown to producers in Europe and Japan, for example. New Zealand food production is increasingly driven by sustainability and globalization considerations; the former can be recognized through the adoption of ISO standards (ibid., p.26), the latter in external investments into the food and food processing industry (ibid., p.29). It is an open question, to what extent the two that are 'increasingly intertwined' (ibid., p.28) can survive together.

What we need to do is to try to repair the damages caused by the globalization process and globality. A positive answer to the threats would be 'deeper and more responsive democracies, nationally and internationally' (Klein 2001), accepting that there are more economic models than just neo-liberalism (ibid.). Beck (2002) offers a few solutions in his chapter on responses to globalization, such as increasing international cooperation on the political level, the creation of a sort of transnational sovereignty in addition to the individual state's sovereignty, new products (organic food) and market orientations (regional labelling, privileging regional economic systems over transnational ones) that rely on local and regional markets, etc. However, he is not terribly optimistic about the necessary adaptations our society ought to make. In other words: a radical change in the dominating values (see Chapter 3) is not imminent. The fashionable understanding of globalization is that it will lead the entire humanity forward on the road to progress. Which progress? To this question, the answer given by the anonymous economic and financial circles usually reads like Western lifestyle and civilization, in other words a standardized and monotonous world. That there are other ways of living, other ways of thinking and that there is the richness of ecological and cultural diversity is something they simply shut off from their reflections. The thinking behind globalization, as the neoliberal defenders understand it, is utterly monocultural (Shiva, 1993). Progress, however, demands not only innovation but creative imagination, and this can only be found if alternatives to simple, standard solutions to problems are accepted, i.e. if the otherness becomes part of thinking (see Massey 2002), if the plurality of ways of thinking is recognized. Humanity has a right to all kinds of diversity, and this right has to be defended by all means (Leimgruber, forthcoming).

And yet, we observe that a global solidarity is gradually emerging, in particular in the field of humanitarian aid. The global interconnectedness makes people in the North aware of the sufferings countries and their populations in the South endure; part of the criticism of globalization and globality must be attributed to the way the North deals with the South. Fighting corruption, eliminating landmines, campaigns against Human Rights' violations, etc. (Florini 2000) indicate that there are positive sides to globality. The Internet and e-mail make instant information about humanitarian catastrophes and violations of Human Rights possible. It becomes increasingly difficult for dictators to silence opponents. Transnational corporations will find themselves accused of environmental pollution and abuse of workers, and their corporate image may suffer. We remember the campaign Greenpeace started against Shell on the question of the Brent Spar oil platform that was to be submerged in the Atlantic. Mass media and electronic information permitted the launching of a boycott of Shell petrol stations in Europe that eventually caused the company to abandon its plan and have the Brent Spar dismantled in Norway. Nowadays part of it serves as a ferry terminal in Mekjarvik harbour (near Stavanger). The case is already a few years old, but it has become a sort of model issue, and information is still available on the web (BBC, 1998).

The emergence of a transnational society may yet be far away, but the simple fact that this idea has been put into the world proves that we have to develop 'governance mechanisms ... to deal with issues with which the nation-state system by itself may be ill equipped to cope.' (Florini and Simmons, 2000, p.11).

Deregulation – or re-regulation?

Another fashion term of the present is deregulation, usually referring to the retreat of the public sector (the state) from an increasing number of domains of public interest. Indeed, the role of the state is undergoing profound transformations. Its traditional domain of setting the framework of social, political and economic life has not been contested as such. However, since criminals and pollution know no boundaries, cooperation and persecution on a supranational scale via international and transnational law has become more and more frequent. Since the end of World War II, state activities have become progressively more outward-oriented, a process manifest in the creation of international communities and free-trade areas (European Union, Organisation of American States, Organization of African Unity/African Union, ASEAN, NAFTA, ECOWAS, NEPAD etc.). On top of these multilateral agreements we have the influence of the United Nations and its numerous sub-organizations. Blake (1998, pp.248 f.) attributes this transformation to a combination of various factors: the increasing globalization of economic activities (which includes the liberalization of the financial sector), the information and communication revolution, the proliferation of international agencies and organizations, the interconnection of national and global security issues, and the growing awareness of the transborder dimension of environmental problems. However, 'the evidence suggests that the nation-state is likely to remain the basic building-block for political associations in the future.' (Blake, 1998, p.250). The state is not dead, it remains one of the agents of society, albeit in a new societal environment and with a re-defined role.

Processes occurring on a global scale take place in an inter-national framework, i.e. between nation-states that are partners in and signatories to the myriad of multinational treaties, conventions, declarations, and protocols that regulate both politics and the economy. This fact alone demonstrates the necessity of the state to survive as an actor in the global system (Holton, 1998, p.80). It will also remain an agent for the simple reason that it is a specific form of social organization. By representing the society and the population as such, it fixes certain rules that are laid down in the constitution, in laws and ordnances. These rules regulate the society and allow it to function according to the prevailing value system and worldview at a given time – they are essential for every society, and they vary according to the evolution of a society. Regulatory regimes are not static, but subject to modifications. Contrary to changes in the economy, however, their dynamics occur in a different time frame. From its sheer size, a state cannot be run like a company, and the political participation of the population requires careful reflections on what and how to change. Regulatory regimes are a necessity and must be maintained. A society without regulation will end up in anarchy. Dicken (1998, p.116) therefore rightly points out that the process of '*de*regulation is really *re*regulation.' (emphasis in the original), i.e. it is a change of the rules of the game towards liberalization. State monopolies are replaced by private competition, and the state will watch that private monopolies do not replace its own that have disappeared. Accordingly, new rules concerning the free market are necessary. Deregulation does not mean to eliminate the rules but to adjust them. The society will continue to function under a new regulatory framework, and its members will adapt to the new situation, because 'regulatory institutions are a constitutive part of the global economy rather than external constraints.' (Holton, 1998, p.68). The state is also essential to 'the transnational monopolies' (Adamo, 2001, p.608) who need it precisely because of the regulatory frame it provides and because of the diversity of regulations around the globe. It may look paradoxical: on the one hand, the transnational companies strive towards homogenization, on the other they profit of the variety of laws and legal practices. And powerful nation-states (such as the USA and their ally Britain) can protect their interests wherever they operate.

To this effect, new regulatory institutions have emerged during the 20th century: trade blocs (the European Economic Community EEC, the North American Free Trade Agreement NAFTA, etc.; see Appendix 2) and the Bretton Woods institutions (World Bank, International Monetary Fund, GATT and its successor, the World Trade Organization WTO) have become additional agents on the world economic scene. Reregulation takes place not only on the national level but increasingly on the supranational scale. This is the opposite of the process that was originally intended, and it seems to create more confusion than before.

Apart from these formal supranational institutions, new agents and decision-makers are also influencing the regulatory framework: on the global level the non-governmental organizations (NGOs), and on the sub-state level the regional and municipal governments who have the right to regulate. Whilst they have a certain freedom of action, they are nevertheless embedded in the global system, and their regulations have to conform to standards developed higher up in the hierarchy.

The aim of deregulation (to stick with this term for the moment) is to promote competition, i.e. to expose the domains concerned to the market where demand and supply determine not only the price but also the degree to which a certain service is offered. Monopolies are thus being replaced by polypolistic conditions (at least in theory), a process most remarkably illustrated by the privatization of the telecommunications sector all over Europe in the past 10 to 15 years. However, deregulation has occurred in a very differentiated way. In many instances, the state did not expose these services to the open market in an uncontrolled way but preferred to stay in business as a shareholder, for example. The service is thus forced to operate under market conditions, but the state maintains some degree of (political) control and at the same time participates in the profits generated.

As the representative of a national society, the state has also the obligation to ensure minimum infrastructure services to the entire population, wherever people live. This duty obviously conflicts with the goal of market liberalization. Deregulation (or re-regulation) has therefore carefully selected from the two options available: complete exposure to the self-regulating market, or the gradual implementation of carefully selected market elements without loss of total control (Neubauer, 1988, p.12 f.). In the telecommunications sector, the first option was applied, whereas in the health service, new rules were implemented which introduced certain market elements without eliminating state intervention. The example of Switzerland will be presented below (pp.30 f.).

With the advent of New Public Management (NPM), the state has adopted market elements into its thinking. Cost-efficiency and flexibility have become guidelines for government agencies who thus had to develop new initiatives and carry more (financial) responsibility. The old civil servant with a life-long job guarantee had to give way to a new employee who was to be motivated and creative, and whose job is far from secure. With the NPM philosophy, the state has lowered itself from the pillar of untouchability and become one agent among others – despite the fact that it still has a considerable public burden to carry. It is hoped, however, that politicians will find more time to go about their real business, whereas government departments and agencies learn to live with their new competences.

Re-regulation implies the entire or partial privatization of domains that were originally seen as falling solely into the public sector's own competence. Market liberalization, the criticism of monopolies, and the profit-maximizing ideology has called for new actors in many services. Infrastructures such as rail, bus and postal services, telecommunications, broadcasting (radio and television), at times also the energy sector and certain social services (medical care, old age pension schemes) have been or are being privatized. Such a deconstruction of the *service public* (public service) has provoked criticism in countries where the population has become used to a functioning public sector at little or no cost. Indeed, the deregulation philosophy is based on profit-making possibilities in an urbanized world, but it is not necessarily applicable to regions where the costs of provision exceed the gains to be expected. Remote areas, even in rich countries, pay the price, unless new government regulations guarantee minimum services.

It is precisely in such regions that state intervention has developed over the past 50 years or so. Its objective is to reduce economic disparities and fight outmigration through the implementation of regional policies. They were originally developed to assist farming areas suffering from unfavourable production conditions: mountain areas (Brugger et al., 1984; Majoral et al., 1996; Matsuo, 1998), boreal regions (Siuruainen 1987; Malinen et al. 1993), and rural areas in general (France, see Parker, 1979, p.33 ff.; Ireland, see Ó Cearbhaill, 1998; Canada, see Troughton, 1978; etc.). When as a consequence of technological transformations in industry and energy, traditional industrial regions (Northeast England, the Walloon region, the Ruhr) were thrown into crisis, they became beneficiaries of regional aid (Keeble, 1976; Parker, 1979; Schrader, 1993). Many regions with a weak or imbalanced economic structure were included in state promoted development programs (Bylund, 1989; Gade, 1989; Wiberg 1994). In Switzerland, a programme of assistance to mountain regions of 1974 (LIM, Law on Investment aid in Mountain areas) was subsequently enlarged to comprise almost the entire country with the exception of the highly urbanized regions (Regio Plus programme of 1997). This kind of intervention aimed at compensating production disadvantages in agriculture (as a consequence of remoteness from the markets and unfavourable topography) or at encouraging investments into new industries or into services in regions with a poor industrial structure. With the Structural Funds programme, the European Union plays a major role in this process as well, assisting 'regions where development is lagging behind (Objectives 1 and 6), and for declining industrial areas (Objective 2) and rural areas (Objective 5b).' (Gourlay, 1998, p.227). Similar programmes have been developed elsewhere, for example in South Africa where the coal mining area of KwaZulu-Natal ran into difficulties in the 1990s, after the end of apartheid had ended the isolation of the country and exposed it to worldwide competition (Nel and Hill, 2001). Other regulation services concerned the levelling of certain costs across an entire country, irrespective of the actual cost differentiation that would normally be the case. Equal prices for basic food (bread, milk), for postal and telecommunication services, or for railway tickets in fact blurred the economic reality between regions with high demand (where basic costs could be distributed on many customers) and regions with little demand. Finally, state monopolies guaranteed infrastructural services, but because of the lack of competition, they often dwelt on their merits and lacked the incentive to modernize and/or improve their offer.

An interesting example of (permanent) re-regulation is furnished by the Swiss health insurance system that has been based on the new law of 1996. Public health has been of concern for a long time, and state regulations can be traced back to the 19th century when the first laws on safety in factories were enacted: 1864 in the highly industrialized Canton of Glarus, and 1877 on a national level. The first law on compulsory sickness insurance was enforced in 1911. This law was very restrictive and, whilst allowing competition between private health insurance companies, made a change of company almost impossible. Traditional schemes were monolithic, offering within the same policy a certain range of provisions, but a person could only be insured with one health insurance company.

This quasi-monopoly was increasingly criticized, and the 1996 revision of the law broke new ground. The idea of compulsory health insurance has been maintained, but the law differentiates between obligatory and complementary insurance schemes and exposes both, to a varying degree, to market forces. The state regulates the contents of the former fairly strictly, whereas the insurance companies are free to define the conditions of admission and the offer of the latter. In either case, the length of the contract is set to one year (tacitly renewable), a change is not possible within this period. The *compulsory* scheme is subjected to competition in that the individual insurance companies calculate the premium every year (according to regional conditions and in cooperation with the department of health insurance), thus offering every insured person the option to chose the cheapest or most efficient company for the coming year according to the personal financial possibilities. The premium in this scheme tends to rise every year, sometimes considerably, because new treatments or drugs may be included. The health insurance companies are obliged to insure everybody for basic treatment under the compulsory scheme, irrespective of their state of health. Because of the annual variations of the premium, switching between companies at the end of the year has become a standard procedure. The *complementary* scheme, on the other hand, does not suffer from excessive fluctuations. The premium tends to be stable over time (it may even be reduced), hence the insured will remain faithful to the company. Because of its complementary character, this scheme does not invite for too many changes: the companies may edict age limits for admission. Elderly people will therefore stick to their company, otherwise they risk being left without insurance cover for specific illnesses that are not covered by the compulsory scheme.

Swiss medical care has always been essentially on a private basis with a relatively minor state intervention, and this low degree of regulation has now even been lowered to allow for more choice and competition, while maintaining the goal of general public health. It contrasts sharply with countries that operate a national health service (the UK, Sweden etc.). However, NPM has also left its traces in the Swiss medical landscape: small regional hospitals have become subject to investigations into their profitability, and there is currently a tendency towards centralization.

What holds good for healthcare in countries of the North does not apply to the South or to the Transition Countries of the former Soviet Union. 'As in the rest of the world, World Bank, IMF and donor policy advice and lending in transition countries has two potentially contradictory directions in relation to social services. On one hand, investment in social services is upheld as essential for building and maintaining a healthy, educated population. On the other, concerns about fiscal deficits lead to strong pressure to reduce public spending ... In Kyrgyzstan, the incidence of tuberculosis – a disease strongly related to poverty – has more than doubled in the period 1991-1998 to affect over 120 per 100,000 of the population.' (Save the Children, 2001). Similar criticism comes from India: 'The World Bank (WB) is privatizing healthcare in Third World countries by funding commercial projects in the name of poverty alleviation. India is an excellent example of this trend. The Punjab Health Systems Corporation (PHSC), a parastatal financed by the World Bank, is facilitating the commercialization of healthcare in Punjab. The result, however, is more expensive and less accessible healthcare for the poor, parallel health infrastructure, and increased corruption.' (Gupta, 2000). The privatization of this service in countries with large segments of poor populations adds to their marginalization.

Not all services can be re-regulated in this same manner. In particular, the restructuring of postal services (letters, parcels) and busses is subject to fierce debates because the fate of marginal (i.e. below-profit) regions is uncertain. Lacking the prospect of profit, they are outside private investors' interest, unless the state (local or regional, hardly central) intervenes to maintain certain minimum services. 'In vulnerable and marginal regions the role of the state in ensuring basic functions is often discussed.' (Persson and Österberg, 1998, p.78). This discussion must include an evaluation of the regional consequences of deregulation and privatization prior to taking a decision. The two authors demonstrate that the deregulation of domestic air transportation in Sweden has reduced the quality of services in the sparsely populated north and promoted direct flights to Stockholm or even abroad over interregional connections (ibid., p.83 f.). Deregulating the mail service, on the other hand, seemed to have had few negative effects (ibid., p.82).

State intervention will usually be in 'lower gear', e.g. through setting minimum service provision standards for infrastructure. It may also support regional initiatives to improve a seemingly desperate situation and in the framework of regional policy offer incentives in the domain of infrastructure. The privatization of services may well improve the situation of a region, but the private sector thinks in a different time-scale from the public one, hence there may be no long-term guarantee for service provision. The privatization of British Rail, for example, and the separation between railway tracks and train services has not had positive effects only, and re-regulations have been necessary (Knowles and Farrington, 1998; Cumbers and Farrington, 2000).

Deregulation is usually justified with the need for more competition, lean structures, the saving of costs and the reduction of bureaucracy. Similarly, so-called transversal subsidies have been criticized: every service in an enterprise ought to be profitable according to its own cost-benefit structure. For example, the profits from the telephone services were used to cover the losses of the postal services. Deregulation has now led to a dramatic drop of telephone costs, whereas postal services tended to become more expensive. The reduction of cost in this sector has to be achieved through rationalization (automatization), which means reducing the number of jobs. The result is regional unemployment and marginalization of the population concerned.

The idea that the privatization of services as a result of deregulation would automatically be beneficial to society is not necessarily appropriate. While the state agencies reduce the number of employees, private companies will create new jobs; while state bureaucracy may decline, private companies will build up an administration. The costs and the bureaucracy are shifted from the public to the private sector, and this may well turn out to be a zero-sum game – Schmähl (1988, p.100 f.) pointed to this problem in the social insurance sector more than a decade ago. The essential difference between public and private enterprise is the principle of solidarity: state agencies are a *service public*, and the state has the obligation to look after all populations, whether they live in attractive urban centres or in remote marginal regions. In this way, disparities can at least be partly reduced. The private sector, on the other hand, is profit-oriented and does not consider solidarity a high value. Most profit can be obtained in central areas where human and pecuniary capitals converge. Peripheral and marginal areas where little or no profit is expected will be losing out. Disparities and marginalization are therefore programmed.

One of the most critical issues of deregulation in our time is the privatization of water supply in the South where water is becoming a scarce resource. Water is central to health, and the supply of clean drinking water (and the evacuation of waste water) has so far been one of the main tasks of the public sector. It must have priority in a world where cities grow at an alarming rate. However, it is precisely due to the galloping process of urbanization that local authorities arrive at the limits of their possibilities to keep up with the demand. Privatization is often seen as the only way out of an almost hopeless situation. Private interests in profit thus replace public responsibility for general supply. For low-income households, this leads to the catastrophic situation that they cannot afford to buy clean water at prices that bear some relation to their income. It has become standard policy of the World Bank and the International Monetary Fund to link the privatization of services to policies of reducing or eliminating debts. 'In the last year, some of the poorest countries in Africa – Mozambique, Tanzania, Cameroon, and Kenya – have privatised their existing water services. This was a condition of gaining access to World Bank, IMF and donor resources.' (Hall, 2000). The poor, already marginalized by their social status, are thus even more marginalized.

Situating marginalization

The chapters in Part II will offer ample opportunity to discuss the three concepts of marginality, marginalization, and marginal regions in greater depth. The preceding pages have set the general frame for our future arguments, and we shall strive to highlight the relations between the three notions in the book subtitle. First allusions in the chapters on globalization and deregulation have demonstrated that it is not difficult to develop interrelationships, in particular if we limit ourselves to the late 20th and early 21st centuries. A look back in history (which may become necessary in the course of the book) may reveal a different picture – unless we are aware of the relative connotation of 'global'. It has been pointed out above that marginalization is not a recent process but has its roots long ago in human history. Rousseau, in his essay on the origin of inequality reminds us of the social origin of differences that go beyond the natural laws of the diversity among humans (Rousseau, 1987, p.88). There have always been people whose social status and economic possibilities were marginal with reference to others (who exploited their superior position), and there have always been spaces that were considered marginal for a variety of reasons. As such, the phenomenon has a history of its own that cannot be traced here, but that may well be worthwhile to be looked into at some detail. A history of marginality might help us to put the present situation into a broader context (Gupta, 2000).

This introduction concludes by enquiring after the role of globalization, globality and deregulation in the marginalization process. *Globalization* and *globality* tend to focus our attention on the small worldwide scale, and on generalization, standardization or homogenization – but they cannot erase all specificities. They may, however, prevent us from paying attention to the variety of phenomena and processes that exist all over the globe at a variety of scales – diversity is a notion that is basically absent from globality and has to be maintained and claimed for by the victims of globalization. *Deregulation* (or re-regulation) leads towards more privilege for the powerful, the rich and the intelligentsia, i.e. for the elites around the globe, thus increasing the gap towards the less fortunate segments of the populations (Seitz, 1995, pp.3-25). As the increase of new poverty in countries of the North illustrates, the North-South metaphor can be applied to every scale around the globe, and to every country. Growing disparities and, as a consequence, growing social unrest are the result.

These two statements shall occupy a prominent position in Chapters 4 and 6 where economic and socio-cultural aspects of marginality are to be discussed. However, they will underlie the argument of the book as such.

Chapter 2

Peripheries and Margins

Following the introductory remarks on marginalization and its relations to globalization and deregulation, we now concentrate on the first of these three terms again. This chapter wants to put it into the perspective of the issues of centrality and peripherality. Two fundamental questions can be asked in this context:

1. What is marginality, how can we characterize marginal regions?
2. How can we differentiate between marginal regions and peripheries?

We shall argue that marginality has to be clearly distinguished from peripherality. The answers will be the basis for definitions of marginality, which are in turn the input for Part II, where various domains of marginality and marginal regions will be discussed at length.

While it is our aim to define marginalization, marginality and marginal regions, the term 'definition' is not used in its etymological meaning (i.e. to outline, to draw a line or limit around something) but rather in the sense of explaining or clarifying. The wide field of meaning of a notion like 'marginality' (as well as of other normative concepts) precludes an exclusive interpretation and calls for an inclusive approach. It has been emphasized in the introductory chapter that marginality depends on scale as well as on perception; we cannot therefore determine with absolute certainty where it begins and where it ends.

This chapter looks first at the relationship between centre, periphery and marginal regions. It evaluates the Core-Periphery model and examines its utility for our arguments. This is followed by first approaches at marginality and possible definitions. It concludes with a presentation of first ideas about the non-material background to marginality, ideas that will be developed in Chapter 3.

Core, periphery and marginal regions

Core and periphery, notions borrowed from geometry, have always been essential concepts in geography (Gottmann, 1980, pp.11 ff), albeit not always by these names, and in many cases implicitly rather than explicitly. At first glance, marginal could be regarded as synonymous with peripheral, as both notions are clearly the opposite of central. Not surprisingly, the first model to spring to mind is the Core/Centre-Periphery model, first proposed by Prebisch (1959), later improved by Friedmann (1973), and further elaborated by Reynaud (1981). Christaller's central place theory of 1933 (Christaller, 1968) is in a way one of its forerunner, but already in the 1920s, Estonian geographers, in particular J.G. Granö and Edgar Kant had been studying the town-hinterland relations and the hierarchy of central places in Estonia (Granö, 1922; Kant, 1926, 1935; Maide, 1931) – the same concept as Christaller's, but based on empirical research and not on economic theory. Indeed, the Estonian geographers preceded Christaller (Kurs, 2002).

Prebisch based his arguments exclusively on the economic contrast between industrialized and 'underdeveloped' countries (a 'two-region model'; Schätzl, 1981, p.141), denominated centre and periphery respectively, and concentrated on the trade flows between them and the resulting terms of trade. His regard on centre and periphery was restricted to the global scale; he did not attempt to seek analogies at national or even subnational levels. This deficit was amended a few years later by Friedmann (1966; 1973). Going beyond purely economic reasoning, he included social, psychological and political elements as well as the temporal scale. 'Human activities and social interactions are considered as forces shaping space and depending on space.' (Schätzl, 1981, p.144; transl. WL). According to Friedmann, core areas possess a potential for innovations from which the periphery may ultimately benefit. The relationships between the two are characterized by dependency and power: the periphery depends on the core, which decides on any actions taken. While sounding extremely simple, this interpretation resembles a one-way street and reduces the power-dependency-relationship to a closed system. We do not agree with this view. A look back into history will demonstrate that power and dependency have always existed in a dynamic equilibrium: no human power is permanent, even the worst case of dependency may come to an end. Even if he does not quote Friedmann in his book, Reynaud (1981) criticizes this unilateral and static view and proposes a dynamic approach. A closer look at Friedmann's work, however, discloses that in his model of economic development stages (1966), the core-periphery relationship corresponds to the second of four stages; it is a transitory phase in the evolution from individual pre-industrial local economies to a post-industrial integrated economic system. However, it is not clear in which way power and dependency will evolve. Friedmann also assumes that an economy can be totally integrated. As a consequence, marginal regions are absent from this model.

One important component of the centre-periphery-model (and inherent in the discourse on marginal regions as well) is the notion of development. According to Friedmann, development is a discontinuous cumulative process, which manifests itself in series of innovations and will eventually 'lead to a structural transformation of the social system.' (Schätzl, 1981, p.144 f.; transl. WL). Development is a process of innovation that includes the economic, social, and cultural, but also the political and even the ecological spheres. Etymologically, development means to unroll, to unfold, to open out. It can also be understood as progressing from one stage to the next. Therefore, we can define it as 'a process leading from an original situation (which is judged as unsatisfactory) to a new situation, judged as better or satisfactory' (Leimgruber, 1994/5, p.392), applicable to economic and social life, politics and culture. Even the environment can find its place in it. This definition has been inspired by the one proposed by the South Commission (1990, p.10) that defined development as 'a process which enables human beings to realize their potential, build self-confidence, and lead lives of dignity and fulfilment.' Development therefore signifies to eliminate or at least reduce regional economic, social and other disparities; this is how we understand it in the present context, and it is a directional and irreversible process. It advocates change, both in attitude in general and from familiar old habits; as such, it is truly innovative. Needless to say that this is a purely qualitative interpretation of the term; development is not synonymous to growth, which is purely quantitative. When looking at the current state of the world (from the perspectives of ecology and society) one can say that the human race as a whole requires some form of (humane) development if it wants to survive (see e.g. Ekins, 1992; MSF, 1997; Seitz, 1995). Development is a normative concept, and there is no final state but rather a continuous progression from less to increasing satisfaction. Incorporating the notion of development into the discussion of marginality and marginal regions allows us to back the idea of reversibility (inversion), introduced by Reynaud (1981) in his dynamic view of the centre-periphery model.

Marginality and marginal regions are opposed to centrality and central regions. Reality, however, is not composed of extremes but of continua between extremes: a region is more or less marginal or central, depending on the point of reference from which we measure a specific quality. From the point of view of natural resources, for example, Switzerland is very marginal, but this drawback is offset by the focus on human capital where the country occupies a fairly central position. In addition, the condition of marginality may be temporary, i.e. while marginalization is basically a unidirectional process that may ultimately lead to misery, it can, in the course of time, be halted and reversed, and a region can re-emerge out of a seemingly hopeless situation. This can be seen with reference to resources: a material will become a resource once its potential has been perceived, but as soon as it can be substituted by another material, it will lose its 'status' and become unimportant. Coal, for example, has been a major driver of the industrial revolution in Britain, and coal-mining areas rose to national importance. However, when better quality coal was imported, when the steel industry moved to coastal locations, and when coal seams were gradually exhausted, these regions lost their strength. Ultimately, coal was largely substituted by oil and electricity, and the fate of the coal-mining regions was sealed. Coal can still be regarded as a potential resource, but in the oil and gas period, it is extremely marginal. A similar fate awaits the oil producing regions around the world in the future.

How to discriminate between marginal regions and the periphery? Although semantically the two words are very close (meaning at the edge, less important, not of central importance), we contend that they have to be kept apart. In fact, the periphery finds itself interacting with the core or centre in a mutual dynamic relationship of varying intensity. Reynaud (1981, p.62) demonstrates this by identifying different types of centres and peripheries (Table 2.1). His typology is based on the flows of people and capital on the one hand, of raw materials on the other, and it describes a stable situation.

Table 2.1 Types of centres and peripheries

Centre	Periphery	People and capital flows	Raw material flows
dominant	dominated	P → C	P → C
hypertrophic	abandoned	P → C	--
dominant	integrated and exploited	C → P	P → C
hypertrophic	integrated and annexed	C → P	--

Source: Reynaud 1981, p.62, adapted from Leimgruber 1994

This typology looks extremely static, almost deterministic or fatalistic; it is this outlook that has prevailed in the modernization theory and has been used to justify specific western models of assistance to the 'underdeveloped' ('developing') countries, or the 'Third World'. This static position was based on the idea that the periphery wanted to catch up with the 'developed' world but did not take into account that the periphery 'does not necessarily want what we of the centre have. Indeed we want what they have – their raw materials, their cheap labour, their potential markets …' (Miller, 1998, p.26). The fallacy of modernization theory is that it perceives the centre-periphery model as rigid. Reynaud, however, is well aware of the long-term reality, of the inherent dynamics of every human construction. The relationship between core and periphery is not stable for eternity; a centre can decline and the periphery may draw advantage of this, the two may gradually become autonomous, and through the process of inversion, their respective status can change – we can even imagine that the former centre declines but does not transmit its dynamism or spirit of enterprise to its periphery (Figure 2.1). Indeed, centre and periphery may possess elements that can be mutually beneficial and thus promote exchange. The difference is valuable, not detrimental. In this way, Reynaud includes the dynamic element (the aspect of time), emphasizing that nothing is final or definitive.

There are numerous examples for this process: traditional industrial regions that were once leading economic areas have become development regions (Northeast England, Wallonia, the Ruhr area, etc.), the Alps were once feared by travellers and indigenous inhabitants, but have since turned into one of the recreationists' core areas (Leimgruber, 1992). The steel town of Consett (Northern England) is a case of inversion in the not too distant past. The town was home to a steel mill, founded in 1848, which in the 1970s offered more than 5,000 jobs, was closed in 1980, following 132 years of operation and only 17 years after it had been updated and enlarged by the construction of an oxygen plant, to say nothing of former modernizations in 1957 (three additional blast furnaces) and 1960 (a plate mill). Its fate was sealed because 'the British Steel Corporation was committed to large integrated coastal plants.' (Else, 1984). To find alternatives for over 5,000 unemployed in an inland location, where people depended for their livelihood entirely on a declining industry, proved impossible, and the future in particular for the young generation looked bleak. To this must be added the psychological factor, the loss of a culture and an identity (the miner and the steelworker), and the feeling of a large community being suddenly a useless member of society. State intervention cannot be the only solution to such a destitute situation, although it may bring some hope – a point to be discussed in Part III.

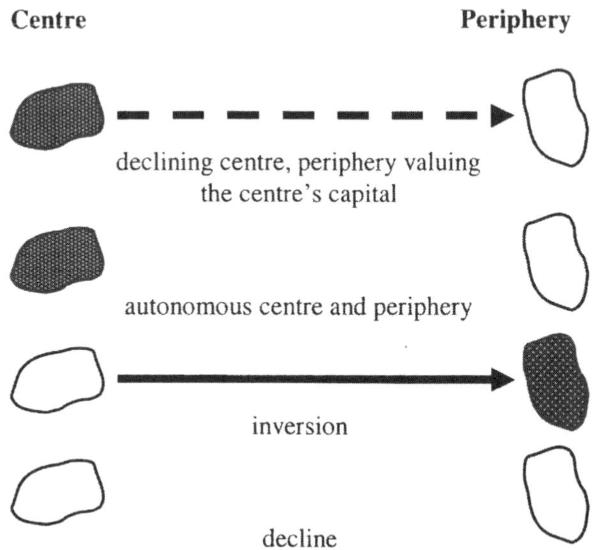

Figure 2.1 The dynamics of centres and peripheries

Source: Reynaud 1981, p.62, modified

It may be tempting to divide the entire world into centres and peripheries, allocating every single place and region to the respective type. In the second part of his book, Reynaud presents case studies of such a regionalization at the global, continental, national and regional level. However, he avoids the temptation of dualistic simplicity and admits that at certain scales not all regions fall into this typology. He singles out three particular cases of regions that cannot be simply classed as centres or peripheries:

1. A place (or a region) can lead a life of its own and yet be closely associated to a major centre (for example the port city of a major centre). Reynaud calls this type of place or region an *associat* (associated place). It maintains close ties with the centre, but there is no unilateral dependency relationship and the *associat* is not well integrated into the surrounding region (p.76).
2. A place (or a region) may have intense internal relations but cultivate little contact with the outside world, thus remaining isolated from it and living according to its own rhythm *(isolat)*, often lacking the will for change (p.84).

3. A place (or a region) can be situated off the major communication flows and lead a very introvert life at the edge of a system, lacking a development potential of its own *(angle mort)*. It is usually considerably more backward than the *isolat* and displays a feeble internal organization (p.87).

Type 1 (the associated place) is neither a centre nor a periphery, and it is not marginal. The *associat* operates in interaction with a (hyper)centre to which it is related in a complementary and not in a dependent way (Figure 2.2). We can therefore exclude such places or regions from our further discussions. However, given the dynamics of centres and peripheries, the decline of the centre might include the marginalization of the *associat*.

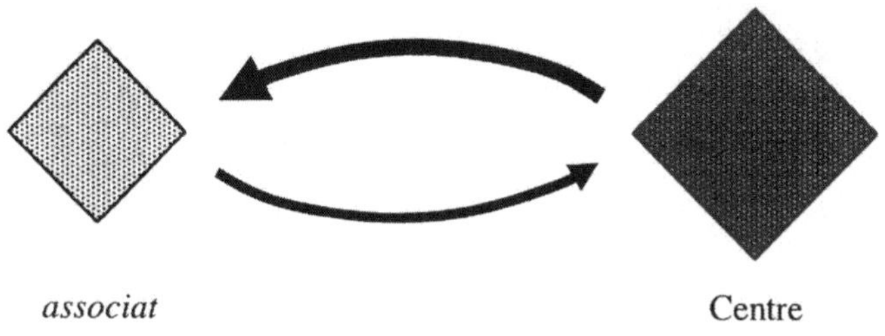

Figure 2.2 The relations between a centre and its *associat*

Source: Reynaud, 1981, p.76 (modified)

Types 2 (isolated region) and 3 ('lost corner'), on the other hand, 'are situated inside the centre or the periphery but cannot really be attributed to them' (Reynaud, 1981, p.83; transl. WL). Both can be regarded as marginal regions as they lead an isolated life. They distinguish themselves from each other by the intensity of the flows (Figure 2.3). Whereas the *isolat* is characterized by strong internal flows, but has weak or no relations at all to the outside world, flows are almost absent in the *angle mort*, where little to nothing happens. These two types of region appear to lie at the margin or even outside given systems.

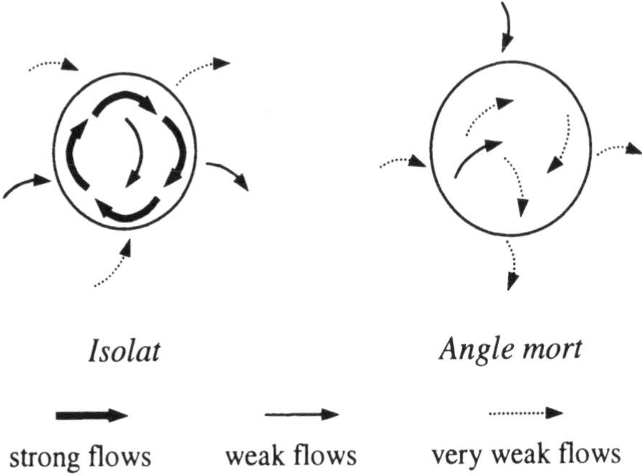

Figure 2.3 ***Isolat*** **and** ***angle mort***

Source: Reynaud, 1981, p.84 (modified)

The centre-periphery model in its original polarized form (i.e. without Reynaud's complements on the dynamics and the exceptions) reveals itself as too narrow in outlook; as Schmidt (1998, p.49) notes, it 'should never be considered in an isolated or absolute way because it depends completely on the space and phenomenon scale under consideration and it is relative to the comparison context and frame where it is inserted.' It can be applied 'at all levels of resolution' (ibid.) – it is a relative concept. The same holds good of marginality and marginal regions: they occur at every scale and in every context.

As geographers, we emphasize the spatial aspects of marginality. However, the term's semantic field is much wider and can be associated with any phenomenon, such as the human body, a community or an entire society, the domain of science, and politics. In some cases, space is of no or little importance; in others it will be an essential element when the geographical differentiation and organization are studied. Socially marginal groups, for example, often congregate in specific locations (as is exemplified by the dwelling areas of the Untouchables in India, the Ghetto for the Jews in medieval European cities, the dwelling areas of crafts workers in Ethiopia (Freeman and Pankhurst, 2001; Pankhurst, forthcoming), or council estates in countries of the North) where social and spatial marginality coincide. The townships in apartheid South Africa furnish an extreme example. Similarly, gated communities (Gmünder et al.

2000; Coy and Pöhler, 2002) are a sort of marginal region, although from a radically different perspective. Affluence creates fear for the security of life and material welfare. Gated communities are supposed to provide shelter from violence and robberies, although this is only one aspect. This topic will be taken up in Chapter 6.

Approaches to marginality

Interest in marginality and marginal regions has appeared after World War II, when successively questions of inequality, spatial disparities, social justice, and welfare began to enter research in human geography. External drivers of this new orientation were the post-war economic boom in countries of the North, the process of decolonization, and the ensuing problem of (economic) underdevelopment in vast parts of the world. While 'underdeveloped' has since been replaced by other notions, such as 'less developed' or 'low income', they are all based on western perceptions of 'development'; they emphasize the material (financial, consumer) situation and are therefore condescending and discriminatory. However, the western standard is by no means the only one valid on the globe, even if the agents of globalization pretend otherwise.

The international perspective dominated at first, but from the early 1960s onwards, we also observed a growing awareness of uneven economic growth within countries, of out-migration from certain areas, and of population concentration in cities with the ensuing problems of housing and quality of life. As a consequence of these processes, regional disparities began to increase, and state intervention was called for as a remedy. Marginality was not verbally an issue, but it was underlying the various regional policy measures proposed. Haggett (1972, p.405) identified three broad fields: a) direct state investment (examples quoted are the Tennessee Valley Authority in the U.S.A., the building of the new capital of Brasilia), b) inducements to private companies (from capital grants to tax relief), and c) inducements to individuals and households (such as land reforms, and compensation payments). Economic concerns were at the core of every measure, disparities were seen from the income perspective only.

With his focus on well-being, needs and wants, quality of life and level of living, David Smith introduced the welfare approach in geography. He defines 'human geography as *the study of who gets what where, and how*,' proposing to analyse 'all human geographical patterns in terms relevant to human life chances' (Smith 1977, p.7; emphasis in the original). Again, marginality is not mentioned as such, but it is implicitly there.

A major problem of regional disparities had been identified in Switzerland immediately after World War II, when a first regional policy measure was introduced in favour of mountain regions in 1948 (Chiffelle, 1983, Leimgruber, 1985). Further policy instruments followed in the 1950s and 1960s. In the course of the 1970s, mountain development received a new impetus through the Law on investments in mountain regions (1974), an instrument of regional policy that combined top-down with bottom-up strategies. Again, marginality was not verbally expressed, but it formed explicitly part of the arguments. We shall discuss this planning tool in greater detail below (Chapter 8).

This book looks at marginality from a wide angle, irrespective of the definition of the term. This broad perspective has two advantages: 1) it is a-spatial, i.e. it can be used in a variety of contexts, and 2) it can be employed at every possible scale. Only in this way, it can be applied universally.

From what has just been said emerges that marginality is a relative concept, set in a normative context. It has to be defined precisely according to circumstances, domains, and spatial and temporal scale. A marginal region is most commonly thought of as a poor or underdeveloped region with no internal growth potential, but such simple and commonplace terms require clarification: what do notions like 'poor' and 'underdeveloped' really mean? Which is the point of reference for measuring poverty (Chapter 6) and underdevelopment? Besides, we usually associate marginal with economic deprivation, although other criteria (such as social exclusion or (geo)political considerations) can be applied as well.

In the previous section, marginal regions have been compared to Reynaud's specific regional types of *isolat* and *angle mort*, regions living a life of their own with few contacts to the outside world. The two terms suggest that such regions have been forgotten or overlooked, that they have not shared the development processes of other regions. However, we must not look at these two types of regions as static and immutable; behind the *isolat* and *angle mort* there are various processes that have led to this specific situation or have prevented a region from changing its status. There is no proof, however, that dynamics is excluded in the future. In fact, Reynaud's concept is flexible, as is demonstrated by the process of inversion where centres and peripheries can exchange role and thereby preserve the equilibrium inherent to our dual worldview. Marginal regions *(isolats, angls morts),* therefore, do stand a chance to step out of their isolation.

Definitions of marginality and marginal regions have usually been economically based. Economists use the term 'marginal' to describe a situation where a firm makes neither loss nor profit (the break-even point), whereas in agricultural terminology, marginal land consists of surfaces whose yield is limited to such an extent that they will be cultivated in emergency situations only. In a profit-oriented economy, this is certainly a negative attribute, although the really negative situation occurs when a firm moves into an extra-marginal situation, i.e. makes losses only (Andreoli, 1994, p.42). Economic activity is a steady process, and a marginal situation may occur in the course of time, even though a firm will try to avoid it. What holds good for firms, however, is not valid for regions: a region is not a firm but a complex spatial unit where losses and profits cannot be measured in monetary terms alone but where elements such as 'quality of life' and 'state of the environment' have to be considered, elements which are ultimately of a qualitative nature. Marginal land, therefore, while yielding very little in economic terms, may excel by a rich biological diversity and hence play a central ecological role (see Chapter 7).

From a sociological perspective, marginality denotes a personality that does not belong to a clearly defined 'mainstream' group (social group, culture, political organization) but positions itself somewhere in between, or even manifests simultaneous adherence to several such groups (Hartfiel and Hillmann, 1982, pp.621 f.). Marginalization can be self-inflicted, resulting from an intentional process undertaken by the individual in order to delimit his or her own sphere from the rest of society, or it can be brought about by the (social) environment (e.g. the marginalization of immigrants, or the choice of locations for old people's homes). The consequence will be isolation within or segregation from the main group, community or society, with the ghetto as its extreme spatial expression. In general, minority groups are often marginalized both socially (discrimination) and spatially (Little Italy, Chinatown). In his study on craftsmen in Ethiopia, Pankhurst (forthcoming) provides an excellent example of this double margianlization.

Definitions

We shall now discuss a number of possible definitions of the marginality concept. Apart from identifying it (discussing the topic) we shall also have to put it into perspective (looking at the scale). The example of the marginal lands mentioned above indicates that marginality is to quite a considerable extent a matter of perspective. What is marginal from an economic point of view may not be so culturally or ecologically, and vice versa; marginal in the 18th century is not the same as in the 20th and 21st centuries, etc. We shall therefore avoid categorical statements about an ultimately correct definition.

In an intermediary report on a research project on marginal agrarian systems, Galante and Sala (1987, p.11) point to the essentially economic understanding of marginality: marginal regions are called by different names, such as 'disadvantaged regions', 'internal regions', or 'difficult regions'. Whatever the denomination, such regions deviate negatively from what can be called 'normal' regions, normalcy being obviously understood as mean agricultural and forest productivity (that has been steadily rising after World War II). This economic below-normal definition has dominated the discussions among Italian scholars (ibid., p.18).

Other researchers in the field have slightly enlarged this fairly narrow outlook. Campus et al. (1987) decided, after a review of the many possible definitions, to identify marginal regions according to four criteria: 1) significantly lower per capita incomes (compared to a reference region such as province, region, or country, or the provincial capital), 2) low infrastructure equipment (concerning both the quality of life and the production process), 3) cultural isolation (having a negative impact on the entrepreneurial spirit), and 4) difficult natural conditions (slope, unstable geology, rough climate) and historical heritage (p.230). The crux of the definition lies with the thresholds or the interpretation of notions such as 'significantly lower'. A simpler approach compared production capacity with market integration and allowed Pieroni and Andreoli (1989, p.3) to discriminate between centre, periphery and marginal region (Figure 2.4). A region can thus be called marginal when it has both a weak production capacity (i.e. a low level of technology and know how, and few alternatives for the solution of problems) and a low market integration (i.e. the production is essentially subsistence and not market oriented). Whilst limited to the economy, the contributions by the Italian scholars permit to situate the problem of marginality and marginal regions.

Miller (1998, p.262), from his more spiritual perspective, characterizes marginal regions as having:

- a cultic relationship to the environment, space, and place;
- a stable body of knowledge, which will include lore and tradition, stemming form the relationship to the environment, space, and place; and
- a means of fostering and maintaining – defending society's stock of tradition, lore, and praxis.

While these three qualities suggest a static situation, marginalized regions do show their own degree of dynamics that is anchored in the relationship to the land and in the difference between themselves and the centre. However, they do not necessarily want to adopt the pace dictated from outside but change or develop according to their own rhythm, using endogenous knowledge and techniques that are in harmony with their culture. Reynaud (1981, pp.77-83) describes this scenario when he writes of the periphery that relies on its own strength. In development theory, it corresponds to the model of endogenous development. From the perspective of the centre (that favours modernization), it may be wrong, but from their point-of-view (difference) it will be perfectly acceptable.

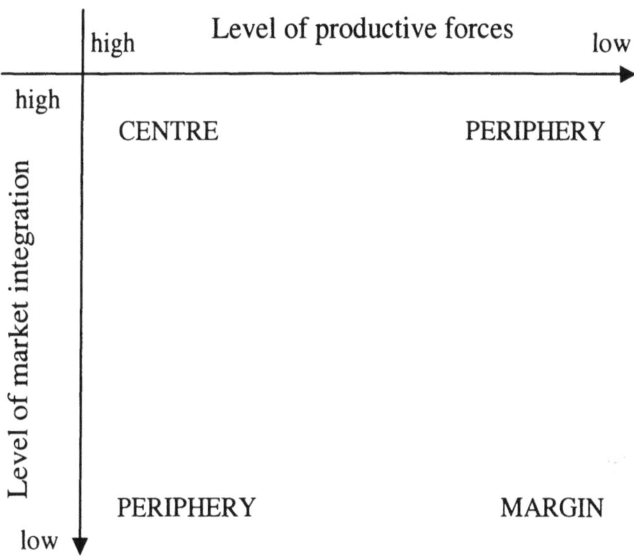

Figure 2.4 Marginal regions as interplay between productive forces and market integration

Source: Pieroni and Andreoli, 1989, p.3 (modified)

The model developed in Figure 2.4 shows that from this particular perspective, marginality and peripherality are not the same, and that from a systemic point of view, marginal regions really lie at the edge of a system or even outside. To determine the complexity of a system, Andreoli (1992) used a multivariate approach, based essentially (but not exclusively) on socio-economic variables. 'In this paper it is assumed that marginality is the situation of an area that is located "at the margin" of a system as regards its socio-economic features' (Andreoli, 1992, p.24). The features she used in the analysis covered the demography, the non-agricultural economy, the agricultural sector, social and cultural aspects, and service supplies and living standard (ibid., p.25; for details see Pieroni and Andreoli, 1989, pp.19-30, where we discover that social and cultural aspects are in fact socio-political ones). It is difficult to follow this analysis for, by combining 26 variables in her analysis, the author puts together factors where a region is really marginal and factors where it need not be. Such a quantitative analysis results in a statistically medium degree of marginality of the various regions studied, which is really meaningless because of the mingling of factors. The subjective choice of variables influences the result and can lead to the statement that a region is marginal because someone (the researcher, local, regional, or national politicians, the business community, the farmers' lobby, etc.) wants it to be marginal. The perception of the population (how do the people themselves feel their situation?) is missing. It is true that emigration and an ageing population can be taken as demographic indicators, but urban areas display similar phenomena. Marginality thus reveals itself 'a relative phenomenon' (Andreoli, 1992, p.42), indeed it is 'on the whole a rather intuitive concept' (Andreoli et al., 1989, p.282) because it depends on the point of view from which it will be studied – marginality is a state of mind (of the people directly concerned, of the politicians, and, last but not least, of the scientific investigators). This holds good just as Rousseau's statement that 'happiness is less an affair of reason than of feelings' (Rousseau, 1987, p.108).

The above reflections therefore do not help us much further. We retain from these first arguments the following elements concerning marginality and marginal regions: marginal regions are situated at the edge of a system, they are relative concepts, they are usually very complex, and there is a strong element of perception associated with them. Besides, the overall association is negative.

The reference to 'mainstream' used above appears to be more promising to understand the concept of marginality. Every society is composed of individuals with a certain range (oscillation or elasticity) of individual behaviour; societies and groups 'are not necessarily unified collectivities' (Giddens, 1984, p.24). The sum of these individual behaviours makes up the 'behavioural oscillation' (we shall call it the 'mainstream') of the society (Figure 2.5). Certain individuals' behaviour, however, is located at the extreme ends of the society's range; it may even reach considerably beyond it. Such persons do no longer belong to the society's 'mainstream' but are marginal persons. This marginalization can occur in two different ways: they may either position themselves intentionally outside the mainstream (self-inflected), or the rest of the society or a group may exclude them from it. This position need not be permanent, i.e. marginalized persons have the possibility to return into the mainstream whenever they want to or are admitted. Again, there are two ways in which this can take place. They can give up the convictions that were at the root of marginalization through an active personal effort, or the mainstream can evolve and widen its oscillation by accepting formerly marginal (i.e. foolish) ideas. In this way, marginal individuals will be absorbed without acting against their convictions; on the contrary, they have contributed to the broadening of the mainstream, to the evolution of the society. While the mainstream is relatively stable over time, marginal individuals can thus exercise a medium to long-term impact on the society in that they may bring about transformations through innovative ideas. Marginal groups are in fact essential elements of every society, as can be read implicitly from the following statement: 'While the continued existence of large collectivities or societies evidently does not depend upon the activities of any particular individual, such collectives or societies manifestly would cease to be if all the agents involved disappeared' (Giddens 1984, p.24). Marginal agents have their place in every society, and the sum of such marginal individuals can be called a marginal group or community.

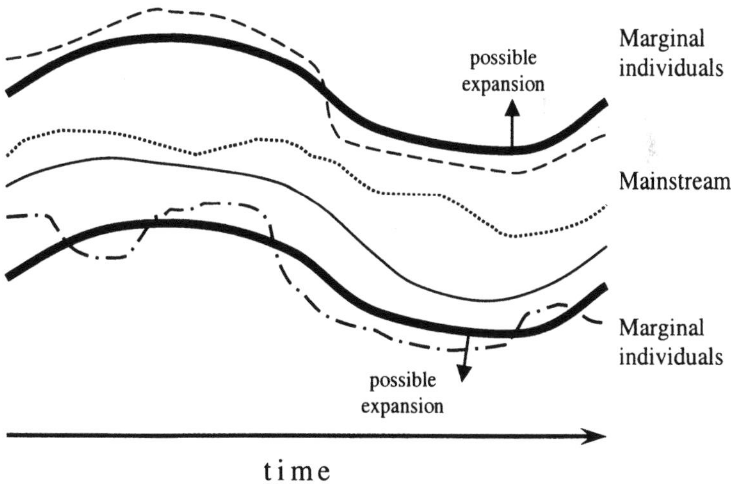

Figure 2.5 Mainstream and marginality

From this dynamic perspective, the notion of marginal loses its purely negative (almost depressive) connotation. To be marginal may signify exclusion from the mainstream, and this situation can be negatively interpreted. However, it also contains a chance towards positive development, if one is ready to throw old habits over board and begin with something new. Indeed, a marginal individual or phenomenon may point the way out of a seemingly hopeless situation, even if inventors and innovators may lead a solitary life and be misunderstood by their generation. Organic farming was considered as queer in the 1950s, but is seen as probably the only long-term solution to the environmental crisis in agriculture (see Chapter 7).

Opening up perspectives is thus the positive side of marginality, an aspect that tends to be overlooked. Douglas (1996, p.91 f.) suggests that marginal subjects (individuals, groups) can be agents of social transformation (we might add: of innovation). She quotes a lengthy passage from Horton (1969) who points to the role marginal spirits can play in a society. Indeed, persons who have brought about substantial changes and exercised long-term influence were often socially marginalized and lived in this way (St. John Baptist, Jesus Christ, St. Francis of Assisi). Even Einstein, in his outer apparel, did not conform to standard social expectations.

The various disciplines have their own interpretations of marginality. In *ethnology,* a marginal area has been characterized 'by its general poverty or old-fashioned character' (Hultkrantz, 1960, p.181) – although there seems to be 'no agreement as to the real significance of the concept.' (ibid.) This is clearly an evolutionist perspective where 'marginal' is equivalent to 'backward' and contains a distinct value judgement that can be used in a denigrating context. Despite many efforts, we have not yet overcome this way of thinking. Carrying this thought further, we discover the roots of development policy: marginal areas and peoples have to be gotten out of their backwardness (which is seen as a negative attribute) and developed, in order to conform to 'civilized' standards (which are naturally presented as positive). No wonder that such ideas emerge in core areas and serve primarily their own interests. This is one of the criticisms of development policy: it is the 'North' that has defined the standards of 'underdeveloped' and of 'poverty' but that at the same time is opposed to efforts that could bring the South to the same technological and consumption level as the North. While lip service is paid to the need for development, environmental stress is claimed to be the official reason, but power and hegemony are the real motives. Marginality is thus considered a global necessity to ensure the primacy of the North.

From a *political* perspective, marginality is usually first associated with political boundaries, which define the limits of a state's authority and the range of the applicability of its legislation. For centuries, state boundaries have been perceived as separations, and regions along the borders were usually considered as risk zones – the military function of the boundary played an important role in decisions on investments at a country's periphery. Many studies of border regions illustrate the relative marginalization along international boundaries. Its visible expression can be the many routes which run parallel to the border rather than cross it or which stop short of it (Wolfe, 1962). In extreme cases, such as was the case with the boundary between the Federal Republic of Germany and the German Democratic Republic (and is still the case between North and South Korea), the military and the ideological functions combine to create a rift in land use and transborder communication, thereby cutting across old established social patterns (Schwind, 1981; Kim, 1990; Ante, 1991). Sometimes natural and/or historic reasons have precluded a border region from developing strong ties with the neighbouring areas or the growth of intense transborder contacts. The Franco-Swiss Jura region offers a good example of a strongly marginalized area, where political and physical factors combine (Gigon, 1991). On a different scale, transborder tourism in the Swiss Alps has begun to promote exchange on a regional scale and reduce marginality to some extent (Leimgruber, 1998b).

Political marginalization occurs also at other scales. Inside a country, it may appear within the party system, and it often characterizes political actions and decisions. Who obtains which political post, which region will be preferred in regional policy, how will the tax system be organized, and so on – these are questions where people, regions, and groups are at stake and where the struggle for power becomes manifest. On the other end of the spectrum, geopolitics is mainly about marginalization. These aspects will be dealt with in greater detail in Chapter 5.

This variety of meanings is the reason why a wider approach to marginality and marginal regions is being proposed, an approach, which may help to revise current thinking about the dichotomy between core areas and marginal regions. The three domains briefly presented above (economy, society, ethnology) have in common that they are limited to humans as social beings. This is not negative in itself, but it falls short of what has been said in the beginning: humans live together in society *and* in interrelationship with nature (the environment), as is suggested by the human ecology triangle (Figure 2.3) which can be seen as a modified version of Schütz' three-world-model.

Such a statement may sound very commonplace, but our industrial and service society has almost forgotten it. It was only in the 1950s that this fact has been rediscovered and widely recognized. The symposium on man's role in changing the face of the earth, held in 1956 (see Chapter 7), has played a major role in this process. Following a gradual shift in perspective, contemporary thinking has become increasingly 'ecological', and 'sustainability' has become a fashion term. As a consequence, most political parties have adopted the 'environment' into their programmes, some explicitly (the 'Greens' that have grouped themselves in most countries of the North), others in a more cautious manner (in particular those parties that are intimately linked to the economy).

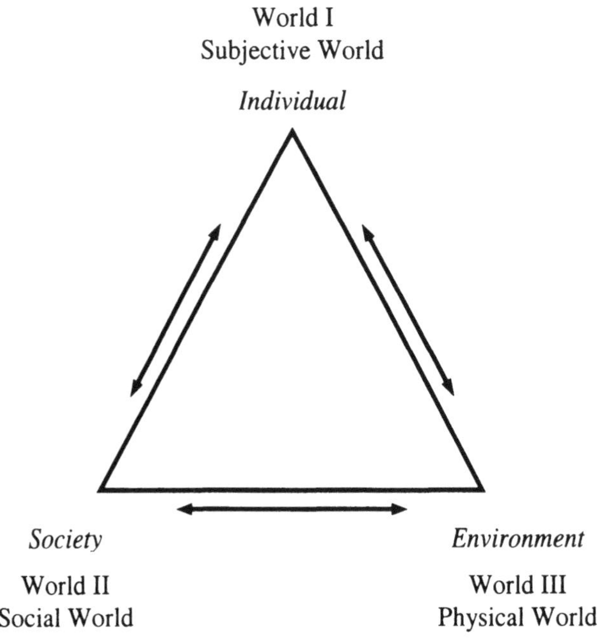

Figure 2.6 The Human Ecology triangle (in bold) and the Three Worlds

Sources: Steiner, 1988; Werlen, 1998

Despite the knowledge about the interactions between humans and the environment, many people pay mere lip service to it and act differently from what they say, i.e. they act just as they did before (business as usual). There still exists a wide gap between knowing and acting (Schaller, 1998), not only among the adults (who have grown up in the society of surplus) but also among the young generation (who were born into the new awareness), and this gap will only be bridged slowly, probably by a new generation that has still to be born. Schaller's results of a cluster analysis (sample size: 723 young people between 14 and 25 years old) might give rise to pessimism: for 55 per cent, environmental problems do exist but can or will be solved by the society. Only 19 per cent display a real concern and demand rapid action, whereas the remaining 26 per cent are relatively indifferent (Schaller, 1998, pp.173-178). The result may look surprising, but it was obtained from a generation that has grown up with parents who were children of the post-World War II 'throw-away-society'. They used to consider the ecosystem as free of charge, causing zero cost for any economic calculation and serving as sink for all waste produced. This attitude is about to change: in many countries legislation has imposed ecologically motivated measures (smoke filters, sewage, noise protection, prohibition of use of dangerous substances etc.), but so far, the general economic boom has helped both economy and population to absorb the extra cost – and during an economic slump, jobs are more important than the ecosystem. The environment has maintained its marginal status in economic thinking, especially as the flexibility of the economic system permits firms to change the location of critical production units to countries which either lack ecological legislation, where existing laws can be ignored, or where corruption makes everything possible. Even on the continental scale of Europe, the international division of labour leads to grotesque forms of behaviour. It favours the transportation of semi-finished products across the continent to be assembled where labour costs are low (usually from the North to the South), and the final products have to be carted back. Current transportation costs seem to justify this procedure, but they are falsified because the costs of air pollution and noise resulting from road traffic are externalized: the environment does not charge us, and the costs of noise protection are borne by the public sector (taxpayers' money). The retail price of the product is therefore not based on real costs. Attempts to solve the pollution problem by dealing with pollution certificates are still a theoretical concept, and it concerns industrial production, not necessarily transportation. This is one aspect of the globalization process that is usually passed over in silence.

Human activity still tends to marginalize the environment, putting human needs and wants (and greed) in first place, even if we know that the natural environment furnishes the fundamentals of our existence. If the ecosystem continues to be marginalized, it will eventually be downgraded or even destroyed to a point where the survival of the human race is at risk (Leimgruber, 2000). The world can continue to exist without human beings, and the ecosystem can recover in the very long run, but we humans, we need the world.

Aspects of marginality

In our first steps on marginality research, we pointed out that marginal regions can be viewed from different angles, depending on the parameters chosen. We distinguished four major possibilities to look at marginal regions: geometrical, ecological, economic, and social, but referred to further possible approaches such as political, cultural, risks/hazards, and perception (Leimgruber, 1994, p.8). In the course of time, these crude ideas have been refined, and in this chapter we are going to discuss three specific aspects which underlie the various thematic sides of marginality, marginalization and marginal regions: geometric, systemic and processual features. They may appear individually or in combination. We have retained the geometric perspective from the original ones, while the others have been transferred to the field of domains (Chapters 4 to 7).

Geometric marginality

As has been pointed out during the discussion of the core-periphery model, marginality can first and foremost be looked at from a geometric perspective. It then becomes a synonym to peripherality. In a way von Thünen (1826), who used marginal location rent as a criteron for his model, has already done this. His circular land use pattern is based on strict economic principles and on rational decisions taken by the farmers. Marginal is the point where profit turns into loss – marginal regions reduce themselves to a geometric line. None of the land use rings is marginal because each is logically defined. The geometric principle lies also behind Christaller's central place theory that is based on a centre-periphery approach. Basically, both are static models, describing functional relationships very much similar to 'old' globalization (Figure 1.2). However, they could be developed into dynamic models if the time dimension were introduced. One has to accept that changes in spatial structure do occur when conditions change. Changes in transportation technology and consumer taste bring about a transformation of the agricultural model. If a service on a specific hierarchical level can be obtained at a lesser cost from a different (new) centre from the one that used to be patronized, the central pace system will readjust itself. Neither von Thünen nor Christaller did take dynamic aspects into account, except that von Thünen later modified his rigid original model to care for lesser centres or specific spatial situations (river transport). As Reynaud has pointed out, inversion has to be taken seriously as a process that occurs over time. Accepting change does not devaluate the fundamental ideas behind the von Thünen and the Christaller models but would approach them to reality.

Accessibility and range of action (i.e. mobility) may therefore be the principal criterion for geometric marginality, at least from a rational economic point of view. However, social aspects must not be neglected, in particular if we change the scale. Various studies on rural areas (Moseley, 1979; Moseley and Packman, 1984; Huigen, 1984; Lois-Gonzales, 2002) have shown that in particular three components have to be considered: the resident population in an area and its regional distribution, the availability of services, and the potential communication links between the people and the opportunities (Bull et al., 1984, p.62). The provision of public services to homes (waste collection, water and electricity supply, milk collection) as well as people's trips towards public and private services (medical care, shops for short to medium term needs) may be crucial to the survival of a rural community. Rationalizing on such basic services may drive rural areas deeper into marginality that they may already be. Certainly, private mobility may satisfy various demands, and other services may be provided by mobile shops, the regular passage of the milk collector, visits by doctors and/or nurses etc. However, such mobile services have an important handicap: as specific provisions they cause extra cost which may not automatically be borne by the community and even less by a private provider. In particular, remote rural areas suffer from this economic disadvantage which reduces their attractivity even further. It is true that modern mobility has opened up many rural areas, but not all inhabitants are mobile; children as well as many elderly and disabled persons have to rely on provisions furnished either by a member of their family (private mobility) or by some mobile service. The costs of mobility weigh heavily, but it is 'not just the often considerable vehicle-related costs, but the opportunity cost of time spent by the service supplier behind the driving wheel: mobile services involve a lot of "dead time"' (Moseley and Packham 1984, p.85). Providing accessibility by public means is therefore a socio-political task which does not pay. Financing is achieved either through direct public subsidies or by transversal subsidies, i.e. the profits from transport in one region cover the losses of another. Apart from major communication routes in urbanized areas, public transportation is not profitable; hence it tends to be sparse (weekdays only), limited to certain hours of the day (early in the morning and in late afternoon), and generally slow because of frequent stops. According to Huigen's study in the Netherlands (1984, p.91), it offers access to only about a quarter of the opportunities a person can reach by using a motorcar; even the bicycle (a very common private vehicle in the Netherlands) is a more efficient means of transportation. From this perspective, rural areas really can be considered as marginal, and this is a sort of geometric marginalization.

A further aspect of geometric marginality can be found in the political domain, a topic to be dealt with in greater detail below (Chapter 5). State boundaries are often paralleled by belts of limited economic activity or low investment in infrastructure for the reason of (military) security. It may be for the same reason that national parks and nature reserves are often found in similar locations, especially when the border runs through land used extensively or regions unsuitable for human activities. It is easier to protect nature when human activity is low or even absent (see Chapter 7).

Systems aspects of marginality

In the chapter on definitions, we have started with Andreoli (1992, p.24) who called marginal regions such areas as are 'located 'at the margin' of a system as regards its socio-economic features'. Both from her studies and from other authors (e.g. Andersson, 1992) one could believe that marginal is equal to rural. If this were truly the case, our discussions would be reduced to the definition of towns and countryside, the former being uniformly central, the latter marginal. This may be a valid statement at a certain degree of spatial resolution, but not at others, and it is beyond the point: it is no use to play one against the other; we have to concentrate on the changing interrelationship between the two. Besides, what is central and what is marginal, is also a point of perspective. For the production of victuals, the urban space is certainly less important than the countryside. Finally, there is no clear border between them, they merge into one another, and the limit is a zone of interference rather than a clear-cut line. By concentrating on socio-economic and socio-cultural elements, Kirk identified 'a number of 'urban' and 'rural' traits representing polarities' (Kirk, 1980, p.12). None of these traits is absolutely dominant in the rural or in the urban space; even cities have parks where some form of horticultural activity takes place, and the local pub is a service that is an important part of the countryside. Rather the two coexist and overlap to a varying degree (Figure 2.7).

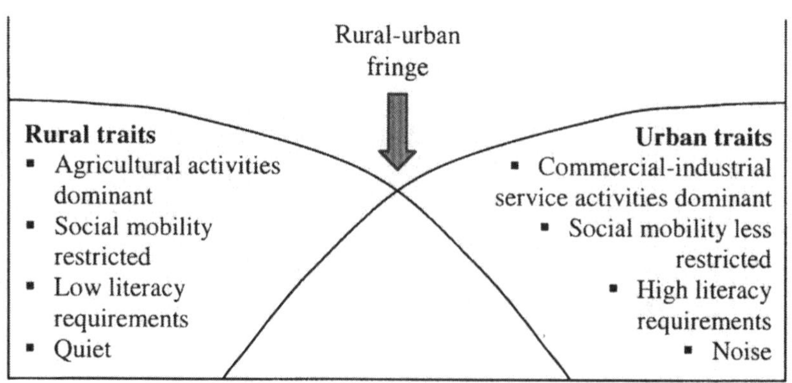

Figure 2.7 The transition between rural and urban spaces

Source: Kirk, 1980, pp.13, 15 (adapted)

It is true that rural areas are very often marginal, and they tend to become even more so in the future, but marginal areas exist in cities as well – it is a question of scale at which we want to study marginality. The countryside, however, tends to become increasingly dependent on the urban centres (at the regional scale), or on centres at a global scale. This is illustrated by the various transformation process in agriculture: the growth of large holdings at the expense of small ones, the emergence of a new class of absentee urban landowners (lawyers, doctors and dentists) who employ agricultural engineers with the objective of profit maximization, the production contracts of farmers with the agro-industry business, and the endeavours of agro-chemical companies to obtain seed and pesticide monopolies. More precisely, persons whose experience lies outside the rural world define agricultural production (the products themselves, the sowing and harvesting period, the production methods, etc.); the new decision makers are totally detached from the producers who are forced to accept the decisions. Under specific contracts, farmers may even be obliged to buy seeds every year instead of using the traditional method of setting aside some of the harvest for sowing in the following year, and to purchase the matching herbicides and pesticides as well. Such macro-scale dependence may guarantee production and marketing (i.e. income) to some extent, but it goes along with a loss of independence at the micro-scale (on the farm level). The case of vining peas in Britain is but one example (Robinson, 1988, p.262 ff.).

This last example demonstrates how macro- and micro-scales are closely interrelated. The micro-scale can be a very important level of analysis when spatial differentiations are to be shown that normally 'disappear' in more convenient macro-scale research, as has been shown by Forsberg (1998). Discussing rural/urban migration in Sweden, he pointed to the need to look behind net population figures. In the 1980s, Sweden had witnessed the counter-urbanization movement as most post-industrial countries. In the case of Sweden, however, not only did young families move to the countryside, there was also the reverse movement of elderly people to cities – which created sometimes the illusion 'that nothing had happened' (Forsberg, 1998, p.20). In this case, micro-scale does not necessarily mean to go down to the municipal level or even below that, but to take the entire migration movement into account, not just the migration balance.

Micro-scale analysis at the spatial level is important to show the dispersion of a specific phenomenon (e.g. the poor) or 'the degree of heterogeneity or internal variance within spatial units of analysis.' (Sommers and Mehretu, 1994, p.45). Such an analysis can only be conducted if the necessary data and the required spatial resolution are available. We shall present a few examples in Chapters 4 and 6 where this limitation will be clearly become visible.

Scale thus is also a methodological instrument in the analysis with a potential to manipulate information. Looking more into details and into micro-space permits to detect the true picture of marginality. Such analysis requires qualitative research, unless statistical data is available at the lowest possible scale. Microregional or microspatial research has received increasing attention in the past ten to fifteen years, and it is steadily developing (see e.g. Blom, 1998; Dahlgren, 1997; Mehretu and Sommers, 1997; Persson, 1998; 1999; Persson and Wiberg, 1995; Sommers, Mehretu and Pigozzi, 1995). Micro-regions are subsystems within macro-systems, and the systems view of marginality is therefore applicable.

Schmidt (1998) interprets the notion of marginal region in the sense Reynaud has used for his two special cases of *isolat* and *angle mort*. Her two assertions, 'the marginal regions which are practically outside the system', and 'marginality implies the character of something not being part of the system' (p.50) go well beyond Andreoli's statement quoted at the outset of this paragraph. Lying outside the system (which can be economic, political or social) means on the one hand that the system can operate without them, and on the other hand that such regions 'stay at the lowest point of their function, covering just the minimum requirements' (Schmidt, 1998, p.50). However, she does not exclude the possibility for marginal regions to emerge from their position and find their way back into the system. Internal or external factors may assist, e.g. the discovery of a new resource, the impact of new technology, a specific crisis. In such circumstances, a marginal region may become a periphery with several prospects: dependent and exploited, or evaluating and developing its own potential (ibid., p.52; Reynaud, 1981, p.77).

An interesting case in point may be the situation of Switzerland within Europe. Doubtlessly, the country is part of Europe, having shared the continent's history and living in the same cultural tradition. The economic ties are very strong (Germany alone is the destination of about 23 per cent of Swiss exports), and both socially and scientifically there are intensive contacts. Yet, Switzerland does not belong to the European Union, she lies outside this politico-economic system. The bilateral treaties concluded in 1999 and enforced in summer 2002 are strengthening the links (e.g through free mobility), but they are only a partial improvement of the situation of a non-member of the EU. Further bilateral negotiations are becoming increasingly difficult because they touch sensitive questions (the banking sector, the Schengen agreement on the elimination of border control). Since many Swiss do not consider EU-membership a priority, the country will keep its marginal position from this perspective. However, Switzerland is not a marginal part of Europe, although it lies outside the EU-system, because both in the capital market and in the European transit it plays a capital role. As a country outside the EU-system, Switzerland has played a pioneering role in European transit road traffic by installing lorry transport on the rails through the Alps – an idea that is gaining ground in other EU countries as well. It is not our intention to trigger off a discussion on Swiss–EU relations, but the example demonstrates the limits of a simplistic approach to the topic of marginality.

The question arises, if a region or a person can really lie completely outside any system. The answer is likely to be 'no', because at the macro or world scale, we have the ecosystem where every region and every individual is one element.

Processual marginality

The geometric and the systems approach are completed by a third view that we may call processual marginality. Mehretu et al. (2002) have developed a new look at marginality by looking at the processes lying behind human activity and its results. These processes can be either intentional (active) decisions, or unintended consequences of decisions and have to be seen in the context of the prevailing production system (in this case in neoliberal capitalism, i.e. Fordist and post-Fordist production and consumption patterns). From these processes result the two primary types of contingent and systemic marginality, while leveraged and collateral marginality are indirect types (p.197).

Contingent marginality is a result of the free market economy where the actors are exposed to full competition. Contingent marginality 'results from *competitive inequality*' (ibid.; emphasis in the original) and is the negative outcome of market competition. A person, region or country may become marginalized, but according to the free market logic, this is an unwanted by-product rather than an end-state; it is always possible to get out of it, given favourable circumstances. Time usually is a key factor in this context, another is free mobility (spatial and/or professional). This view corresponds to our conviction that nothing is definitive or static, and it takes up Reynaud's idea of inversion.

Systemic marginality (not to be confused with systems marginality, discussed above), on the other hand, is inherent to a system where 'hegemonic forces of a political and economic system ... produce(s) *inequities* in the distribution of social, political and economic benefits' (ibid., p.198; emphasis in the original). Totalitarian governments provide excellent examples for this kind of marginality: privileged groups or regions are favoured in the distribution of goods and services, the political roles are allocated within the ruling group irrespective of the individuals' competence, etc. The majority-minority cleavage opens wide when it comes to education, regional assistance, and social benefits. Systemic marginality is not self-correcting but tends to persist over a lengthy period of time, until the dominating forces at the origin of marginality are replaced.

Collateral marginality can be described as an unintended 'by-product' of a process. It is a sort of neighbourhood effect where a member of a majority might be inadvertently marginalized simply because of his or her presence within a marginalized minority neighbourhood or of his or her proximity to a vulnerable community (ibid.).

Leveraged marginality is to some extent intentional. The pressure exerted on individuals and groups by the economic actors (who demand higher profits and less costs) inevitably leads to marginalization. Mehretu et al. (2002) illustrate it by reference to the competition industrial labourers in the North face with regard to low-pay countries in the South (p.199). The phenomenon of 'new poverty' (see Chapter 6) may be its most vivid expression, but more visible is probably unemployment (documented by official statistics) that to some extent results from the delocalization of firms, but also from large-scale mergers.

The division of labour, which has been assuming a global dimension at the end of the 20th century and the second millennium, stems from the

Fordist production and consumption model patterns. Mehretu et al. (2002, pp.195 f.) point to an interesting evolution in the course of this century: whereas the Fordist model helped workers in the automotive industry to become, for instance, car-owners (those who produced the product could also afford to purchase it), in the Post-Fordist era, production and consumption model have been mismatched: those who produce (e.g. the car) cannot necessarily afford to buy it. Most of the world's purchasing power is concentrated in countries of the North (where the decision-centres are located), but low-cost mass production has been shifted to the South where large segments of the population have to survive on minimum salaries. While the Fordist process, to some extent, demarginalized labour, the Post-Fordist era is putting pressure on it. The scale of operation has changed, and so has the relationship between decision-makers and executing groups.

Chapter 3

World-views and Values

Before embarking on the discussion of various domains of marginality it is necessary to consider backgrounds that are not frequently mentioned in the scientific discourse, although they are – at least in our opinion – crucial for the debate. Ordinarily, outer manifestations of marginal situations are highlighted, such as segregation and deprivation, and the human and social drivers behind are identified, such as market capitalism or a dictatorial regime that wants to maintain its position of power. Sometimes, natural factors are added to those responsible, such as remoteness, mountain or desert situation, or natural catastrophes. There is no reason to doubt the validity of such factors to marginalization and marginality, but we believe that they are missing the essential point, i.e. the deeper roots, which are of a non-material or inner nature. Such factors are seldom in the forefront of explanation. It is as if a general fear of being ridiculed dictated our thinking when one speaks about spirituality and the supernatural world. The scientific world-view has indeed marginalized so-called pre- or non-scientific outlooks and relegated them into the realm of fantasy or even superstition.

In recent years, the wall that has separated the ways of thinking in the natural sciences from those in the humanities since the Age of Enlightenment has begun to crumble. Until the late 20th century, reason triumphed over emotion, rational explanations were sought for every phenomenon and every process, and the European world-view gradually saw man as capable of achieving everything. As a result, we are caught in a fatal process of messing about with the Creation and its mysteries, and nobody can really tell what the consequences will be. Religion has lost its significance, mainly because the Churches have degenerated to institutions where worldly prestige and power have become more important than spiritual guidance, at least with many representatives of the upper layers in the hierarchy. The Enlightenment meant that science was liberated from a narrow dogma that limited investigations, but it shot beyond its aim and destroyed belief in the supernatural.

The pendulum is now swinging back, but the new tendencies are no return to a pre-Enlightenment situation, rather they intend to redress the balance. Whilst the difference between religion and science will be maintained, it has dawned even to 'hard' scientists that there may be overlaps between the two. Whatever exists in the material world has its correspondence in the non-material or spiritual realm. With its almost heretical title, the book by Fox and Sheldrake (1996), a theologian and a biologist, breaks new ground. Angels are not physical beings; they 'have no place in a mechanistic world, except perhaps as psychological phenomena, existing only within our imaginations' (Fox and Sheldrake, 1996, p.7). They can be understood as bearers of energy (ibid., p.141). Maybe angels represent a different state of consciousness compared to our modernist perception; we can infer this from the 'amazing rupture and perversion [that] have occurred in human consciousness in the last few centuries as we have attempted to divorce ourselves from our relationship to angels and spirits.' (ibid., p.26). They are as invisible as are radio waves, they have neither mass nor body – just like photons (ibid., p.21). To believe in angels is a considerable challenge to modernist thinking, it means 'to recognize the objective existence of nonhuman intelligences, and that's the challenge that faces us now.' (ibid., p.27).

The same can be said of God. Most of the time He has been personified, but in reality He is a supreme universal force, that force scientists are trying desperately to discover but will never find because of the limits to science. While this is a challenge to science, many scientists agree (maybe not in these words) that some sort of mystery is a necessity to keep their investigation going.

Accepting the limits to human scientific investigation means also to recognize the limits to growth (Meadows et al., 1972). Nobody can tell, where these limits lie, but they do exist, and we may well have transgressed them without taking notice of an eventual warning sign (Meadows et al., 1992). However, the sheer awareness of limits has lead to the idea of sustainable development and hence to organic (systemic) thinking that is much closer to reality than simple linear thinking (Wood, 1987), promoted by the Cartesian tradition. To consider interrelationships and to accept invisible forces as existing have become the central way out of current problems.

The evolution of the western world-view

Throughout history, humans have held a certain image of the world, constituted of the three essential components God (the supreme force), humanity, and the rest of the Creation (animals, plants, the a-biotic environment). They were usually organized in a strict hierarchy with God at the top and humans above the rest of the Creation. Christianity, following the Jewish tradition, bases its world-view on the Old Testament where is written 'replenish the earth and subdue it' (Genesis 1:28). The interpretation of this saying is that humans stand below God but above the rest of the Earth. In the course of time, this has been taken too literally: to subdue was equalled to exploitation. The result was the gradual destruction of our environment, in particular once the technical means had been invented to work in a rational way. However, to subdue does not exclude management, i.e. care.

The Christian world-view has undergone several transformations (Figure 3.1). Throughout the Middle Ages, the traditional hierarchy was upheld until in the 16th century. God occupied His supreme place, and mankind stood above all other parts of the Creation (Lea, 1994). Following the new thinking in the Age of Enlightenment, humanity began raise itself to God's level, while the rest of the Creation kept its low position. God was thus gradually demystified. With the new technological possibilities of the Industrial Revolution, He was relegated to a lower position than man, standing only above the remainder of the Creation.

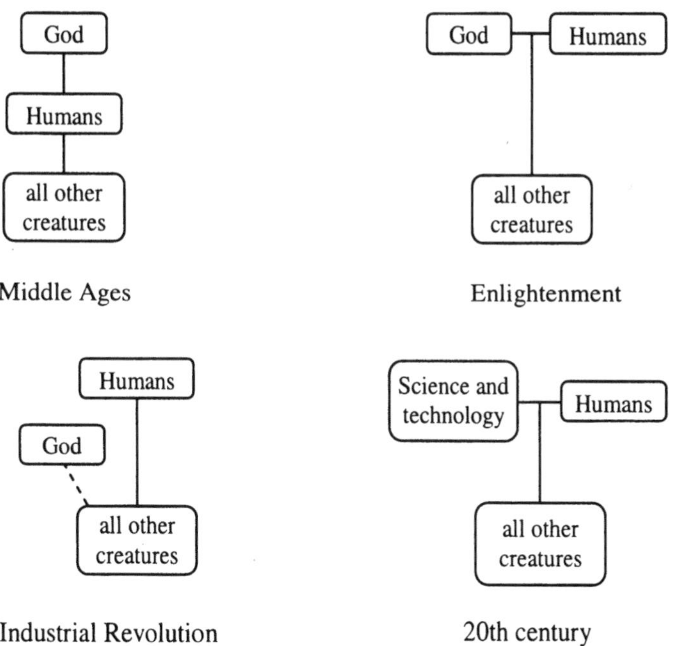

Figure 3.1 Christian world-views from the Middle Ages to the 20th century

Source: Leimgruber, 2003c (forthcoming), after Lea, 1994

A new God has emerged as a result of technological progress: science and technology have taken the place of God on the same level as the humans – this is the 20th century when all progress is seen exclusively as lying in the material world. Mysticism and with it God has completely disappeared to the profit of (seemingly) rational thinking. Simple solutions are being sought for complex phenomena, and the complexity is broken down (segmented) into individual parts that are studied individually, in a monocultural way (Shiva, 1993), with no regard to the entity they originally came from: matchboxes have replaced systems.

The question poses itself as to the future evolution of our world-views. Figure 3.2 presents three possible scenarios. They are all possible continuations of what has gone on before. The first scenario sees humanity relegated to a position below the New God science and technology. A second possibility is that man is eclipsed by science and technology, finishing as a subdued part of the entire creation, parallel to animals, plant and the a-biotic world. Science and technology would govern everything – this is in a way the basis for science fiction novels and films.

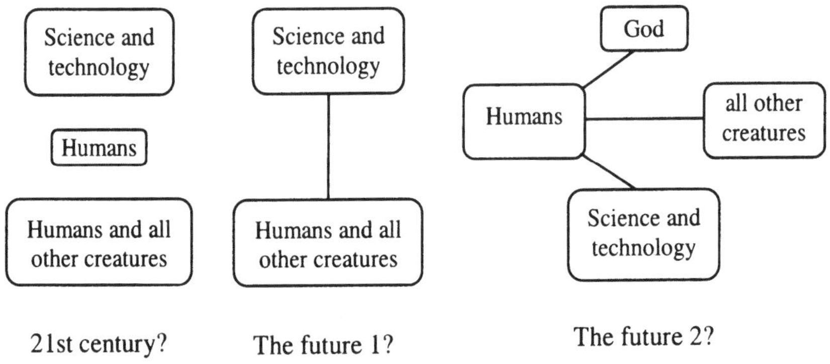

Figure 3.2 Possible future world-views

There is, however, a third possibility, and this is the one that seems to delineate itself gradually. It is a future world-view where God as the Supreme Force returns to His position, where humans begin to understand that the rest of the Creation is not simply below them but has to be recognized as a partner in the ecosystem, and where man controls science and technology. God would then return from the marginal position He has taken for some time, and the human creations would be put in their right place, i.e. subdued to human control to avoid negative influences on the rest of the Creation.

The first two of these alternative scenarios represent a rather gloomy future, such as would probably result if technological progress continued in the same unchallenged manner as it does at present. Instead of declaring that warnings are obstacles to progress, pronounced by uninformed pessimists, they ought to be taken seriously. All too often, scientists who had cautioned about the way science and technology might lead us, have been marginalized, as they did not follow mainstream ideas. Politicians, who have to take decisions, tend to follow the latter and ignore solitary warnings. Yet no member of the 'hard' scientific community who advocates this kind of progress knows where we are heading to, what the ultimate consequences will be. The fathers of the nuclear bomb at first admired the miracles of technology; when they realized the destructive potential of the new weapon and saw the consequences of its use, it was too late to stop further research. They knew what was going to happen, but they were powerless against a self-propelling process, promoted by huge funding from the military who in turn were supported by politicians. We may face a similar awakening and powerless scientists in other fields where long-term experience is still lacking (biotechnology, genetic engineering) – and when the negative consequences will eventually become visible, it may be too late. Science will further be centralized, and the richness of local knowledge will be lost (Shiva, 1993, p.9). Thinking in alternatives, that might cast shadows of doubt over traditional mainstream research is unpopular because it threatens the power of scientists who, in turn, have influence on politicians.

Elements of the third alternative are about to appear, and they even make their way into the economy, although not so much into politics. The growing awareness of the interdependence between the human system and the ecosystem, which has been heightened by the many natural catastrophes of the late 1990s and the early 21st century and by the many protests against 'globalization' (rather: the neoliberal market paradigm), has resulted in many reflections on the way humanity treats the environment and how it could better be taken into account in the future. This has, however, not yet been applied to the fellow human creatures: Human Rights' activists still have to denounce violations all over the world, and the way workers and employees are sacked because of mismanagement and reorganizations is also a sign that the 'social ecosystem' is not taken seriously. Maybe decision-makers have become afraid of the natural forces but still do not respect human beings as what they are, the image of God.

All this may sound very moralistic and out of place in a scientific book. However, scientists, politicians, entrepreneurs, generals, etc. are human beings who hold a particular world-view and follow certain values.

It is imperative to recall these foundations of our decisions and actions before it is too late.

Values as guides to decisions

The world-view is an essential part of human philosophy, and it inspires us to respect, or to despise. The evolution of the Christian world-views as described above mirrors the secularisation of thought and action over time: references to the invisible and the sacred have been replaced by the belief in the unlimited opportunities of modern technology, and taboos have been increasingly ignored.

This process is based on the value system inherent in society and individuals. Every society, even every individual lives and acts according to a set of norms and values that have been passed on from the past through parents and the elders, people usually respected for their experience. Values and norms are by no means static concepts that remain immutable throughout history. They evolve and change constantly, at times over short, at times over long periods, and the transformation of values is usually brought about by some external impetus: immigration, conquest, return migration, input via tales, books, the media, the Internet. The value system is an open system that requires external energy, just as every system depends on such outside sources. The history of mankind is characterized by such shifts in values.

Values are central to human decisions and actions. They are the sources of an individual's behaviour and of a society's norms, concerning the relationships to both people and nature. They are general principles and can be described as a complex of concepts which guide human thinking and actions, serving as fundamental guidelines that legitimate the rules of human behaviour, which itself is defined by norms (Chazel, 1988, p.125). Max Weber (1980, p.12 f.) calls them the elementary motives of human activity, the desired objectives or end-states of a process, standards and criteria of orientation. They direct the behaviour of the members of a social group (Hillmann, 1989, p.53 f). It is important to notice that values are not objective qualities but defined by human beings (Beattie, 1964, p.73). All individuals hold their own personal value system, even if they will ultimately accept the set of values adopted by the various groups (family, workplace, school, etc.) and the society to which they belong (O'Brien and Guerrier, 1995, p.xiii). Values underlie our varying interests and guide our knowledge systems, social behaviour, consumption habits, attitudes to work, etc. The bearers of values are the individuals, but every individual is part of a society to which it transfers his/her values which, in turn, will become a property of the respective society. It is imperative, however, to underline their subjective character. With their personality, certain individuals are able to transmit or even impose their particular value system on a group or even an entire society. In such cases, the desires and the interests of an individual actor are very important relative to the value he/she defends (cf. Livet, 2001).

Values exist as a duality or couple, i.e. they are characterized by fixed terms; we can imagine them as two extremes lying at either end of a continuum. Becker (1959, quoted in Hillmann, 1989, p.141 ff.) uses the two terms of 'sacred' and 'secular' to describe a comprehensive value system (Figure 3.3). These two terms must, however, not be interpreted as religious notions in the sense of a particular institutionalized religion, but rather as general *spiritual* terms, referring to specific attitudes and modes of acting. A society dominated by sacred values will experience problems when dealing with innovations, as it tends towards the conservation of inherited ways of life, whereas a secular society will have difficulties in understanding the requests of conservationists, because it embraces change and considers it an indispensable elements of social progress.

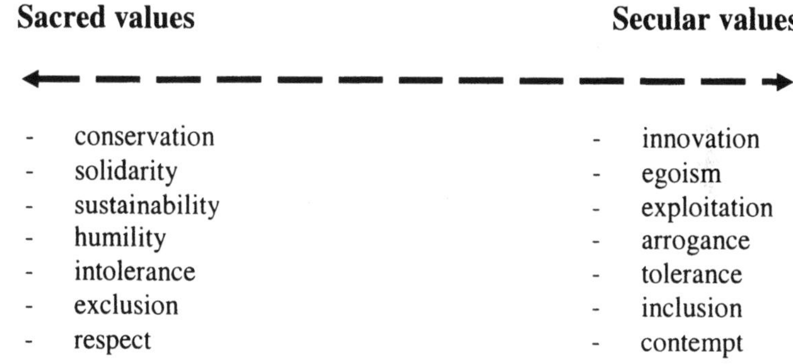

Figure 3.3 The value continuum

The same phenomenon can be described in different terms. Cunha (1988), for example, prefers to speak of the existentialist (sacred) and the productivist (secular) paradigm, whereas Fernandes and Carvalho Tomás (2001) use the couple 'ecocentrism' and 'technocentrism'. From yet another philosophical perspective, secular values correspond to the male or *yang* view, sacred values, for their part, to the female or *yin* perspective. Whatever terminology we apply, the two extremes are complementary (the *yin-yang* symbol is a perfect illustration of this fact) and represent extreme positions that are never fully attained; yet both have to be taken into consideration. Neither is good or bad, but imbalance in either direction can be detrimental: too much reliance on sacred values leads to stiffness and to stagnation, too much confidence in secular values results in fragmentation, restlessness and uncertainty. The evolution of a system can only take place if both sides interact harmoniously towards a dynamic equilibrium between the two. Hence every society, every human being needs elements of both extremes in order to survive.

Every individual and every society, but also every action can be placed somewhere along this continuum, and nobody remains on the same position through life. The value system is given by the cultural background, which depends on many internal and external factors (such as age, education, social environment etc.). A change of attitude (towards another person or towards nature) mirrors a shift along the value continuum.

The history of mankind is characterized by a steady shift from sacred to secular values (Figure 3.4), although in between, oscillations have occurred. The Romans, for example, operated much more according to secular values than the Germanic invaders after the end of the Roman Empire. From predator, man has in the long run gradually become an exploiter, while, currently, he ought to be a curator.

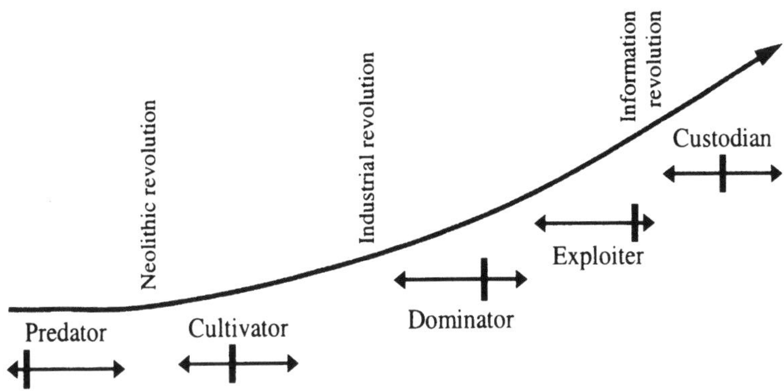

Figure 3.4 Changing value systems in the relations between humans and the ecosystem

A vertical line marks the dominating values: sacred values to the left, secular values to the right

Source: Leimgruber (forthcoming)

Technical innovations (the plough, the wheel etc.) have always meant a passage towards secular values and a greater impact of man on the ecosystem. Given the restricted number of people who applied them in daily life and the low efficiency of these tools, the impact of early innovations on the ecosystem and on societies was limited. When the philosophical basis for the natural sciences were laid in the Age of Enlightenment (since the 17th century), Man increased his grip on nature, and the achievements of the Industrial Revolution promoted the negative impacts on the environment. In his novel 'Hard Times', Charles Dickens describes 'the hardest working part of Coketown, ... the innermost fortifications of that ugly citadel, where Nature was as strongly bricked out as killing airs and gases were bricked in', thus offering a glimpse of air pollution in 19th century England (Dickens, 1961, p.72). George Orwell seconds him in 'The Road to Wigan Pier' where he writes of 'the monstrous scenery of slag-heaps, chimneys, piled scrap-iron, foul canals, paths of cindery mud criss-crossed by the prints of clogs.' (Orwell 1975, p.16). Domination was reinforced by the technological breakthrough of the 20th century and resulted in reckless exploitation of resources, including the pollution of air and water and the degradation of the soils. The pendulum had swung right towards the secular values that came to govern human actions. The ecosystem was thus viewed as a domain at the exclusive service of mankind, as possessing a simple use value: ecocentrism was replaced by technocentrism (Fernandes & Carvalho Tomás, 2001).

The new thinking of the 17th and 18th centuries has been accepted without much criticism because it promised progress and freedom from hard work. Apart from this achievement, however, it was accompanied by one important drawback: by dividing problems into singular sub-elements, it favoured thinking in segregated entities (exclusion) rather than propagating an inclusive (systemic) approach. The health sector offers a telling example. Conventional medicine tries to cure an illness by attacking the symptoms through the well-focused use of drugs, but it neglects eventual negative side effects that could be cured by the administration of complementary drugs, and it does not search for the deeper causes that cannot be dealt with by the application of a medicament. The human being is seen as a mass of individual organs, each in its matchbox, rather than as a system, a view that is the by-product of the segregation of the individual who was 'chopped up' into body, soul and spirit. Instead of looking at the three as a unity, each of the three was to be treated separately.

The much needed return towards the sacred side of the value continuum was begun in the 1970s, when successively the Stockholm conference on the Environment (1972) and the oil crisis following the Yom Kippur War in

Palestine (1973) exposed the vulnerability of humanity to uncontrollable natural and not so easily controllable human events. The explicit request for sustainable development, postulated by the Brundtland Report (World Commission on Environment and Development, 1987) and the alarm signals sent out by the Rio Conference (1992) have emphasized this process. It is true that not all political agents can follow these arguments, that it is possible to find scientists who still minimize the human impact on the environment – yet the problem had been exposed to the scientific community as early as the 1950s (Thomas, 1956).

The idea of sacred values has not been completely forgotten. When conservation areas are sometimes called 'sanctuaries', one remembers that sacred values include elements such as respect and protection for the weak. This recalls a pre-scientific world-view, when animals, plants and mountains were sacred, i.e. untouchable. To enforce the idea of protection, nature was said to be inhabited by spirits, elves, goblins, etc. who were not to be disturbed unless people were prepared to bear the negative consequences. This sounds very much like superstition, something no longer in line with enlightened and rational thinking. However, since humanity has replaced this irrational perspective on nature by the rational one, we face tremendous ecological problems. Nobody would pollute a pond if he or she believed that a spirit inhabited it. Spirits, of course, are immaterial, usually invisible, although they manifest themselves to certain people. More important than a personified elf, however, was the belief in its existence and the awareness that by overexploitation considerable harm could be done to him/her and hence to the pond. This apparently superstitious attitude was in fact a non-political measure of environmental protection, sanctioned by the entire society.

World-views, values and marginality

What does this chapter mean for the topic of this book? I have opened the Prologue with a quotation by the Brazilian writer and poet Paolo Coelho, drawn from a book that has procured him global fame. The intention behind this quotation is that we must not content ourselves with trying to identify marginality (and with it the processes of globalization and re-regulation) through obvious and visible indicators but that we have to look behind it and detect what are the hidden motives of the process, the situation and the region in question.

This short chapter sets the spiritual tenor of our argumentation: whenever we try to prove something by quoting rational arguments, we must bear in mind that there is some irrational background. Migrations, for example, can be easily explained by external factors such as overpopulation, high birth-rates, survival problems, ideological and political pressure, but behind every single factor lies the internal world-view of the migrants and of those who may force people out of their homes and country, can be detected the respective values. Why did many people attempt to flee the Communist regimes in Eastern Europe, why did others opt to stay? Why do some people leave an earthquake zone, but return soon afterwards and reconstruct their villages, whereas others prefer to emigrate for good? There is no simple answer to this question; every decision was taken by weighing individual factors that led to a decision: either to the attempt to leave permanently or the will to stay. Even environmental threats do not automatically lead to mass migration, and the wish to return home to the familiar surroundings may be stronger than the rational insight that an area is a potential threat.

Behind this chapter lies an urgent appeal that may help humanity to overcome the negative aspects of marginality at least to some extent, more precisely to come away from the flagrant inequalities of our time, reduce marginality to the unavoidable disparities, and emphasize its positive sides. It is an appeal to rethink our attitudes and behaviour towards the ecosystem and our fellow humans that has repeatedly been made in recent years. Thus, 'the value orientations within culture, accruing from the ways a culture solves its fundamental problems economically, politically, socially, and traditionally through the use of the past, are of vital importance' (Miller, 1998, p.271). There is not only one western (Christian, or White Anglo-Saxon Protestant) value system that can help to solve all problems, but there are many, different from one culture to another. The problems themselves are inherent to a society, a culture, and must be solved according to its own value system. From this perspective it is wrong to qualify the Amish society, for example, as backward simply because it refuses to consistently use modern technology. The Amish are not a society in a closed system but – and there is a difference – a relatively stable closed society (Miller, 1997, p.9), who live their life in an open society and are well aware of the achievements of modernity (Miller, 1992). However, they make deliberate choices, preferring to 'avoid certain specified types of technology', to stay out of the social security system, and 'to have their own schools, with their own teachers providing the appropriate education needed for the life of an Amish farmer' (ibid., p.146). Their resistance to the combustion engine is not dictated by the wish to remain a backward society forever 'but rather a spatial principle of not providing too much mobility for the group' (Miller, 1997, p.9). The Amish live in a 'traditional landscape' that is forged by their 'level of technology' (ibid., p.8) and mirrors their world view. To modernize them by force would be equal to killing their tradition and their way of thinking, in short to deprive them of their identity. As everybody, the Amish have the right – a Human Right – to their own values. This is a classical example of the conflict on the value continuum: tolerance (a secular value) is required that a group (the Amish in this case, but other groups or subcultures must be included) can maintain its orientation towards sacred values. While '[r]adically critical reflection on values and taken-for-granted folkways within academia therefore remains a perennial task' (Buttimer, 2001, p.13), certain values must not be imposed on other cultures simply because one specific society cultivates a feeling of superiority. We shall come back to this point in Chapter 8.

PART II
DOMAINS OF MARGINALITY

Chapter 4

The Economic Perspective

The primacy of economic thinking

'The love for profit dominates the world' – this sentence allegedly written by Aristophanes (c. 450–388 BC) has kept its validity until the 21st century AD. Since the Age of Enlightenment, thinking in economic terms has increasingly dominated the life of mankind in a fetish-like manner. Activities in daily life are first and foremost considered from the perspectives of profit, cost and benefit, utility, etc., i.e. essentially from a material angle. It is true that the economy occupies an important position in human life, but there are other elements besides, and it should certainly not narrow down thinking to the extent it has done since modern economics appeared.

A second critical point is that the theoreticians from the late 18th to the 20th century have always reflected in terms of the rational economic actor, the *homo oeconomicus*, whose decisions are based on perfect information and knowledge. The economy was considered almost an exact science, where deterministic laws and mechanical forces operated in society (Brodbeck, 1998, p.45). No attention was paid to the huge variety of factors that influence the economy and that cannot be grasped in simple mechanistic laws: world politics and natural catastrophes were excluded; besides, the individuality of every single agent, and the fact that the human being is far from rational and never disposes of the total information required for perfectly rational decisions were simply ignored. The human brain is far too complex to allow for complete rationality (which would be simple linear logic). It is composed of three basic types that have distinct functions and direct our thinking in combination. The reptilian heritage 'seems to play a primary role in instinctually determined functions', the limbic cortex of mammals 'plays an important role in emotional, endocrine and viscero-somatic functions', and finally there is the 'highly differentiated cortex ... of the higher mammals ... which culminates in man to become the rational brain of calculation and symbolic thinking' (Tuan, 1974, p.14). Looking at this heritage, one can understand that purely rational thinking is not possible. Man operates with a bounded rationality, a concept that has been developed by Simon (1957), based on ideas in psychology, in particular of the *Gestalttheorie,* and Piaget's 1948 study (Piaget and Inhelder, 1975). It has strongly influenced the work of behavioural geographers (Downs R.M. and Stea D., 1973, p.3 f). In his studies on Swedish agriculture and on the migration decision, Wolpert (1964; 1965) has demonstrated the validity of this approach.

Interestingly enough, economic theorists have hardly if at all looked at these findings as far as economic decisions are concerned, and the entire world of the economy continues to follow the traditional narrow outlook. The only field where psychology was extensively applied was in marketing: creating needs among customers is first and foremost a psychological task, and so is public policy designed to attract firms to invest in a certain region or locality (Leimgruber, 1990). Once this mechanism has been started, it becomes a sort of self-propelled chain of progress-constraints leading to permanent change (Brodbeck, 1998, p.236). Customers are accustomed to buy always the latest (i.e. best?) product and to dispose of the old (i.e. bad?) one. As a consequence, resources are wasted and garbage is produced, garbage that can only partially be considered a resource for recycling. What is modern has become central to our thinking, what is old is therefore discarded as marginal, uninteresting.

A number of globally active TNCs have, however, discovered that consumers are not standard persons and that tastes and demands vary across space, cultures, and time (Holton, 1998, p.50). The variety of consumer habits characterizes every field of consumption, and in the food sector it may also vary according to gender (Bernhard, 2002). Fortunately, a global consumption culture dictated by a few TNCs is not in sight, and it will probably never become reality. To live with diversity is an essential component of human life, and this holds good for our eating habits as well – Barberis (2002) praises precisely this aspect of the food that has a local connotation and is part of the landscape feeling. One can even go one step further: 'From standardisation comes boredom which in turn leads to desperation' (Pitte, 2002, p.23). From this results that the diet 'can rediscover an infinite variety on the surface of the Earth, in order to better satisfy the physiological and cultural needs of humanity' (ibid.). This is a powerful argument against genetically modified food. The tendency to simplify the food we eat seems to have reached its apex and return to a more balanced situation where the quality of food and its influence on health are taken into account. This, at least, can be derived from many small-scale bottom-up initiatives that have appeared around the globe since the 1970s (see Ekins, 1992 for examples from Costa Rica on pp.78 ff., Sri Lanka on pp.100 ff., and from Kenya on p.151 f.). The 'workshops for the future' that were organized during the national exhibition EXPO 2002 in

Switzerland also pointed in this direction. The soaring costs of the health service and the reorientation of agricultural policy towards environmentally friendly farming (organic production; see Leimgruber et al., 1997) demonstrate that the link between nutrition and health is increasingly taken into account. Warnings about forgetting traditional cultivation systems and local food practices have been repeatedly uttered (Garine, 1991, p.23). For example, the farmers in the middle Senegal valley in Mauritania did not like the taste of the type of rice that was forced upon them during agricultural development campaigns; 'they sold it in order to procure themselves broken rice that corresponded much more to their taste' (Garine, 1991, p.37). Food and health are inextricably linked throughout human life. Breastfeeding is the essential start into life for newly born babies. The battle against Nestlé's publicity for baby food and in particular milk powder to replace breastfeeding, which has been going on since 1977, demonstrates the awareness for this link (König, 2001).

Apart from the field of publicity, the consumer is not seen as an individual human being who enjoys the free will to decide and act according to his or her own (bounded) rationality, but rather as a machine that works along the predetermined logical lines of economic (seemingly objective) rationality. The free will has been marginalized, individuals who want to decide for themselves are clearly off the mainstream. The neo-rurals, who after 1968 left the comfortable urban life of their parents and retreated to abandoned mountain regions to lead their own life, usually as farmers, were a marginal group (Keller, 1999). Not all of them, however, managed to survive under simple conditions for a long time. It required a radical break with the past and the firm intention to forsake the comfort of the modern society to resist the temptation to leave.

Conventional belief is that market forces will automatically level out inequalities and lead to development. The condition for such an automatism is the free working of the economic forces, similar to natural forces, i.e. under no political, social, cultural or ecological constraints. A completely free economy according this theory would be entirely deregulated. Whilst this looks good in theory, it would result in complete anarchy, because it would presuppose a society where regulation is not necessary, every member behaving rationally as well. This is, of course, not the case. It is therefore futile to build an economy on such assumptions; the reality has shown that despite efforts to reduce or re-orient the regulatory framework, inequalities have not disappeared but are increasing.

This chapter is not destined to condemn the economy and show the many fallacies of its thinking. We realize that many leaders, both in politics and in the global economy, are aware of the fact that the economic system does not work like a machine; the problem is that they still hope to make it work like one in the future. On the other hand, it has also been recognized that profit alone cannot be the ultimate goal of an enterprise; entrepreneurs know that both the natural and the social environment are important elements in the economy, from the locality to the region, the state, supranational systems, and the entire globe. We shall therefore attempt to discuss in this chapter various sides of marginalization in connection with the economy, selecting examples on various scales.

Economic marginality: the global scale

The best illustration of economic marginality on the scale of the globe is to adopt a dualistic centre – periphery perspective and look at the Triad, the tripolar macroregional structure dominating world finance, trade and manufacturing (Figure 4.1). With the exception of Singapore, all members of this superstructure are located to the north of Rufin's (1991) new *limes* (see Figure 1.1). Among themselves, these three pillars of the world economy produce and exchange the majority of manufactured goods. 'In 1994, these three macroregions contained 87 per cent of total world manufacturing output and generated 80 per cent of world merchandise exports.' (Dicken, 1998, p.60). Their share has been steadily increasing since 1980 when it amounted to 'only' 76 per cent and 71 per cent respectively. The economic power of the Triad is further demonstrated by the fact that, in 1997, 48 per cent of world exports took place inside it, as 'internal' trade, and this figure has not substantially altered since. The concentration of economic decisions-makers in the countries belonging to the Triad is such that these people will often produce leveraged marginality in certain areas of the world because of their global outlook and activity space, and for being generally well informed about economic and political trends and processes. However, even if they are usually better informed than other economic actors, they are not free from erroneous judgements and personal aspirations, as the scandals around many firms and the breakdown of large enterprises in 2002 alone (Swissair, Enron, Worldcom, to name but three spectacular cases) demonstrate.

The Triad also dominates foreign direct investment (FDI), but in this field it is not so much the flow of capital inside the three major players but the fact that they dominate the entire FDI pattern. As Poon et al (2000) have shown, there has been a remarkable increase of FDI in areas outside the Triad between 1985 and 1995, and also a shift within the countries of origin of the capital. The authors put the Triad into perspective and demonstrate that a structure, identified at one moment in time, need not continue indefinitely into the future and must not be cemented in our minds.

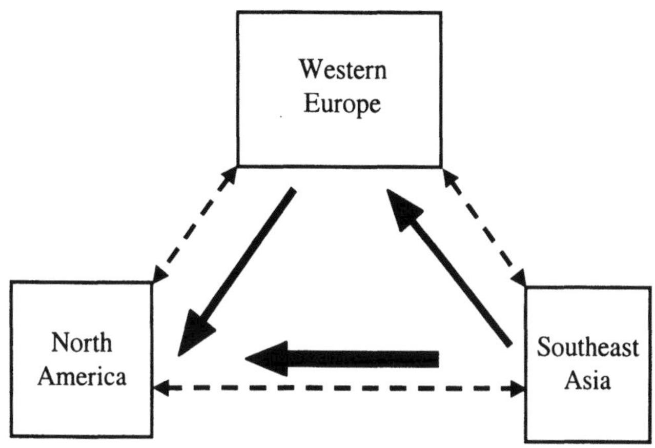

Figure 4.1 The Triad

The size of the rectangles corresponds to the trade volume (imports plus exports) of the three partners in 2001, the black arrows represent the trade balance between them in the same year with the arrow head pointing towards the partner with a negative balance (Europe in one case, North America in two cases)

Source: WTO statistics, 2002

On this global scale, the marginal regions are composed of those countries that lie outside the major information and merchandise flows between the three core areas of North America, Europe and East/Southeast Asia; they are the majority of countries and populations in the world, the 'new barbarians' (Rufin, 1991). The most marginal continent in this respect is Africa, extremely rich in mineral resources, but recklessly exploited by the colonial powers first and multinational mining companies since independence (see the following section). This marginalization is one of the great concerns of the New Partnership for Africa's Development (NEPAD, 2001).

From an organizational perspective, the Triad (which is an example for domination and exploitation) contains the potential for the successful control of the world market and of the world as such. The bulk of immaterial resources in the fields of management, research and development, and financing is concentrated in the North (Figure 4.2); although most of the production capacities have been delocalized towards countries of the South, it continues to preserve a certain production potential. The South is entirely dependent on the potential of the North, and from this perspective, the Triad retains its dominant position.

Decisions are taken in the North, which disposes of both capital and know-how, and where also the vast majority of consumers are located who can afford to purchase the goods put on the market, although little more than 20 per cent of the world population live in the North. The South, on the contrary, has been attributed the role of dependent producer. There is little to no room for independent initiatives. The northern management methods are also exported to and adopted by independent firms in the South in order to gain access to the world market and become competitive (Trap, 2002). Under the conditions of the old globalization, the old industrial division of labour (OIDL), the South played an important role in the export of raw-materials and manpower (slaves); with the advent of the new globalization, the new industrial division of labour (NIDL), however, raw-materials have become less important; emphasis lies on factors such as cheap labour and legislation favourable for northern investors (in matters of environment, working conditions, and capital transfer).

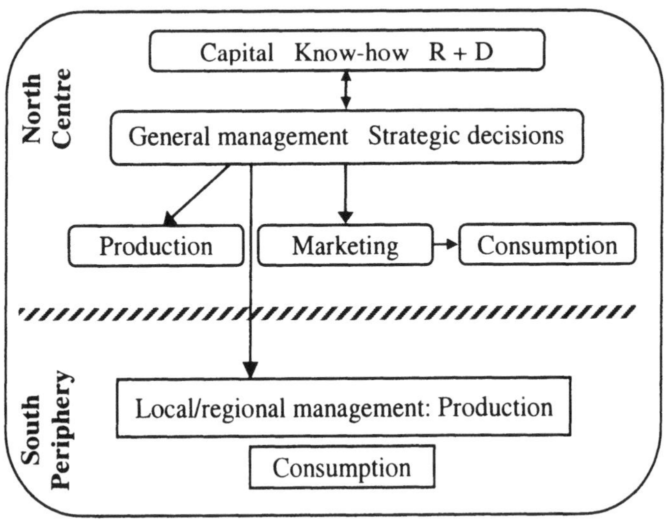

Figure 4.2 The North-South divide

Looking more in detail at the structure of the Triad, one sees that in fact the real decision-making centres (the cores of the core) are limited to certain areas within the three pillars (Figure 4.3). In North America, it is essentially the Northeast of the US and the adjacent part of Canada (parts of Québec and Ontario), parts of Texas, and California; in Europe, it is the 'European banana', stretching from Southeast England to Northern Italy with an offspring along the Mediterranean to Catalonia, and in East/Southeast Asia it is part of Japan, South Korea, Taiwan, and Singapore. There are cores within cores, just as there are marginal regions within marginal regions. Divisions run right across countries.

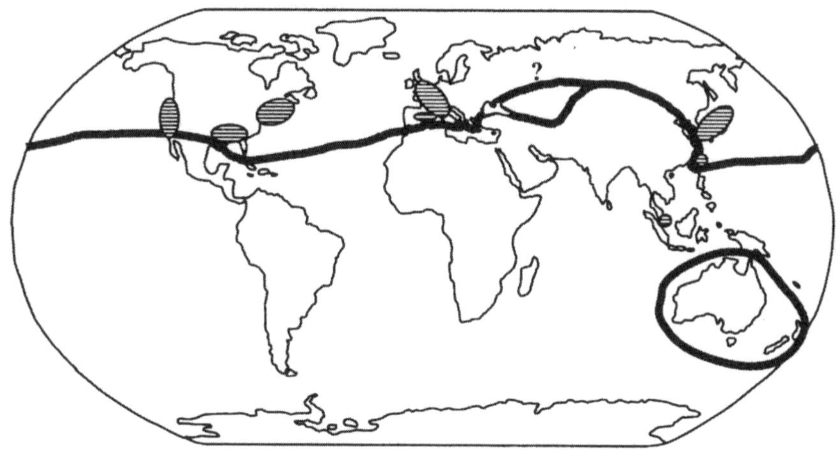

Figure 4.3 The true core areas of the world economic system and the contemporary *limes* according to Rufin (1991)

The example of the Triad illustrates in very general terms the relativity of the marginality concept and its link to scale. At this point it does not make sense to dwell on details of the North-South contrast; a lot has been published about it, very much will still be written on this topic because the gulf is at any rate widening.

North and South have become a metaphor for rich and poor, although these two terms do not mirror the geographical reality. In England, for example, the (poor) South is in the north and the (rich) North in the south. They stand, however, for asymmetric relations between better-off and less well-off regions or countries; in such spatial contexts, the geographic denomination behind vanishes. The map (Figure 4.3) is in a way misleading because it suggests that all the other areas are marginal, whether they are situated on the northern or on the southern side of the new *limes*. Within the countries of the South, there are important centres, just as there are considerable marginal regions in the North. Besides, the entire spectre of peripheries (integrated, dominated, exploited, or abandoned, see Table 2.1) can be encountered around the globe.

In a more precise way than the simple division suggested by Rufin (1991), membership in the OECD reflects another (and somewhat more realistic) dimension of the North-South division. Founded in 1961, the OECD (Organisation for Economic Cooperation and Development) at present comprises 30 members, all belonging to the North (Figure 4.4), albeit with considerable differences between them (Appendix 1). Contrary to the simple partition of the world in Figures 1.1 and 4.3, the true division is therefore far more complicated. The former Soviet Union and a number of her ex-satellite states have so far not been admitted into this 'exclusive club' of countries, nor has China. Most of the OECD members are economically advanced and relatively well off. Others are present for strategic geopolitical reasons. The European Union cannot be a member, but it is represented through its member states, and the EU Commission participates in the work of OECD. Africa, Central and South America, the Arab World and South and Southeast Asia are absent. This puts the global marginality into a new light.

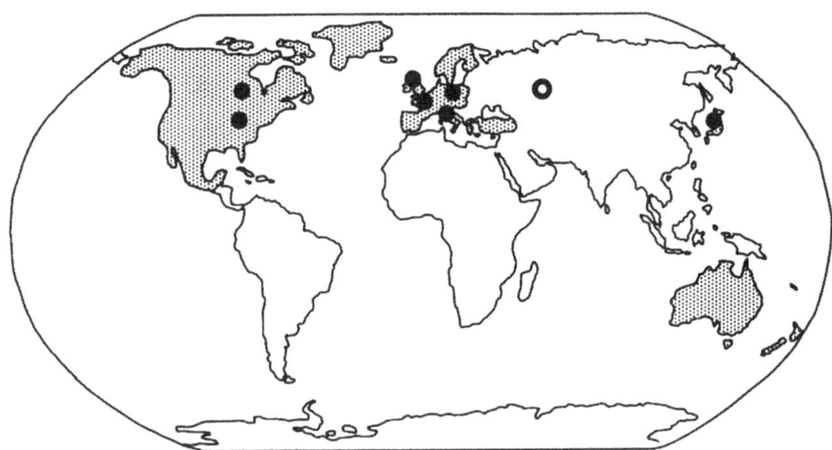

Figure 4.4 The OECD and the G-7 (G-8)

Even more restricted is the G-7 (G-8 with Russia), which comprises the seven most important industrial countries of the world (Figure 4.4). However, the G-7 is a forum for the exchange of ideas rather than a formal organization. Despite its informal character, it looks as if the largest part of the globe was even more marginalized than through the OECD, as seven countries determine the future course of the economy. Even though many of the decisions aim at an improvement of the economic and social situation in countries of the South, the G-7 practise a distinct top-down approach that is heavily criticized.

A shift from the global perspective to a larger scale takes us to international trade blocs. At first sight, they look like a sort of response to the simple division of the world into 'rich' and 'poor' as reflected in the Triad and in the OECD. Trade blocks, however, represent regional realities; they are a sort of 'transborder political economic spaces', usually composed of contiguous member states that conclude an agreement on economic cooperation and the reduction of tariffs and other trade barriers. Their aims are usually to promote economic growth and full employment, to facilitate trade and technology transfer, to reduce barriers to capital flow, to coordinate national economies – in short, they attempt to create a climate of peace and security through the promotion of the regional international economy. In some cases, however, hegemonic aspirations are behind such initiatives. A large number of states belong to one or even more regional trade blocs (Appendix 2). The engagement of the United States in both NAFTA and APEC demonstrates their key-position as a cornerstone of the Triad. This subdivision of the world political map into economic preferential zones can also be regarded as a sort of (leveraged) marginalization: countries that have little to offer or are viewed as politically non-acceptable can be excluded from such agreements. Again, this is a general view of the reality – rather: it is one of the many realities. A connection to geopolitical models (Chapter 5) can be detected.

By continuing the economic view down to the regional and local scales, we discover that the dualistic worldview is present at every scale and dominates the study of regional disparities. Inequalities in cities as the largest scale to be discussed in this context will be treated in Chapter 6 under the heading of urban poverty. Poverty is first of all measured in economic (monetary) terms, but to be privileged or underprivileged has also to be seen in the social and even in the psychological context.

Africa – the most exploited and marginalized continent

The initiatives taken by the African political leaders in 2001, to create the African Union (AU) as a follow-up organization of the OAU (Organization of African Unity) and promote the New Partnership for Africa's Development (NEPAD) is a reminder of the marginal position this continent holds in the global context. It can also be understood as a cry of despair, an attempt not to be entirely forgotten, an aspiration to endogenous development. Ever since Europeans have begun to penetrate the African continent, its riches have attracted both political and economic agents.

With their geopolitical vision of the *mare nostrum*, the Romans incorporated the southern shore of the Mediterranean into their empire. They spread their culture, the culture of the dominating and thus superior power and created a first global culture with the city as one of its primary characteristics (Rostovtseff, 1970, p.333 f.). The African provinces (Northern Africa) were considered the granary of the Empire and supplied most of the grain required – but little more (ibid., p.237). Although the Roman influence was limited to a narrow strip of land in the very north of Africa, it represents the first example of Europe-based exploitation of the continent.

Following the first exploratory voyages to find the sea route from Europe to India, the discovery of African peoples and resources has occupied the European rulers. The power situation on the Old Continent largely determined the way the outer world was explored, conquered, and appropriated. The Spanish and Portuguese were the first to establish coastal bases from which they developed trade relations with the interior. They were followed by the Dutch, the French, and the English. The characteristics of this early colonization were the construction of port cities and the gradual development of territorial corridors along the navigable rivers towards the interior (Taaffe et al., 1963). Model relics of this kind of strip development are the West African states of Gambia, Benin, and Togo (Harrison Church, 1981, p.263 f.).

Colonies were seen both as strategic possessions and sources of raw materials. Gold and spices were the early drivers for the risky voyages, but later, other goods were added. The emphasis was on the export of minerals and food products and the import of manufactured goods that arrived from the North. This pattern has survived until today. Very early, however, the South-North trade was supplemented by a sort of South-South trade when colonial powers introduced new food crops from the New World into Africa (e.g. manioc, maize and peanut; Hellie, 2002). Among the resources drawn from Africa we find slaves who were 'exported' from black Africa since the 7th century, mainly to Islamic countries north of the Sahara and across the Indian Ocean; up to the early 20th century, there may have been as many as 18 million for this trade alone (ibid.). The European colonial powers traded between seven and ten million from the 15th century to the formal end of slavery in 1867 (ibid.). The Europeans began to export Africans to Latin America when the representatives of the Church opposed the use of Indians (who were Christians and subjects of the Spanish crown) as slaves. The Franciscan (later Dominican) friar and bishop Bartolomé de las Casas was one of the most ardent opponents to this practice, and when the Spanish settlers, who wanted to keep their cheap labour, threatened him, he recommended the use of black slaves in the Spanish colonies (Berthe, 2002; Berthe et al., 2002).

The gradual penetration of Africa and the continuing exploration by natural scientists slowly revealed the richness of this continent in raw materials of all kinds, not only slaves for manpower but also minerals and elements of the vegetable and animal realm. Nobody has an idea about the quantity of material resources (excluding people) that have been drawn from Africa since the beginning of colonial exploitation. The fact is that the profits went exclusively to the colonial powers and their trading and exploitation companies – companies that have now been replaced by or have themselves mutated into integrated transnational enterprises where trade figures on specific raw materials are difficult (or even impossible) to be obtained. This pattern still holds today, even if a number of countries export industrial goods such as textiles and clothing or processed agricultural products. In every respect, Africa is a minor partner in the economy, for example in manufacturing growth, world trade, and foreign direct investment (Dicken, 1998, pp.29; 32; 47 f.). In particular, sub-Saharan Africa is in an extremely and increasingly weak position as her 'already tiny 0.6 per cent share [of manufacturing trade; WL] in 1970 has been halved' (Watkins, 1997) during the last decades of the 20th century. Very little value is added in Africa herself, despite the richness of the continent in both plant and mineral resources, and the profit reverts to the countries of the North, essentially Europe and North America. It is difficult to avoid the term colonialism in this context, but the current situation resembles it anyhow. The WTO statistics on merchandise trade in 2001 we read that Africa imported goods for a value of 127 billion US$ and exported for 141 billion US$. This is a tiny fraction of the total world exports of 5,984 billion US$. 56.7 per cent of the exports consisted of mining products; manufacturing goods amounted to 25.3 per cent, and agricultural products accounted for 14.9 per cent (WTO, 2002).

The extreme marginal situation of the entire continent can be seen from the Human Development Index (HDI). Of the 173 countries included in the 2002 Human Development Report (UNDP, 2002b), only one African country (Seychelles) figures among those with a high human development (HDI 0.811). 21 African states have a medium human development (one quarter of the 84 countries in this group), and 29 display a low human development, i.e. four out of five in this category. African countries occupy the ranks from 147 to 173. Civil-war-torn Sierra Leone ranks last with an

HDI of barely 0.275. The north and the south of the continent and a few countries in between are in a better situation than most other countries (Figure 4.5). Just to put things into perspective: the country with the highest human development, Norway, arrives at an index value of 0.942.

The HDI is based on three indices: life expectancy at birth (representing health), adult literacy rate and combined primary, secondary and tertiary gross enrolment ratio (education), and Gross Domestic Product per capita (economic performance). It offers a simplified overview but has been completed with a variety of other indices that allow us to draw a more detailed picture.

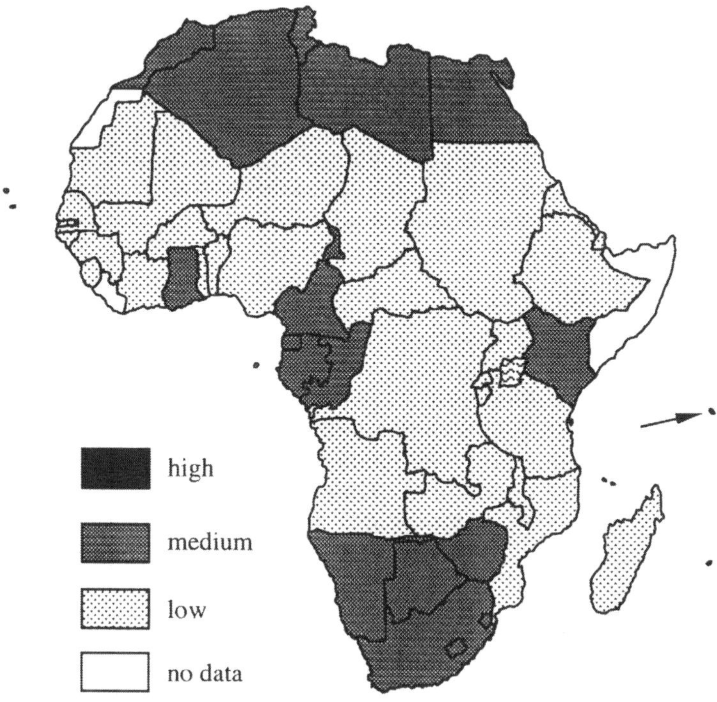

Figure 4.5 Human development in Africa: the HDI for 2000

The Human Development Index (HDI) is composed of the life expectancy, the education and the GDP indices. Index values between 1 and 0.800 signify high human development, values between 0.799 and 0.500 medium human development, and values below low human development. The Seychelles are marked with an arrow

Source: UNDP, 2002b

The map clearly shows that sub-Saharan Africa is the most disadvantaged part of the continent. In particular, life expectancy lies below 50 years in 22 countries, in Mozambique and Sierra Leone even below 40 years. Disease (including HIV/AIDS), insufficient nutrition (both quantitatively and qualitatively), and lack of clean drinking water combine with civil wars, accessibility problems and bad governance to hamper widespread improvement of living conditions. Even if the people were ready and willing to work towards this goal, their efforts are often ruined by external circumstances. Droughts destroying harvests and killing animals are natural causes that cannot be avoided, but belligerent action is a man-made reason that could be averted. War actions destroy harvests and kill animals, and they prevent the farmers from tilling their fields; in the aftermath of a war, minefields constitute a long-term danger and render large areas unusable for agriculture. Most African countries have been and are still ravaged by wars (Smith and Braein, 2003, p.60 f; 84 f.) – the result of European colonization, exploitation and abrupt decolonization after 1955 (ibid., p.86).

Against the background of this tragedy of an entire continent, we can understand the efforts within Africa to remedy such a desolate situation. In fact, the HDI has not changed dramatically over the past 25 years (Appendix 3), and the efforts to attain the Millennium Development Goals take a long time to show the needed effects – this, at least, can be read from a comparison of the HDI in Appendix 3 and the report on progress in Appendix 4. To obtain a better place in the international community and draw benefits of the globalization of the economy is therefore legitimate and figures as the most prominent goal of the African Union. By adopting the resolution on the NEPAD, the heads of states and governments of the member states of the AU pledged themselves to advance on a new road to development and seize the initiative themselves. For a long time, Africa had in fact enjoyed a very bad reputation, in particular as concerns the power struggles and politics in general. A new generation of statesmen has, however, emerged, and the will to work together seems to have become predominant. As a consequence, the condescending attitude of the North will gradually have to give way to a treatment on equal level: Africa is trying hard to reform herself and become a partner and no longer remain a subject. This point has been made quite clearly by the president of Burkina Faso, Blaise Compaoré, in an interview in the UNDP Magazine 'Choices', June 2002: 'The first point to make clear is that for us the development of Africa is primarily a matter of Africans.' Help by the international community is required, in particular to achieve political stability and overcome the internal conflicts that still haunt many countries.

Development is in this context understood as a broad concept, although economic objectives occupy a very prominent position throughout the document that details the ideas, wishes and actions to be taken. In particular, the spirit of collaboration is strongly evoked, both within Africa and with the developed world. Collaboration signifies that both partners – the North and Africa – take each other seriously, and that the mutual relationship is based on sharing and not on appropriation. Despite this optimism, the legacy of the past is still present. The exploitation of the mineral wealth, for example, still figures as a prominent goal in the document.

The NEPAD initiative has since received support from UNDP who opened a credit of assistance amounting to 1.9 million US$ towards the goal 'to raise living standards in the region through improved local governance, better global trading and financing terms and intensified attention to priorities such as education and health care' (UNDP, 2003a). While this amount may look important at first sight, if broken down, it amounts to an average of 38,000 US$ per country (irrespective of the size of population), or 0.2 Cents (US) per inhabitant of the entire continent – a sum that must be seen as an encouragement rather than effective assistance. Even the fact that 'the contribution is part of a broader $3.5 million project expected to draw funding from additional international partners' (ibid.) does not make things better. Real help, a point stressed in the NEPAD brochure, will require far greater financial involvement: 'The new long-term vision will require massive and heavy investment to bridge existing gaps.' (NEPAD, 2001, p.28, point 66). This money can flow out of public or private sources; given the history of Northern exploitation of the continent, it would in fact be the (moral) obligation of those countries that profited from this disequilibrium – not only the colonizers (!) – to contribute substantially towards the financing required to attain the goals defined and help the populations to escape from the distress they are currently experiencing.

Given the richness of the continent, both in terms of natural resources and indigenous culture, Africa could be an attractive region for investors. The authors of the NEPAD programme are, however, realistic enough to recognize that internal measures must be taken to enhance this attractivity, measures to encourage and secure domestic savings on the one hand, and 'conditions that promote private sector investments by both domestic and foreign investors' (ibid., p.50, point 146) on the other – obviously political and legal frameworks that encourage indigenous entrepreneurs and attract non-African capital and offer sufficient guarantees for a long-term engagement (ibid., p.52 f., point 155; p.61, point 169). More considerable support is likely to emerge from the Tokyo International Conference on African Development (TICAD) in September 2003 (UNDP, 2003b).

The NEPAD is an initiative to take Africa out of its marginality. It is based on the political will by a new generation of leaders, people who feel committed to their continent and are determined to reverse the process of marginalization that has prevailed so far. This will be a time-consuming and painful process, accompanied by setbacks every now and then, but the longer one waits, the more painful it will be. In particular we must not overlook the fact that the rest of the World moves on as well; the challenge, therefore, is not only to try to keep up but to reduce the ever widening gap.

Regional disparities within countries: the case of Switzerland

The continental perspective discussed in the case of Africa must not deceive us: at a different scale, every country is characterized by economic differentiation and unequal distribution of resources, wealth, jobs, training opportunities and other infrastructures, to say nothing of the unequal distribution of the population. This is a matter of principle and holds good for the rich and highly industrialized countries (such as Switzerland, to be discussed below) as well as for the poor and mainly rural parts of the world. The types and degree of disparities and their perception, however, vary from one country to another. People may accept such differences as a normal phenomenon, and in many cases there is little they can do about it. Emigration is the habitual way out of distress, but this road is not open to everyone. The decision to migrate is not only based on rational considerations (such as jobs or income) but is an individual appreciation of its positive and negative sides (Wolpert, 1965). Emigration, besides, does not solve the problem for the society concerned. Rather does it increase a marginal situation. The brain drain will reduce the growth and development potential. Maybe the individual migrant will not find the paradise he was looking for and will feel marginalized in his new surroundings. Far too often, migrants are victims to accidents on their way to the 'paradise', succumbing to the hardship of the journey or to inadequate means of transportation (unsafe boats crossing the Mediterranean, for example). Much more constructive responses would be to develop an active political approach (regional policy) or to encourage local and/or regional initiatives (grassroots movements) aiming at improving the situation. This aspect will be dealt with in Part III.

Most countries contain areas that are of limited economic significance, be it because of their nature (mountains, wetlands), because of difficulties of access (remoteness, topography), because their resources are exhausted, for cultural reasons (inhabited by indigenous people), etc. Such inequalities are often perceived as negative, and measures may be taken to relieve them. However, they can also be seen as a chance for a change. The perception, at any rate, is guided by the speculative stance an outside observer takes. If he arguments from a neoliberal position, he would see them as regions that have not yet profited from the trickle-down effects of the market economy. By using the polarization theory, on the other hand, he would explain the disparities with the self-fuelling accumulation in the centres that drain resources from the periphery. Which of the two theories is correct cannot be proven because no society is entirely deregulated and allows the market forces to operate freely.

Regional disparities must not be looked at with despair but as a challenge and a chance. A marginal region may offer sudden prospects that have not been thought of before. The history of tourism in the Alps, for example, is also the history of a changing perception of mountain regions (Leimgruber, 1992b). Mountains impressed man in quite different ways; they were at times sacred, later looked upon as terrifying, then as useful scientific and tourist objects, while in our time they are complex ecosystems facing degradation from the multiple menaces of the late 20th century lifestyle (see Figure 7.1).

The Swiss political system

Before entering into details, it may be useful to briefly sketch the political background of the country. Switzerland is a federal state with a highly decentralized three-tier political system: the Confederation as the central state, 26 cantons (regional states; see Appendix 5), and about 2,900 municipalities (communes), the local states. Political life takes place at every level. This is possible because three essential elements are decentralized: taxes, police, and education. The Confederation obtains its money essentially from indirect taxes; the direct federal tax (levied by the cantons) yields about 20 per cent of the total income. The so-called federal police is responsible for terrorism and crimes against state security. In the field of education, the Confederation runs the two polytechnics in Zürich and Lausanne, whereas the universities are run by the cantons. The central state has essentially a coordinating task. This is perhaps best visible in the field of planning. The federal planning law sets the general framework of the planning process, but the Confederation can only ensure harmonious coordination among the cantons. Concrete planning measures are taken by the cantons, whose directing plans have to be accepted by the Confederation. These plans are the basis for local planning where various zones (building zones, agricultural zones, and protected areas) are defined in a democratic process.

Political life is much more intense in the cantons and in the municipalities. The cantons collect direct taxes (on income and wealth), organize the police, and are responsible for the entire school system. As a result, Switzerland comprises 26 fiscal, police and school systems. This complicated structure can only be understood from the historical background; it is not very efficient in economic terms, but important for the regional identity. The municipalities also levy income and wealth taxes, according to the needs of their budget, and often based on the cantonal tax; they hold some responsibility for kindergarten and primary school, and they may also have a local police force. The cantons, however, regulate the degree of autonomy of the municipalities, hence there is not such diversity inside the cantons as one could suppose. However, with their fiscal autonomy, the communes can conduct an independent economic locational policy, as long as they respect their own local planning regulations.

The macroeconomic perspective

The common image of Switzerland is that of a rich country with a strong economy, resistant to the vagaries of the global economy. This seems to be confirmed by many economic indicators, for example within the context of her neighbour states (Table 4.1).

With 388,568 million Swiss francs (approximately 260,000 million Euros) in 1998, the GDP looks respectable, but it is only a fraction (about 15 per cent) of neighbouring Germany, for example. The idea of regional disparities will therefore sound strange at first sight. Poverty and deprivation seem to be notions that would apply for the rest of the world, but not for a country in the heart of Europe. Indeed, very much depends on the definition of such terms – a point further discussed in Chapter 6.

The very nature of Switzerland, however, calls for economic differentiations: common knowledge tells us that mountains are not endowed in the same way as lowlands, and rural areas are less well off than cities. It is not this obvious fact that lies behind disparities, however; it is the way human agents operate under certain circumstances. Regional variations exist within mountain areas and in rural space, and cities are not simply islands of affluence. The cultural diversity is another factor of differentiation: we must take language areas and religious adherence into account, a fact to be discussed in greater detail later in this chapter.

Table 4.1 Economic indicators for Switzerland and her neighbouring countries

Indicator Year	Unit	CH	D	F	I	A
GDP per capita 2000	OECD = 100	126	105	97	103	109
Industrial production index 1999	1995 = 100	112.2	110.2	112.4	102.9	123.2
Active population 1999	per cent	67.8	57.9	55.8	48.1	59.0
Unemployed 1999	per cent	3.1	8.9	12.1	11.7	4.7
Strikes 1999	Days	2,675	78,785	705,120	909,100	0
Consumption of households 1998	EU = 100	132	106	104	88	106

Source: Statistical Yearbook of Switzerland, 2002

Because of the negative connotation of the term 'marginal', one prefers to speak of disparities and thereby avoids to divide the country into central and marginal areas. Regional disparities point to differences rather than inequalities. They can be measured by a variety of indicators, such as employment, economic performance, economic growth, regional aid, economic structure, etc., and every indicator will yield a slightly different picture. This can be illustrated in the present context by the use of two different indicators on the cantonal level, the per caput national income, and the yield of the Swiss direct federal tax, a tax that is levied across the country according to the same principle, calculated every two years. The former informs about the economic strength of the individual cantons, the latter is an indicator for the income of the taxpayers. In either case the data are comparable on a national scale. In order to supplement the coarse picture of the cantons, we have selected a third indicator, the personal disposable income, represented on a regional scale, although the latest available data refer to 1980.

National income per head is the only figure that is available on a regional (i.e. cantonal) basis. It is of a very general nature, and the cantonal level does not necessarily mirror the sub-regional reality of the cantons with a large proportion of mountain areas. The disparities can be viewed from two angles: as a static situation (Figure 4.6) and as the evolution over a certain lapse of time (Figure 4.7). They will show that the two views are different, that the rich do not automatically become richer and the poor poorer.

The static perspective (Figure 4.6) is based on the deviations from the national mean and shows a 'classic' picture with the cantons of Basle (+ 69.3 per cent), Zug (+ 58.4 per cent), and Zurich (+ 29.5 per cent) dominating the country. The reasons for their economic strength are the chemical and pharmaceutical industry in Basle (nowadays largely consisting of R&D activities and administration), favourable tax regulations in the case of Zug, and the combination of a strong finance sector and machine industry in the canton of Zurich. The considerable overspill to neighbouring cantons can only be seen in part as it manifests itself predominantly on the sub-regional (municipal and district) level. Geneva as an international city with UN regional headquarters and many other international organizations as well as a strong position in private banking lies somewhat isolated from the rest of the country.

The picture is much more differentiated among the less fortunate cantons. While parts of Aargau and Vaud profit from their strong neighbours Zurich and Geneva, the largest part of Switzerland ranks fairly low. Two cantons (Jura and Obwalden) have more than 30 per cent less national income per capita than average (–30.9 and –34.6 per cent respectively). Although both lie in mountainous regions, this factor alone does not suffice to explain their marginal situation. Both cantons lack good accessibility and are situated outside major communication routes, but such overt factors again are only of partial explanatory value. Fribourg is essentially a canton of the Swiss Plateau and finds itself in the same category (–21.2 per cent). And the Valais, while lying entirely in the Alps, has some industry and a fair amount of tourism that should help bolster national income, but with a per caput income 27.7 per cent below the national mean, the reality is different.

The causes for these disparities must therefore be more complex and cover both external economic influences and internal conditions that are in part historically founded. Tax regulations can explain a lot, but regional economic history (traditional inclination for agriculture) and the influence of the general economic situation have to be added.

Jura is a relatively new canton; it was carved out of the canton of Berne in 1979 and has since been struggling to survive economically. The main industries (watch making, machines) have been in crisis for decades, and the location in a region with difficulties of access does not favour competitiveness. Although it is bordering France, there are few synergies from this seemingly favourable situation: the railway line that is connecting the canton to the industrial region of Belfort-Montbéliard and the major axis to Paris has never been electrified on the French side. A motorway is about to be constructed, but being of secondary importance from a Swiss perspective, its progress has been very slow. There is some tourism, but the region lies off the major tourist flows.

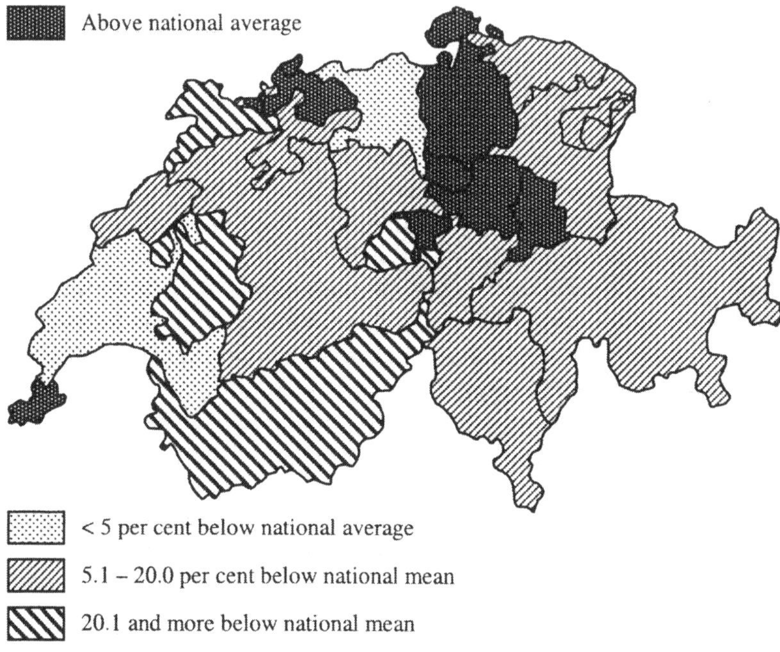

Figure 4.6 National income per inhabitant, Swiss cantons 1999: deviations from the national mean (CHF 46,620)

For the names of the cantons, see Appendix 5

Source: Statistical Yearbook of Switzerland, p.260

Obwalden consists of two portions of territory. The main part lies on a route that connects Lucerne to the Bernese Oberland, whereas the exclave of Engelberg is the upper part of a valley tributary to Lake Lucerne. Engelberg is known for its Benedictine monastery and as a tourist centre, whereas the other territory is characterized essentially by agriculture and small industries, bypassed by the most important communications routes.

Fribourg has developed an active economic policy in the 1960s only and has since tried and succeeded to attract a variety of industries. However, the canton finds itself in a vulnerable position, as the dynamics of the 1990s show (Figure 4.7) and has not been able to catch up. Even the canton's location on the linguistic 'border' between French and German (a cultural advantage over other parts of the country) does not help it, nor do the two major motorways and the main east-west railway line that transverse it.

The Valais is best known for its tourist centres (Zermatt with the Matterhorn, but also Montana and Crans, and Verbier, to name but a few); it has a reputation as a fruit and vegetable as well as wine growing region, is home to some major hydroelectric power stations, and also houses industries that are based on energy (electro-chemicals, aluminium). Decisions concerning all these economic activities, however, are taken outside the Valais: tourism depends on preferences and the economic situation in the regions of origin of tourists, the electricity produced is delivered to the main centres of the country and to the export market, and the industries are production units of companies located in Basle or Zurich and in competition with other units of the same enterprises. The Valais is a classical resource periphery with little decision potential of its own.

Figure 4.7 The dynamics of the national income in Switzerland, 1990-1999: cantons above mean growth of 22.1 per cent

Source: Statistical Yearbook of Switzerland, p.260

The dynamics of the national income per inhabitant in the 1990s put the static picture into perspective (Figure 4.7). It may not be astonishing that cantons like the two Basle, Zurich and Zug have grown considerably; what looks surprising at first sight is the evolution of the two Appenzell, the Grisons, and Neuchâtel.

Comparing the dynamics 1990-99 with the point of departure (1990) confirms both the strength of the two Basle cantons, Zug, Zürich and Nidwalden, and the continuous weakness of Jura, Valais, Fribourg and Obwalden (Figure 4.8). Geneva, on the other hand, falls short of the expectations suggested by the image of 1999. Neuchâtel, on the other hand, demonstrates considerable dynamism. This is due to the growth of the microelectronic industry that builds on and continues the tradition of watch making in that part of the Jura. A substantial part of the Swiss Plateau, particularly in the west, seems to be characterized by a relatively weak economic performance.

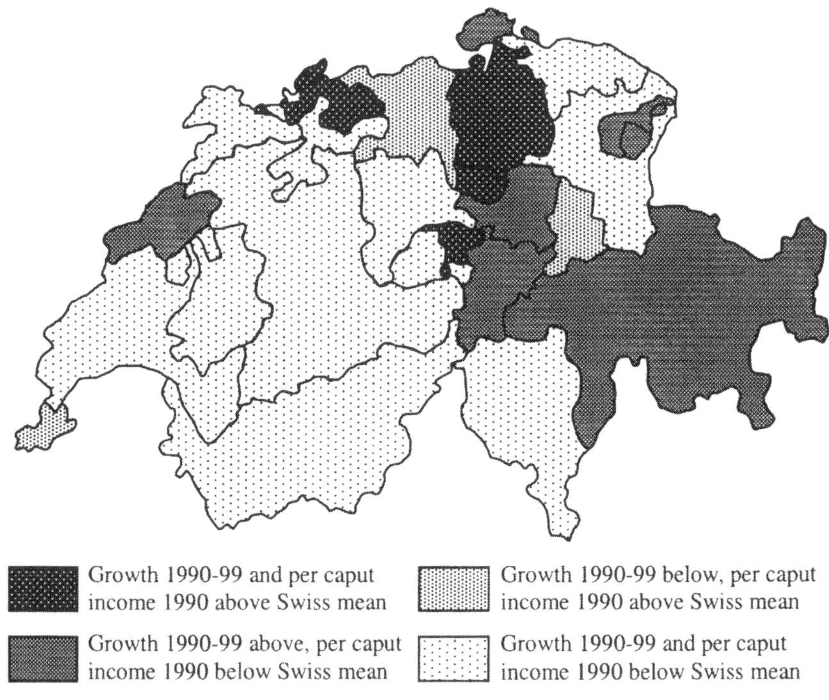

Figure 4.8 The economic dynamics of the Swiss cantons, 1990-1999

For the names of the cantons, see Appendix 5

However, the overall figures used for this analysis reflect only a partial reality. Wachter (1995, p.141) reminds us that the official database for the study of disparities is extremely narrow: the cantonal scale can mask considerable internal differences, and the fiscal autonomy of the cantons renders the comparison problematic. Cantons with a particular tax regime for holding-societies tend to rank high on the national income scale, although the economic activity is limited to some office space. The very high per caput-income in such cantons does not mean that the population is particularly wealthy. Besides, the overall figures do not take regional differences in price and income levels into account. A large canton like Berne, for example, is characterized by considerable differences between the capital region and certain areas in the hinterland (Figure 4.9).

Figure 4.9 Regional differentiation of the per capita Gross Domestic Product in the Canton of Bern, 1999

1 Jura; 2 Bienne and Seeland; 3 Bern Midland; 4 Oberaargau; 5 Emmental; 6 Bernese Oberland

Source: Cantonal Statistical Office, Bern

The canton of Berne reaches from the Jura mountains across the Swiss Plateau to the Alps and is the second canton of Switzerland with a surface of almost 6,000 square kilometres and 950,000 inhabitants in 2000. It is not a particularly important canton for industry and private services; the main activities are public administration (as the capital of Switzerland, Bern houses the federal government and many government agencies) and tourism, particularly in the Bernese Oberland.

The Capital region with the Bern conurbation is the most wealthy region of the canton; its per capita GDP in 1999 (CHF 46,800) is slightly above the national mean of CHF 46,549. Even in 1998 and in 2000 (according to provisional figures), the Bern and Midland region's GDP per person exceeded the Swiss average. All other regions fare less well, with the Jura (CHF 30,800) and the Emmental (30,100) at the bottom of the scale. The latter region appears thus as particularly marginal, a tendency that will be confirmed by the data on personal income (Figure 4.10) – a marginality that has a long-term character. Tourism, on the other hand, supplies the Bernese Oberland with sufficient income to occupy the third rank among the six regions, although the per capita income (CHF 32,100) is not significantly above the Jura. While the canton as such is not an economically strong part of the country, internal differences show that the weakest region is situated in the hilly to mountainous prealpine zone (Emmental) and in the Jura rather than in the Bernese Alps.

Personal disposable income: the consumption potential of the population

The view from the national economy masks the reality of the population that may not necessarily profit from the money earned by the 'big' economy. While the relationships between the performance of large firms and the income of the 'ordinary citizen' are not to be questioned, the GDP-figures do not tell much about the income and the level of living of the man in the street. It is, however, difficult to obtain statistical data on a spatial resolution that allows us to detect disparities on the regional or even on the local scale. Official statistics limit themselves to rather general levels – in Switzerland to the country or at best to the individual cantons. Large cantons with highly different living conditions are thus amalgamated to a general figure that tells little about the reality.

The detailed regional pattern of disparities has been studied in the 1980s by an economist research team, who undertook the difficult task to calculate the disposable income per inhabitant in 106 regions (Fischer et al., 1983; Fischer, 1985), using what is called MS-Regions (MS standing for *mobilité spatiale*, spatial mobility and representing essentially commuter regions). This indicator is not directly comparable to the national income used above: the personal disposable income is the amount of money available to a household or a person once direct taxes and the social insurance premium have been paid. It can, however, be considered as a measure of consumption potential. Unfortunately, these data have only been calculated for the period 1970-1980, hence we can only offer a historic picture (Figure 4.10). However, we may safely assume that the situation has not changed significantly during the 1990s.

Regional differences are striking, in particular between cities and the countryside. The poorest areas are found in parts of the central Alps, and along the limit between the Alps and the Plateau. They are rural regions, characterized by difficult production conditions in agriculture and problems of accessibility. An exception to this is the southernmost region of Mendrisiotto, bordering the economically strong area of northern Italy. The economy of this district entertains intense relations with neighbouring Lombardy; a considerable part of the workforce is composed of border commuters – 43.5 per cent in 1980 (Leimgruber, 1987, p.134), 42.2 per cent in 1998 (Demeter and Pancera, 2000, p.45), 13,499 or 47 per cent in 2002 (personal communication, Ticino Statistical Office). Usually they work for relatively low salaries, a fact that weighs heavily on the general wage level. Border commuters area a sort of buffer workforce, and their number varies from one year to another. In the entire canton Ticino, for example, they were 29,757 in 1980, 40,707 in 1990, but only 31,290 in 1996 (Torricelli et al., 1997, p.80).

The presence of wealthy areas in the (rural) Alps of the Grisons is due to tourism in places like Davos and St. Moritz and their surroundings. However, this economic activity is no guarantee for high incomes. Both the cantons of Valais and Ticino are important tourist destinations, but the mean disposable income per inhabitant lies below average in all regions – even the financial centre of Lugano does not stand out particularly. A further contrast can be detected between the western and the eastern part of the Plateau. The western part is economically weaker than the eastern part; the dividing line approximately follows an old cultural boundary (the Brünig-Napf-Reuss line) that has been identified by the ethnologist Richard Weiss, following his extensive studies on Swiss traditions (Weiss, 1947).

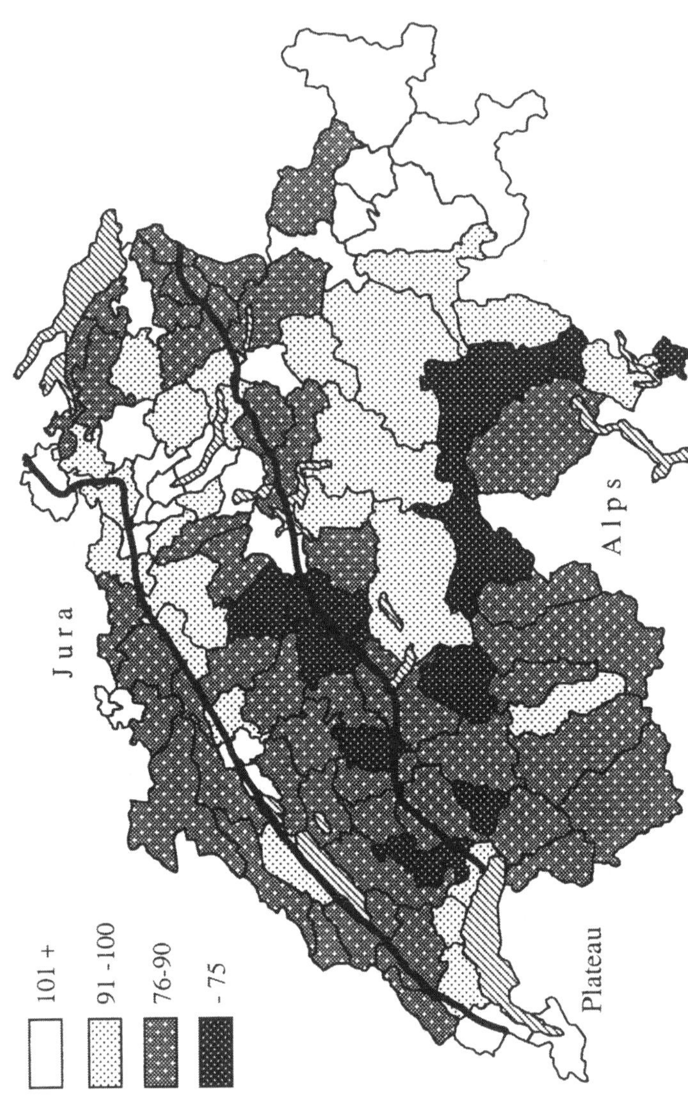

Figure 4.10 Personal disposable income in Switzerland per inhabitant, 1980 by MS-Regions (Index figures Switzerland: Fr. 17,506 = 100)

Source: Fischer, 1985

Tax revenue as an indicator of disparities

National income is a macroeconomic indicator; personal disposable income refers to the financial potential in a region for private consumption. Both represent the private sector, which is profit-oriented. The public sector, on the other hand, draws part of its income from indirect and direct taxes, used to finance the general basic infrastructural services in a large variety of domains. Indirect taxes are related to consumption patterns, while direct taxes mirror the income situation of a population. We can therefore use the latter as an indicator of regional disparities.

Public expenditure reflects the regional differences in infrastructural needs according to particular situations – in the Cantons of Geneva and Zurich, for example, police and fire-fighter expenditure is particularly high because of the security efforts in and around the airports (Leimgruber, 1995, p.19). On the income side, taxes are a data source that permits us to gain another insight into regional disparities. Due to the specific political organization of the country, Switzerland offers two distinct perspectives. On the one hand, we can look at the differences between the cantonal tax systems, on the other, we can study the income from the direct federal tax as an indicator of disparities. Cantonal (and municipal) taxes usually have fairly constant tax rates, but adjustments will occur from time to time, decided democratically through municipal assemblies or local and cantonal parliaments, with the option of a popular referendum. The situation in the year 2000 (Figure 4.11) thus looks somewhat different from the one elaborated for 1990 (ibid., p.18). There is thus considerable inequality in fiscal policy from one canton to another, but it mirrors what has been illustrated above with reference to national income.

It would be wrong to conclude from this map that the higher the taxes the more 'marginal' a canton. A comparison with Figure 4.6 reveals that the pattern is more complex. The case of Basle (the city canton) is very illustrative in this context. Basle is a rich and economically important canton (as can be seen from Figures 4.6 – 4.8), but the index of the tax rate on income and fortune amounts to 110.9, whereas the neighbouring countryside canton of Baselland has an index of 87.4. The city of Basle has always promoted a fiscal policy that favours low incomes; as a consequence, people with high incomes chose to move to Baselland.

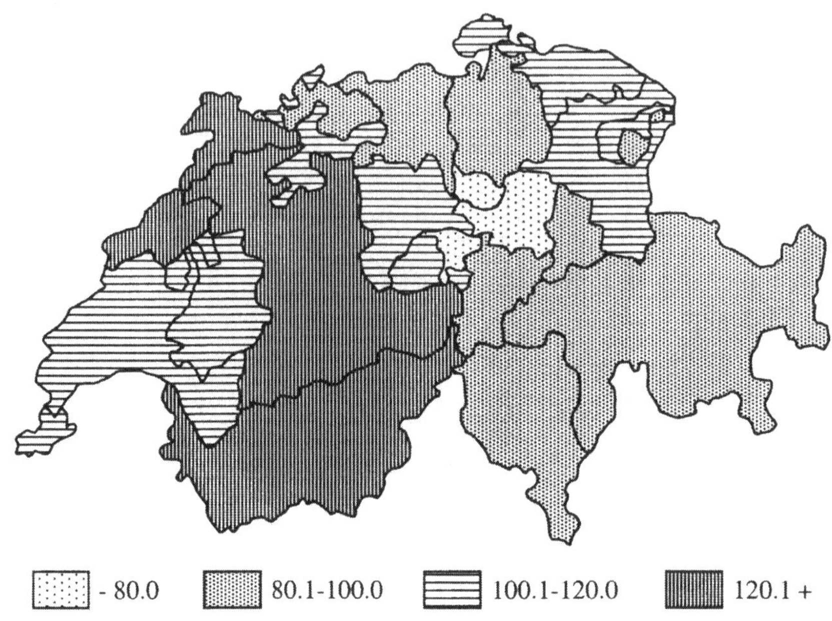

Figure 4.11 Index of taxes (total fiscal charge) by canton in Switzerland, 2000 (Swiss mean = 100)

For the names of the cantons, see Appendix 5

Source: Statistical Yearbook of Switzerland, 2002, p.823

In addition to the cantonal and local taxes, the federal government levies a direct federal tax that concerns both individual taxpayers and corporate bodies. This tax is calculated every two years, and its results are published for the individual municipalities and cantons. Despite limitations contained in the system itself, this is the only data source that allows us to compare relative welfare on the local, district and cantonal level and thus demonstrate regional imbalances that go beyond the coarse pattern of the cantons. Based on the same principle of calculation, the direct federal tax permits a countrywide comparison beyond regional legal particularities. The data allow us to look into inequalities on the large scale and hence discover rural-urban disparities within cantons such as Berne or Zurich that cannot be shown in Figure 4.11 where the individual tax regimes forbear a comparison.

The most recent federal tax data available cover the period 1997-1998 (Figure 4.12). The map once more displays considerable regional differences, and it confirms the information obtained from the private sector. Being a uniform indicator for the entire country, it can be compared to Figure 4.5. Not surprisingly, deviations are not very significant and most cantons can be found in the same category with either indicator, i.e. above or below average. There are, however, four exceptions that may retain our attention: Glarus and Schaffhausen have an above average national income per inhabitant, but a below national direct federal tax revenue, whereas Outer Appenzell and Vaud find themselves in the reverse situation.

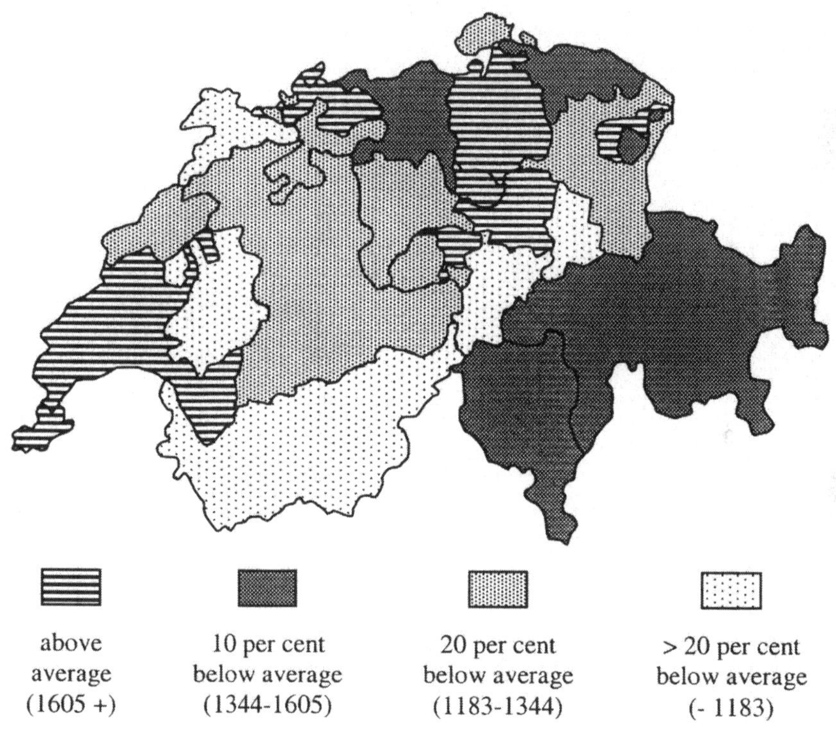

Figure 4.12 Direct federal tax in Switzerland: mean annual tax income per taxpayer by canton during the fiscal period 1997-1998 (in Swiss francs)

For the names of the cantons, see Appendix 5

Source: Administration fédérale des contributions, 2002

The first two cases may find an explanation in the presence of industry, which produces added value for the cantonal economy, whereas the latter two are related to the presence of a certain category of taxpayers that increases revenue. The cantonal tax law may attract persons with high revenue, but equally or even more important will be the quality of residence: Vaud offers unparalleled scenery and a mild climate along Lake Geneva, an attraction not only for the affluent business class and nobility, but also for diplomats and senior officials of international organizations who work in nearby Geneva. Outer Appenzell is close to the city of St. Gallen and offers similarly unmatched scenery in amiable hilly and mountainous surroundings. In either case, the cantonal tax rate and the revenue of the federal tax are above Swiss average, a sign that the (official) tax rate is not necessarily a determining factor for residence. The affluent have a set of preferences of their own.

The cantonal map suggests an image of homogeneity that can only be corrected by a change of scale. We have already pointed to this problem in the case of the GDP when we examined the canton of Bern (Figure 4.9). In the present context, we prefer to look at the example of the Zürich conurbation in order to exemplify the disparities on the large scale of the municipalities. We use three profiles (Figures 4.13 and 4.14) that allow us to detect the variety in tax income and the complex spatial differentiation in the wider Zürich region (see Table 4.2 for details and the explanation of the numbering).

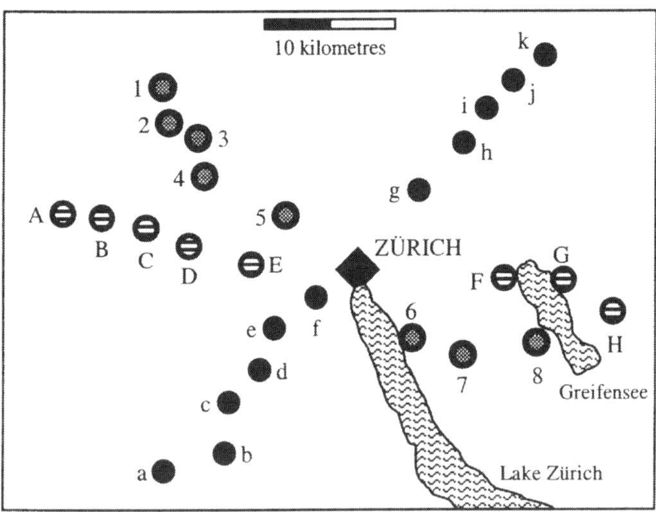

Figure 4.13 Greater Zürich and the three direct federal tax profiles (Figure 4.14 and Table 4.2)

From whatever direction we approach Zürich, the city always emerges as the municipality with the lowest per caput tax income or the most unfavourable (the 'poorest') place in the conurbation. The average tax revenue per taxpayer is situated 40 per cent below the Swiss mean, which places the town into the lowest category in the national context shown on Figure 4.12. Although tax revenues do not exactly mirror personal income, one can nevertheless find in these data the confirmation of the relative low mean standard of living of the population in the core of the conurbation. The same phenomenon appears at the eastern end of the west – east profile with the regional centre of Uster (H). Relatively low incomes are characteristic for towns, and even in the Swiss context poverty has become an issue (see Chapter 6). From this perspective (and from this one alone), towns tend to become marginalized.

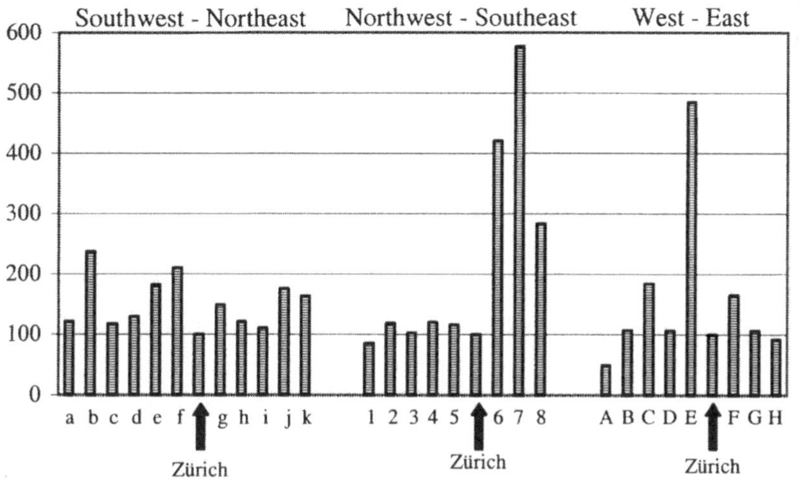

Figure 4.14 Direct federal tax: revenue per taxpayer in the Zürich region, 1997-1998, by municipalities (index values, Zürich = 100)

For the names of the communes, see Table 4.2

Source: Administration fédérale des contributions, 2002

Parallel to the negative result for the core of the conurbation, the profile shows where the wealthy taxpayers reside. They live in municipalities in the suburbs, either on Lake Zürich (second profile, Nr. 6, 7 and to some extent also 8) or on the western slope of the hill to the west of Zürich (Ütliberg; third profile, E). It is not by accident that the eastern shore of Lake Zürich has been nicknamed 'Goldcoast'.

The profiles demonstrate once more that 'marginality' is a matter of scale. Inside one of the richest cantons of Switzerland, such data (Table 4.2) allow us to document the remarkable differences that occur within an area of only about 30 kilometres in diameter and that are masked by the cantonal data used in Figure 4.12. Similar dissimilarities can be detected throughout the country.

Table 4.2 Direct federal tax: revenue per taxpayer in the Zürich region, 1997-1998

	SW-NE			NW-SE			W-E	
Nr.	Municipality	Tax	Nr.	Municipality	Tax	Nr.	Municipality	Tax
a	Ottenbach	1,178	1	Otelfingen	825	A	Künten	490
b	Affoltern/A.	2,301	2	Dänikon	1,156	B	Bellikon	1,035
c	Hedingen	1,150	3	Dällikon	1,003	C	Bergdietikon	1,792
d	Bonstetten	1,254	4	Weiningen	1,168	D	Urdorf	1,029
e	Wettswil/A.	1,769	5	Oberengstringen	1,124	E	Uitikon	4,717
f	Stallikon	2,044		Zürich	972		Zürich	972
	Zürich	972	6	Zollikon	4,094	F	Fällanden	1,600
g	Wallisellen	1,446	7	Zumikon	5,611	G	Greifensee	1,032
h	Dietlikon	1,179	8	Maur	2,748	H	Uster	891
i	Bassersdorf	1,076						
j	Nürensdorf	1,713						
k	Brütten	1,596						

Source: Administration fédérale des contributions, 2002

I do not want to be misunderstood: marginality is a normative concept, and in this example, its use is tied to the specific case of a rich country. The comparison is only possible within this context, not, however, beyond the national borders.

Unemployment – an unsolved problem

Unknown since the end of World War II, the spectre of unemployment has been haunting Switzerland for the past 30 years. The oil crisis in the 1970s brought the economic boom to a sudden end, and the firms found themselves confronted with overcapacities and surplus manpower. The first reaction was to cut down immigrated labourers, people who had been recruited during the 1950s and 1960s on a temporary basis. When their work permits were not renewed, about 200,000 persons had to leave

Switzerland for good (Leimgruber, 1992a, p.65) – a sort of unemployment in disguise (Figure 4.15). This concerned in particular workers with annual permits, whereas the transborder commuters were less involved. As regional manpower, they were an ordinary element on the labour market, and the boundary did not prevent them from keeping their jobs. Subsequently, also resident workers (natives and resident immigrants) started to lose their jobs. There was a call for an improved unemployment scheme, because the existing insurance had so far been voluntary until the slump of the early 1970s changed people's minds. In 1976 the unemployment benefit scheme was declared compulsory, and the law was passed by Parliament in 1982. Due to the persistence of unemployment, an amendment had to be made in 2003.

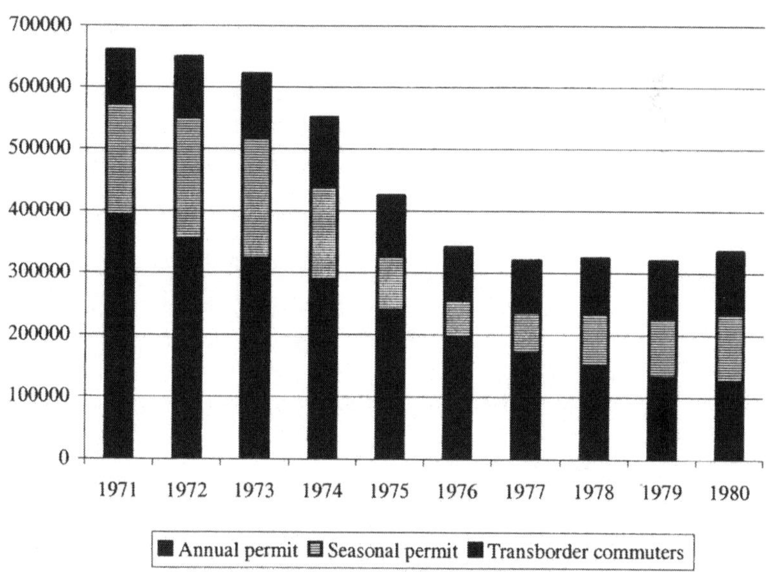

Figure 4.15 Immigrant labour force in Switzerland with limited work permits, 1971-1980

Source: Leimgruber, 1987, p.197

Unemployment depends on many factors. There is the seasonal effect (touching in particular construction and tourism), which usually results in rising figures from December to February; the economic situation in a country can lead to the reduction of manpower in certain firms; the influence of the global economy must not be underrated, in particular for firms and subcontractors that are active on both the domestic and the international market, but also for others; the quality of a firm's management may decide on employment or unemployment, etc. These factors usually interact with each other, and it will be difficult to identify one single cause for mass unemployment.

Despite first discomforting signals in the 1970s, the issue was not taken too serious at first. The foreign workers acted as a sort of buffer, and the proverbial strength of the Swiss economy ensured employment. However, the relative prosperity of the 1980s did not continue forever, and unemployment became a real problem in the 1990s. 'A real problem' means that the rate of jobless gradually rose from two per cent at the beginning of the decade to a record level of 5.2 per cent in 1997, but dropped again to two per cent in 2000. Compared to 8.4 per cent in the European Union in the year 2000 (with 14 per cent in Spain as maximum value), one can hardly speak of a real problem – it is a problem viewed in relation to the period of (almost) full employment before 1973 and during the 1980s, but every jobless person is one too many.

The overall figure tells us that Switzerland is certainly not marginalized within Europe. However, the regional picture within the country illustrates once more the economic disparities that were presented above. A further element in explaining these differences is the cultural diversity, in particular in language and religion. There is a considerable contrast between Italian and French speaking cantons on the one hand and the German speaking part on the other, but also within the latter the differences are at times not negligible (Figure 4.16). Unemployment figures fluctuate from one year to another, as can be observed from the confrontation of two situations in figure 4.15. 1997 was the year of record unemployment, whereas 2000 was much closer to what the Swiss were used to in the past. Within three years, the average unemployment rate had dropped from 5.2 to 2.0 per cent. The major regional difference (French and Italian speaking parts versus German speaking cantons) shows up quite distinctly on the 1997 map, whereas in the year 2000, the Canton of Jura (6.6 per cent in 1997) was among those with least unemployment (1.9 per cent in 2000). This drop, however, does not mean that the economy is booming; according to the law, unemployment benefits are paid for a limited period only (520 days according to the old law, 400 days according to the new law enforced in 2003, with an exception for persons above 55 years of age who lose their job: they will continue to benefit for 520 days); persons who remain unemployed after this period are supported by the social service and disappear from the unemployment statistics.

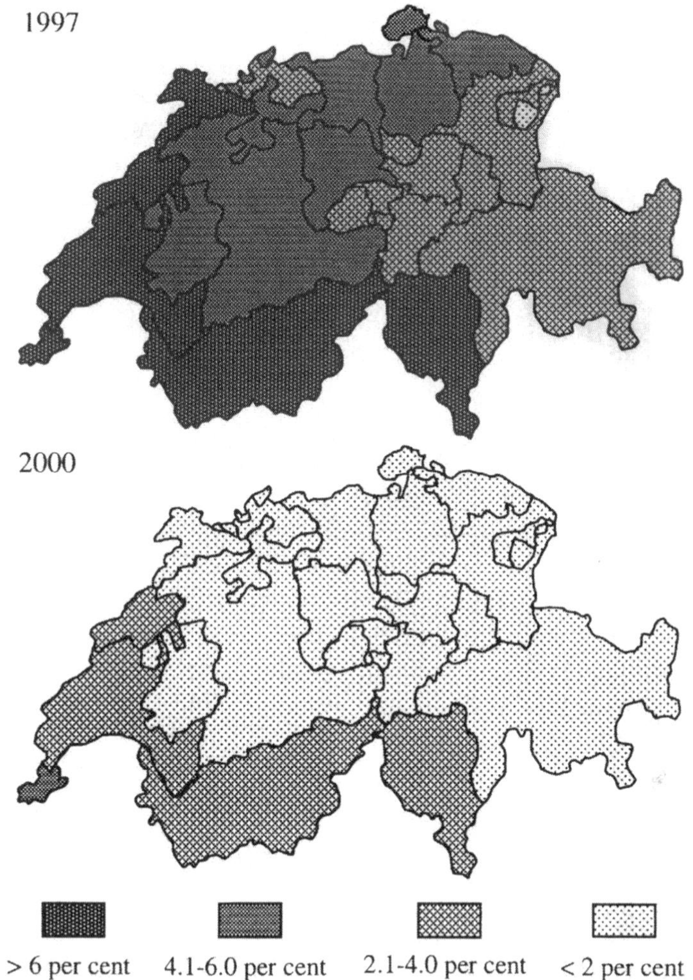

Figure 4.16 Unemployment rate in Switzerland by canton, 1997 and 2000

Source: Statistical Yearbook of Switzerland, 2002

An interesting case is the small cantons of Appenzell, in particular Inner Appenzell. With 1.9 per cent in 1997 and 0.3 per cent in 2000, this small territory sets a record. The economy in this part of the country is essentially dominated by small and medium enterprises in manufacturing and services (no firm with more than 200 full time jobs), and agriculture is still an important activity (one fifth of the active population). Not unexpectedly, the national income per inhabitant is therefore below national average (Figure 4.6). However, both Appenzell rank among the most dynamic cantons of the country (Figures 4.7 and 4.8). The fiscal charge lies below the Swiss mean in Inner Appenzell (Figure 4.11), and the revenue from the direct federal tax is above average in Outer and only slightly below in Inner Appenzell. Although both cantons profit from the proximity to the city of St. Gallen (with a university and some industry) and are close to the Austrian border (to offer jobs for Austrian transborder commuters), their economy is not particularly strong and thus less vulnerable, less exposed to the whims of the global market.

The role of the cultural factor

Despite her high rank in the world economy, Switzerland is not spared the problem of regional disparities. Whichever indicator we select, they will not disappear. As a consequence, there are marginal regions in Switzerland, as the comparison of various indicators has shown, but there is not really a vast region that can really be called depressed. However, poverty is also an issue in Switzerland, despite the low rate of unemployment. Just like marginal, poor is also a normative term that requires some reflection. To be poor in Switzerland is different from Nepal or Senegal. This question will be discussed in Chapter 6.

We must not underrate, however, the cultural factors that play an important role in the discussion of regional disparities. The political system that has been outlined at the beginning of this section, is intimately tied to the history of the country and, in particular, of the cantons. The political subdivision of the country is crisscrossed by many other cleavages, such as the rural-urban one, the language barrier, religious differences, or folk traditions. As a result, Switzerland resembles a jigsaw puzzle. Language will be dealt with in Chapter 6, but a few words about religion and traditions may be appropriate.

After the Reformation in the 1520s and 1530s, Switzerland was divided into catholic and protestant cantons, according to the principle of *Cuius regio, eius religio*. This rigid separation, which was the cause of several civil wars in the 16th, 18th and 19th centuries, continued almost uninterrupted until the 20th century. As long as people did not move around very much, the religious structure of the country was clear. When the rural exodus started in the 20th century, the religious orientation of the destination was not necessarily an important element in the decision to migrate. To find a new workplace and a residence was more essential than to live in a catholic or protestant area. As a consequence, there has been considerable mixing of religions, favoured by the freedom of religion guaranteed by the Constitution and a slowly emerging tolerance. While Catholics and Protestants do not share the same ideas about the economy, Max Weber's thesis (Weber, 1962) does no longer apply in the case of Switzerland. The differences that might have existed between the Catholics and the Protestants in their economic efforts have largely disappeared; what has remained, however, is the regime of public holidays, which are much more frequent in catholic cantons than in protestant ones. Besides, the religious element has also entered politics. A catholic and a protestant party (both with a rather conservative orientation) have emerged, but the latter is of very limited significance in national politics, whereas the former plays a major role. The explanation for this phenomenon can be seen in the strength of the non-clerical parties that defend the interests of the economy – they are largely dominated by Protestants who prefer secular liberal political affiliations to religious conservative ones.

While it is commonly accepted that the language border between French and German is a major cultural division, the cultural anthropologist Richard Weiss (1947) has shown that there is a cultural limit that is far more important. Based on the material gathered for the Swiss folklore atlas, he defined a line that runs roughly from the Brünig Pass (south of Lake Lucerne) along the cantonal boundary between (prostestant) Berne and (catholic) Lucerne. The *Brünig-Napf-Reuss* line is in fact the old boundary of the state of Bern prior to the French invasion in 1798. It divides the Plateau into two major cultural regions with a number of different practices. The cattle bred to the west of this line were traditionally spotted brands, whereas to the east it used to be brown cattle. In the late 20th century, the introduction of new breeds has changed the pattern of cattle throughout the country, and this characteristic has almost entirely disappeared. Other traditions have also vanished, such as the custom to eat potato soup as a rural Saturday dish (to the west of the line only) or the use of the yoke for oxen (to the east only). Their disappearance is a natural sign of cultural change.

One element has, however, survived and still reminds us of this particular cultural border: this line separates the type of playing cards used by the population. The so-called French cards (spades, hearts, diamonds, and clubs) are used to the west of the *Brünig-Napf-Reuss* line, whereas to the east people use the so-called German cards (acorns, flags, flowers, and bells). This is a small relict of what was once an important border, but it is still present in the perception of the population in the region. In a study of this particular dividing line, Schaller (1993) found that the mental map across the *Brünig-Napf-Reuss* line was much more distorted than parallel to it. This is an invisible border effect that mirrors the mental distance separating Protestants and Catholics even in our time. Space perception and playing cards are not very conspicuous and as such not recognized by mainstream research on cleavages in Switzerland (such as the study by Joye et al., 1992).

Concluding remarks

Marginality is related to disparities. While the latter describe various kinds of difference (economic, social, political), the former refers to the position of an individual, a society or a region within a hierarchy. The process of marginalization therefore stands for the loss of this position within a given hierarchy, viz. the downward movement from a higher to a lower status. It can be seen as the antithesis to development: an individual, a society or a region evolves towards a non-desirable state of affairs. Unless the people concerned feel hopeless, this situation need not be permanent and may even be considered as positive. Marginality can also be a matter of degree rather than of principle. Our western society of the early 1950s, for example, survived with less material goods and did not feel marginalized at all.

To some extent, therefore, marginality is a normative phenomenon, defined from a particular perspective (at present the one of Europe and North America) and measured via material goods. It is considered a negative attribute that has to be repaired. However, the northern perspective of the economy, to stick to the topic of the present chapter, is not the only one possible. It has been diffused across the world through the colonization process and enhanced by the globalization of the economy, and it is based on a specific lifestyle that has evolved in Europe and a number of her colonies (North America, Australia, New Zealand) and is propagated as the only lifestyle that leads to personal satisfaction and happiness. It centres on the individual and excludes the idea of solidarity within the community. It represents an extreme shift towards secular values. Removing regulations that are barriers to exchange and diffusion is the major task in this process: free trade areas and the new rules of the World Trade Organization concerning the cultural sector facilitate the propagation across the entire globe of one standard culture, one standard way of living – without, however, creating the necessary conditions that all citizens around the world will be able to profit from it.

This process can be observed on all scales; it is part of the idea of globalization, which includes the control of the economy from a few selected centres in the North, the unification of consumer habits across the world and the creation of a unified world culture (for which the term *McDonaldization* may be used, although it reaches far beyond the fast food business; Ritzer, 2000), the political domination under the leadership of the currently only remaining superpower – in short, the idea of *divide et impera* where one agent dominates the others by fragmenting them into individuals whose happiness lies in the hedonistic satisfaction of needs, both essential and manipulated. The paradox of our time is that on the one hand billions of people lack access to clean water and essential food, while on the other hand they are confronted via TV with the glamour and wastefulness of the northern society, and that this lifestyle is presented as a (theoretical) target of development they will never be able to attain.

The future may look bleak, but there is one comfort, however: human processes throughout history have been cyclical, characterized by the rise and fall of power and hegemony. Just as the economy has experienced booms and slumps at more or less regular intervals (the Kondratieff cycles; see Taylor, 1993, pp.13 ff.), political dominance of one nation-state will come to an end at a certain moment in time. No superpower will rule for eternity, although it will not possible to say when its decline will set in and be completed. All human cycles are probabilistic in nature; they do not respect a clear periodicity but oscillate according to context and social evolution.

Examining the long waves of history demonstrates that great empires have risen, maintained themselves for varying periods of time, and disappeared after some time. The 'hub of the universe' may have existed for many generations, but it has been succeeded by another political, economic and social core. Such waves may look like deterministic laws, but they are not because the duration of such a wave depends on the attitudes of ruling classes and the societies of an empire, and it may experience its ups and downs without collapsing. The story of the Roman Empire is a very good example: rising slowly from the 6th century BC, it reached various apices throughout its more than one thousand years of continuous dominance; it always recovered from temporary declines until at a certain moment in time the society had definitely 'overshot' and was surpassed by the events (the invasion by the barbarians). However, the cultural achievements of the Romans have lasted until the present. The Socialist Empire under the leadership of the USSR collapsed in 1989, and the Community of Independent States (CIS) is the almost helpless attempt to rescue a shadow of the former political and economic empire into an uncertain future. The economic outlooks of its member states, however, are directed towards the entire world. Yet, certain elements of social life established during the Soviet period would merit being maintained. The British Empire ended in the 20th century, but in the surviving organization, the Commonwealth of Nations, it has retained a faint reminder of its former splendour. The US Empire, about to be constructed with missionary zeal on the basis of economic and political factors (the carrot and stick approach), will thus come to an end as well, although we may not be able to foretell the date. The globalization of the neoclassical economy is still propagated as the only code to universal happiness, but the lack of solidarity, which goes along with it, begins to worry people to an increasing degree. Maybe the economy will have to turn back to the local and regional scale and come closer to the citizens. The dropouts on all social levels are an indicator for this need.

Chapter 5

Marginality and Politics

After the discussions of regional disparities and first references to regional policy it may make sense to carry on into the field of politics. This chapter is destined to look at marginality and marginal regions in political systems, using different scales, from global to regional and local. The political perspective is particularly linked to the economic sphere, as could be guessed from the section on Africa and on Switzerland.

The political perspective refers to the public sector, the views and ideas held by agents who are acting for the public good. While there is no doubt that politicians have good intentions, they are not isolated persons taking the best decision, but they are part of a network of people who support them and who in turn want their own interests to be satisfied as well. The art of the politician, therefore, consists in balancing interests, demands and possibilities without doing himself too much harm and thus met in peril his/her re-election – or more generally to endanger his/her power.

This chapter looks first at geopolitics and the changing global power game. Its second focus is on political boundaries and transborder relations. It concludes with reflections on war and its consequences.

Geopolitics: between dominance and control

Conceptual transformations of the geopolitical world map

Centres and peripheries are the two essential components of any geopolitical model that attempts to explain global and continental power relationships. Geopolitics can be described as the outward oriented component of a state's policy, based on power and security considerations, and aiming at securing influence and hegemony. Taylor (1993, p.91) calls it 'the set of strategic assumptions that a government makes about other states in forming its foreign policy'. The protagonists of geopolitics have thus always been the nation-states. Since Kjellén coined the notion in 1916, it has always been used in this particular spatial context. German geopolitics in the late 1920s and in the 1930s, that led along the road to World War II, was a telling example for this kind of world-view, but other powers have practiced geopolitics before. The colonial expansion of Europe (including Russia), for example, did not occur without a clear definition of political interests (both internal and external). Securing an adequate *Lebensraum* (living space) for its population was, according to Ratzel, one of the tasks of the nation-state; its political actions were thus guided as much by the worry for the survival of the people and the state territory as by reflections on hegemony.

Until World War II, the Ratzel-Kjellén idea of geopolitics could be regarded as valid. As long as it remained a theoretical concept, it mirrored the prevalent way of thinking, rooted in Ratzel's Darwinism (Boesler, 1983, p.37). The extreme interpretation undertaken by Karl Haushofer and its implementation by the German Nazis must be excluded from our discussion because they exploited a concept that was basically value-free and not ideologically bound.

World War II and its aftermath radically changed the outlook of geopolitics for the second half of the 20th century. The lessons from the war and the growth of international relations resulted in a gradual erosion of the position of the nation-state, a process that had been formally started with the creation of the Bretton Woods institutions in 1944. Subsequently, two new categories of agents entered the world political stage: top-down international organizations (the United Nations, the GATT and its successor, the World Trade Organization, etc.) and bottom-up non-governmental organizations (NGOs, such as the Amnesty International, Doctors without Borders). The former are aiming at a sort of world-wide governance regime (Newman and Kliot, 1999, p.4), promising to do better than the predecessor of the United Nations, the League of Nations, and the latter could follow the model set by the oldest institution in this field, the Red Cross (see Chapter 9). The number and character of the geopolitical actors have thus considerably increased. The system of international politics has changed from interstate relations to a complex network of public and non-governmental agents (Holton, 1998, pp.131 f.), including sub-national entities and even individuals. This complex interplay has been made possible and is constantly facilitated by the advances in communications and information technology. As a consequence, conflicts of interest began to take a new dimension: the traditional state-to-state antagonism (A versus B with their respective allies) was completed (not replaced) by the intervention of these new agents who try to mediate between the parties (Figure 5.1). Their role is an indicator of the growing importance of international law, which however deprives the nation-states of part of their sovereignty. By adhering to and respecting international law, the members of the international community agreed to share and defend common values for the benefit of humanity. Peaceful cooperation was to replace the violent resolution of conflicts. Its only weakness roots in the lack of an independent enforcement mechanism – the discussions around interventions in specific conflicts (such as Iraq, 2002 and 2003) illustrate the dominance of the old conflict paradigm (the law of the jungle) that still prevails over the idea of international law. The New World Order that might have resulted from the entry of the transnational agents is thus still characterized by considerable uncertainty. Taylor's vision (with a question mark) of a new hegemonic cycle is the reality of the future (Taylor, 1993, p.75), although humanity would like it different and would also deserve better.

Traditional conflict pattern

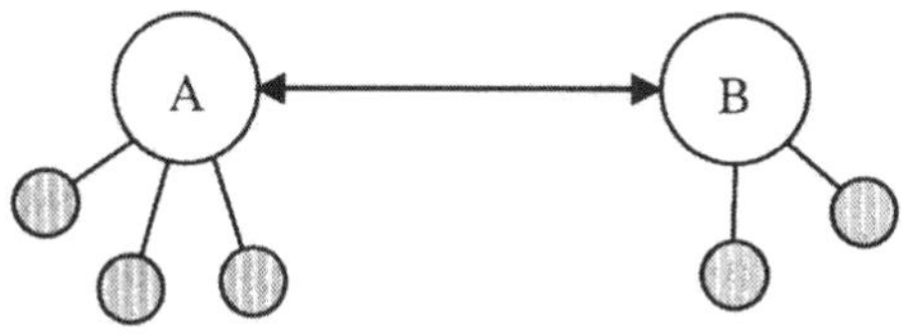

Post-WWI conflict pattern

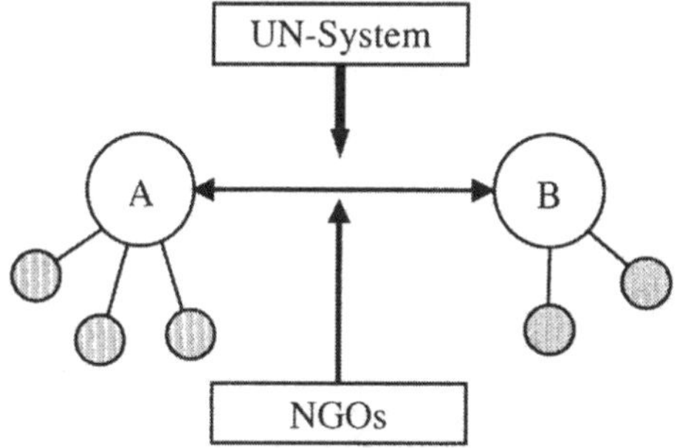

Figure 5.1 Traditional and modern conflict patterns

Source: Own elaboration

Discussing geopolitics in the present context thus looks like cultivating an historical discourse, which would be of little value to the future. What Ratzel, Mackinder and Kjellén wrote at the end of the 19th and at the beginning of the 20th century might be of little avail in the 21st century. We should be forward looking and accept the new world order based on international law as the norm for the future. The idea of hegemony of one power over the others would thus be part of history, and humanity would be united in one global band of solidarity. The only global geopolitical actor left would be the United Nations that would become a sort of world government.

Whilst this sounds good, it does not correspond to the reality in these first few years of the new millennium. When it comes to a critical point, humans are the same as they have always been, and in a certain sense, history repeats itself. The claims to hegemony by the world's only superpower left after 1989 (Blum, 2002) have thus not given way to global solidarity and respect for international law yet. The missionary zeal displayed by the United States parallels that shown by religious communities who are convinced that their faith (and only theirs) is the only valid doctrine. Automatically, therefore, those who do not embrace it will be marginalized and pursued, for the sake of what is deemed right for mankind. The phrase from the New Testament 'He that is not with me is against me; ...' (Matthew 12:30) echoes in President Bush's appeal after the 9-11 terrorist attacks on the World Trade Centre in New York and the Pentagon in Washington DC ('Who is not with us is with the terrorists'). G.W. Bush was in fact paraphrasing Adolf Hitler ... Geopolitical models based on the nation-state idea thus continue to be a reality – we are forced to go on discussing them.

Geopolitical marginalizations

All geopolitical models are based on the hegemony of some nation-states over others, i.e. on the centre-periphery model. Based on the idea of hegemony, they do not take inversion into consideration but display the short-term vision of the politicians. However, unexpected processes take place and result in geopolitical transitions. They have occurred throughout human history, even within the relatively short period from the late 18th to the late 20th century (Taylor, 1993, p.75).

Geopolitical models are comprehensive: every country is incorporated, none is excluded. Exclusion from a geopolitical perspective would result in a power vacuum that had to be filled immediately. Despite the process of economic globalization and the retreat of the state from many of its activities, the nation-state still maintains its role as a political actor on the global political stage. In the United Nations system, the US have tried to dominate ever since their major antagonist, the USSR, has disappeared from the scene and been replaced by the relatively weak Russian Federation, and they are the entire world as one large pan-region, naturally under their influence. They have used instruments such as withholding payments, suspending membership in organizations, refusing to adhere to an organ, or exercising pressure on other member states. The creation of international trade blocs (Appendix 2) has not eliminated the dominance of one particular state who sets the tune for the entire association: the US clearly dominate NAFTA, France and Germany are the main drivers of the European Union, and Japan cannot be ignored in the south-east Asian and Pacific world. It is not possible to separate the political from the economic domain; the two systems support and complete each other, and economic sanctions are very often used as political instruments to exercise pressure on states (South Africa during apartheid, Iraq). Therefore, trade blocs (Chapter 4) can also be regarded as a special case of geopolitical models.

Consciously overlooking a country is not characteristic for geopolitical considerations as this would result in the total geopolitical marginalization of an element of the international system. If by chance one peripheral component were forgotten or neglected – despite the comprehensive character of these models – it would not remain outside very long; either it would be reabsorbed by the system to which it belongs, or a competing system would try to get hold of it. On the other hand, a country (or the ruling class) can also feel neglected or marginalized and strive to find a new partner. The way certain African countries switched between allegiance to the USSR and the West during the Cold War is an eloquent example. In the long run, however, they have understood that they were exchanging one dependency for another. Marginalization can also be an active geopolitical instrument (e.g. via the use of food or military aid to ensure a positive attitude or even compliance with the rules dictated. It can, however, also be a double-edged sword: the US treated Iraq as a friendly country and contributed massively to its rearmament during the 1980s, but outlawed it after the invasion of Kuwait in 1990 (Pilger, 2002, p.65 f.).

Taylor (1993, p.50-102) offers a lengthy discussion of geopolitical models. To repeat it here would add nothing new to the argument. The only exception is the perspective of the pan-regions, a macro scale view of the world that goes back to ideas developed by the German geopolitical school of Karl Haushofer. 'A pan-region is a large functional area linking core states to resource peripheries and cutting across latitudinally distributed environmental zones.' (O'Loughlin and van der Wusten, 1990, p.1 f.). Various models of pan-regions have been developed, and the core is always in the North with the South attached as the resource periphery (see the various perspectives in ibid., pp.6 f; similarly Taylor, 1993, p.55, 62). We prefer a different view (Figure 5.2, based on Galtung's (1979) ten 'super-states' or spatial macro units, four to the North and six to the South of the *Limes*, and arranged from west to east (Taylor, 1993, p.86 f):

- United States plus Canada
- European Community plus the rest of Western Europe
- Soviet Union plus Eastern Europe
- Japan
- Africa
- Middle East
- China
- India plus the rest of South Asia
- South-East Asia with Oceania
- Latin America

The model requires to be updated in one important point: Russia and the Community of Independent States (CIS) have replaced the Soviet Union, but they do no longer represent the power displayed at the time by the URSS.

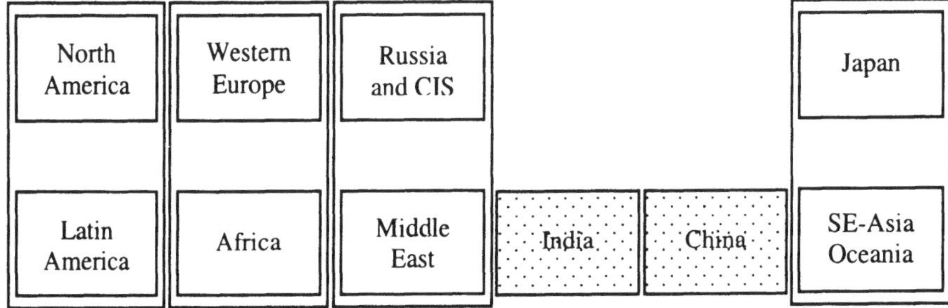

Figure 5.2 Geopolitical pan-regions

Source: Taylor, 1993, p.87 (updated and modified)

This scenario is explicit about North-South-relationships, showing quite clearly the intimate relations that have been built up over time between the two Americas (as a consequence of the Monroe doctrine) and between Western Europe and Africa (the former colonial relations). Japan is the dominant northern power over the rest of Southeast Asia and Oceania, although Australia may try to contest this inferior position in the model. The interests of Russia and the CIS (after the collapse of the Soviet Union) in the Middle East are obvious because this region is one option of opening Russia to the world, the old dream since Tsar Peter the Great. The claims have been reinstated over and over again, as the contacts with the countries in the region have shown. The oil in the region, however, leads to a conflict of interest with the western pan-region whose prime interests would lie in the western hemisphere. The US engagement in Afghanistan and the deployment of troops in former Soviet Republics obviously nourish a latent conflict between these two pan-regions.

The two southern super-states India and China are a case apart. They both occupy an intermediary position in the model, intending to affirm themselves as regional or even as future global powers. With over one billion inhabitants each they present a high demographic potential, and their forced technological modernization shows their will to catch up with the United States and Western Europe. Both are nuclear powers, and India is on the forefront in information technology, while China is a seemingly insatiable market for western products.

The model looks quite convincing and mirrors Rufin's (1991) new *limes* (Figure 1.1). However, it remains a theoretical construct and must not be viewed as static. Recent processes in Africa, for example, demonstrate a readiness of the countries of the South to liberate themselves at least partly from the yoke of the North, to break this vision of Eurafrica as 'order through dominance' (O'Loughlin and van der Wusten, 1990, p.18). Such liberation is a time-consuming process. For too long a time have the Africans been exposed to European perceptions and ideas, has their own culture been considered as inferior to others, and have they been prevented from bringing their own values and experiences into daily practice. Conflict solution through the traditional method of the palaver takes time and requires the readiness to compromise, to find common ground, but a palaver may offer the only lasting solution to a conflict. Is the North prepared to offer the Africans the time required to achieve true independence?

Experience shows that fragmentation is more common than the suggested unity (the *divide et impera* principle). However, unity (or uniformity) is also a reality. The adjusted Galtung model can be challenged from this perspective that has revealed itself still valid during the discussions about the invasion in Iraq in 2002 and 2003. I call it Orwell's geopolitical scenario (Figure 5.3), a model that George Orwell has used in his novel 'Nineteen eighty-four' (Orwell, 1954). In it, the real separation between Europe and North America runs through the Channel, a small but significant deviation from the axis of the Atlantic. We may recall that already Mackinder separated the British Isles from the rest of Europe and included them in the Outer Crescent (see the map in Taylor, 1993, p.55). Although both the UK and Ireland belong to the European Union, at least the UK is on the side of North America (the US, to be precise): the Channel is the boundary between the American and the European pan-region. A maritime power that once dominated the seas cannot become continentally minded within two generations.

Figure 5.3 Oceania, Eurasia, and Eastasia: George Orwell's geopolitical scenario

Source: Own elaboration

Oceania, Eurasia and Eastasia are permanently at war with each other, fighting for world hegemony. Britain is part of Oceania – this choice not only reflects the old bonds that have crossed the Atlantic despite US independence but also Orwell's roots in tradition. Throughout history, Britain has always taken a particular attitude towards the Continent by using its balance of power policy to influence events in Europe.

The US-UK coalition against Iraq in 2002/2003 is thus not a new scenario but continues an old tradition. The Channel is the boundary between the British Isles and the Continent, and it continues to make itself felt, despite British membership in the European Union. Throughout history, Britain has always viewed the Continent from a distance, intervening whenever her interests were endangered – not matter whether the ally was France or Prussia. The united front of the two major continental powers, France and Germany, does thus not constitute a particular surprise, as it offers France a chance to reaffirm her position as an important regional player. What may sometimes be amazing is how old structures survive temporary changes and resurface despite seemingly radical social transformations.

It has been said above that marginalization is not a geopolitical issue, unless one was prepared to take a power vacuum into consideration. The dominant powers in the North carefully watch how their 'subordinates' in the South act, and they will adjust their policies accordingly. Political measures may in many cases be replaced by economic pressure, given the supremacy of the North in this field (see Figure 4.2). Economic sanctions, usually a multilateral measure decided by the United Nations, are the most frequent means to urge countries to behave politically in a way that is compatible with the values shared by the international community. Unfortunately, the individual countries are not unanimous about the best values, and sanctions often fall short of their goal. The marginalization of Iraq through sanctions from 1990 onwards has in fact secured this country a place in the centre of global attention. Maybe in the case of South Africa, the sanctions (imposed in the 1970s only) helped the anti-apartheid struggle, but due to the vague nature of these measures and the geopolitical interests of the North in the southernmost region of Africa, the country was not totally marginalized.

Taiwan and Cuba – examples of political marginalization

Within the geopolitical considerations of the two major powers, the United States and China, the island states of Cuba and Taiwan occupy a singular position. Both are thorns in the flesh of their respective powerful neighbours, both have been able to maintain their independence (Cuba) or autonomy (Taiwan) for more than four decades, and both have developed successful social (Cuba) and economic (Taiwan) models. Although they are minor players on the global chessboard, their cases merit attention as they show how one can survive without being part of the mainstream.

Taiwan was the refuge of the Chinese nationalists that lost the war against Mao Zedong's Red Army in 1949. Chiang Kai-Shek retreated to the island and set up the Republic of China, claiming to be the true representatives of the Chinese people. As such, the Nationalists with substantial US support continued to occupy the Chinese seat in the UN Security Council. The international community (with the exception of a few countries), led by the US, largely ignored the existence of the People's Republic of China until the late 1960s. When eventually the US began to take a somewhat more relaxed attitude towards Beijing in the late 1960s, the fate of Taiwan was sealed. In 1971, the People's Republic replaced the Republic of China in the UN, in 1972, President Nixon visited Beijing, and in 1973, first diplomatic relations with the US were established (Winckler, 2002). These events demarginalized Mainland China from the World Political Map and at the same time marginalized Taiwan. They also enabled the People's Republic to become an important player in the global economy, a process triggered off by the reforms by Deng Xiaoping after 1978 (Wilbur et al., 2002). After having obtained recognition on the international level in 1971, Beijing became an important element both in world politics and in the global economy, whereas Taiwan's role was reduced to the economic field.

Mao Zedong had been unable to incorporate the island in 1949, postponing its conquest for a later moment (Winckler, 2002). Later, he must have regretted this 'omission' because it subsequently created an important source of conflict. The US had integrated the island into their geopolitical system from 1953 onwards (Cohen et al, 2002). From 1949 onwards, Taiwan was ignored by a number of countries who had recognized the Beijing government from the outset (the UK, the Scandinavian countries, Switzerland, the Netherlands), but most western countries supported the US attitude and ignored the People's Republic – the China-Taiwan issue divided the Western World. In particular, newly independent African countries were grateful for the technical assistance they received from Taipei and supported the Nationalist government through diplomatic relations and in the United Nations. By 1971, however, the international situation looked different. France, Italy, Canada and others had established full diplomatic relations with Beijing and broken with Taipei, and the US had undertaken the first steps towards full recognition. The power relations favoured the entrance of the People's Republic into the UN (ibid.). When President Carter closed the US embassy in Taipei and established full diplomatic relations with Taiwan in 1978, the island found itself in a grave situation.

Politically isolated, Taiwan remained internationally present thanks to its export-oriented economy. It was this domain that enabled the Taiwanese entrepreneurs to establish informal relations with their mainland counterparts and eventually arrive at some form of economic cooperation between Beijing and Taipei. While politically, the People's Republic continues to demand the incorporation of Taiwan (according to the formula 'one country, two systems', as applied in the case of Hong Kong; ibid.), economically there is a two-way relationship, and 'Taiwan has also become an important trading partner' (DeWoskin, 2002). The process of normalization is long and requires sacrifices and concessions from both sides.

The case of *Cuba* lies on a different level. The international relations of the island had since the 1959 revolution been coined by the antagonism between the US and the regime of President Fidel Castro. The Cold War period was particularly prone to cultivate a mutually hostile attitude: Cuba towards the US because of the desire to free herself from capitalist dominance, the US towards Cuba because according to the Monroe doctrine, the western hemisphere was under US protection and should not be open to European colonization. Socialism is not an American ideology, hence Fidel Castro's pro-socialist ideology could be interpreted as an European intervention. It is a matter of interpretation in how far the overthrow of the Battista regime could be called a colonizing effort by Europe, as President Monroe had called it in 1823.

The political and economic reality before 1959 was that Cuba was solidly anchored in the US economic and political system, nominally independent, but in fact almost a colony; the regime was a reliable ally, and the sugar production was controlled by US capital: Cuba was a dominated periphery. The revolution of 1959 attempted to free the country from US domination and become an autonomous socialist system. The US replied with sanctions and drove the regime into isolation. As a consequence, Cuba turned to the Soviet Union, the 'ideological fatherland' for support. The autonomy was thus short-lived, and the island became a sort of integrated periphery. Relations went across the Atlantic to the Soviet bloc and, increasingly, to countries of the Third World that were still colonies. The people in many colonies saw Fidel Castro as a model, and he lent support to the decolonization process. He was also an important figure in the movement of the non-aligned countries.

Cuba was marginalized by the dominant power of the capitalist world. The US passed an embargo in 1960, prohibiting the sale of US goods (including medical drugs) to Cuba, the import of goods from Cuba, any other commercial activity between the two countries, journeys to Cuba for the 'ordinary citizen'. This embargo more or less pushed Fidel Castro to seek support with the USSR who subsequently bought the entire sugar production of the island (Barthelemy et al., 2002). The sanctions have been maintained since their inception, albeit with temporary alleviations (Parolini, 2001, p.11). In the 1990s they were internationalized: the Cuban Democracy Act (1992) extended the embargo to non-US countries that were engaged in trade with Cuba, and the Helms-Burton Act in 1996 reinforced the 1992 measures, threatening countries that did not comply with unilateral economic sanctions. To increase pressure and thus promote marginalization seemed possible after the collapse of the Soviet Union.

The new geopolitical situation after 1989 had left Cuba as a remainder from the Cold War, but this relic has had the stamina to resist pressure from the mighty neighbour to the north. However tense the relations between the two countries, even Washington could not ignore the existence of Cuba. The 1970s witnessed a temporary ease of the embargo (ibid., p.12), and the case of Elían González in 1999-2000 re-launched the Cuban question in the US (Jatar, 2000). The visit by Pope John Paul II to Cuba in January 1998 and his subsequent discussions with President Clinton had led to somewhat more relaxed relations (Cameron, 2002), but the fundamental problem remains.

The complete and unilateral isolation of a country seems to be less and less likely in a globalized world. Arrangements on a diplomatic and peaceful basis are always possible, provided each party is ready to make concessions and accept the otherness of its opponent. 'To make peace with an enemy, one must work with that enemy, and that enemy becomes your partner.' (Mandela, 1995, pp.734 f.).

However marginal they are, Taiwan and Cuba are no white spots on the World political map. Their neighbouring powers carefully watch them, trying to take advantage from every new situation, and yet preferring the status quo.

Border regions: marginal regions?

Boundaries and border zones

The global scale is one particular perspective, but political marginalization occurs on the national and regional scale as well. Particularly prone to it – and mirroring the core-periphery dichotomy with regard to the nation-state – are areas on state borders. The boundary as a line separating two countries from each other has not necessarily been a place that attracted particular attention from the political class or from investors. The prime interest lay with the military that saw it as the defence line of their country. Traditionally, therefore, fortifications were a characteristic feature of the border landscape (the Roman *limes,* the Great Wall in China, the French Maginot line, the German Siegfried line, to name but a few), and they still retain the myth of bygone days. When it came to emphasize the ideological function of the western boundary of the Soviet bloc, the military investment on the GDR-side of the border dividing Germany was particularly important (Ante, 1991, p.65). According to Schwind's typology, it is an extreme case of a superimposed boundary (*Zerreissungsgrenze*), tearing two territories apart that actually belong together (Schwind, 1972, p.122; 1981, p.152). To dismantle a fortification is almost equivalent to opening a country to the enemy, apart from the physical efforts and the cost incurred (tearing down the German wall is an exception to this rule).

Apart from serving military and ideological purposes, boundaries hold at least three more important practical functions (Guichonnet and Raffestin, 1974, pp.49):

- The legal function delimits the perimeter within which norms and laws of a society can be applied. It is also the point of departure for norms that are effective towards the outside world (e.g. immigration laws).

- The fiscal function roots in economic interests and the protection of the home market. The state can demand duties and special taxes, or limit the quantities of determined goods.
- The control function arises out of the need to protect the population and to keep an eye on the people and the goods entering the country.

Political boundaries have a centripetal aspect: they are a necessary element to preserve or protect state territories (Figure 5.4). By separating political systems from each other they can impede transborder contacts. In this case, the region along the boundary will be (or become) a marginal region.

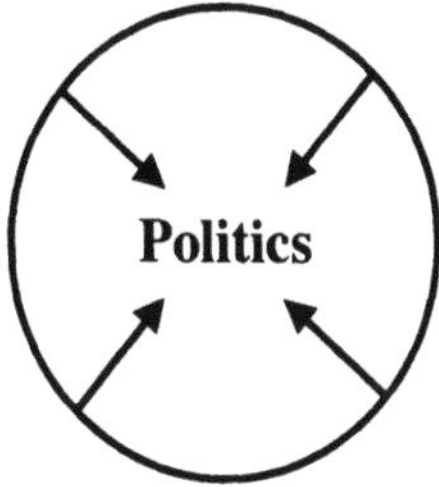

Figure 5.4 The centripetal aspect of politics (containment)

Source: Leimgruber, 1999, p.199

Under the impact of centripetal thinking, border areas have often been considered as vulnerable and not apt for large-scale economic investment; except for military purposes, investments in public infrastructures were limited. Neighbouring countries turned their backs on each other (and continue to do so), as is mirrored in roads or railways that stop short of the boundary or run parallel to it (Wolfe, 1962). Border crossings used to be army strongholds (the armed border guards are a reminder of this past), and the passage of people across the boundary used to be strictly controlled. Such measures discouraged investments in regions close to the state boundaries. The most extreme example of the total marginalization of a border zone has been furnished by the Iron Curtain, where not only the passage was limited to a few selected points but where a large belt of land was used by the German Democratic Republic for security installations. Agricultural activities in this zone were strictly controlled and access was reserved to the people traditionally living in it (and, of course, the security forces; Ante 1991).

One must be blind to believe that these tales belong to the past. Iron curtains are still a reality, e.g. between the U.S. and Mexico, on Cyprus, and across the Korean peninsula. A new iron curtain is emerging along the eastern border of the expanded European Union (Leuthardt, 1999, pp.173 ff.). Western Europe is that part of the world where the political boundaries have lost their role as barriers and have become regions of encounter. This process started with the creation of the European Economic Community and culminated in the Schengen Treaty of 1995, which facilitates mobility inside most of the European Union. In the late 1990s, border installations at crossing points have progressively been dismantled and the systematic control of people has disappeared. Instead, police controls on the roads have been intensified, and the increasingly sophisticated information technologies make 'invisible' controls via closed circuit television and computer databanks possible. The control function of the political boundary may have disappeared as a physical phenomenon, but as virtual reality it is increasing in efficiency. On the other hand, the emergence of new political boundaries, such as in former Yugoslavia or in the former USSR (Holdar, 1994), or as a consequence of the break-up of former Czechoslovakia may create new problem areas. Existing relations will suddenly be disrupted or at least complicated when the former provincial boundary becomes a state boundary (Gosar and Klemencic, 1994). This has most dramatically been shown by the partition of India in 1947 (Markovits, 2000-2001; Menon, 2000-2001).

On the other hand, the European Union in its expanded form (EU 25) will comprise member states that border on the Russian Federation, on Byelorussia and on Ukraine. One of the chief objectives on this outer border is to keep illegal migrants out. To this effect, strict border controls and (partly barbed wire) fences are already in place, heavily funded by Brussels. The future external boundary of the EU passes through regions where the local populations used to travel to and fro without problems (Leuthardt, 1999, p.177 for the Slovak and Hungarian border with Ukraine; p.253 for the Lithuanian boundary with Byelorussia). Now it is becoming a barrier difficult to surmount, as the example of the Seto minority in the borderlands between Estonia and Russia demonstrates (Saar, 2003).

Apart from being a line of division, the border is also a line of opportunities. By separating two political, judicial, and economic systems, it can be a place of advantage: different wage and price levels may give rise to economic dynamism, and places on the border may develop into growth poles. When the centrifugal economic perspective (Figure 5.5) dominates over the centripetal political view, a border region can lose its marginal position.

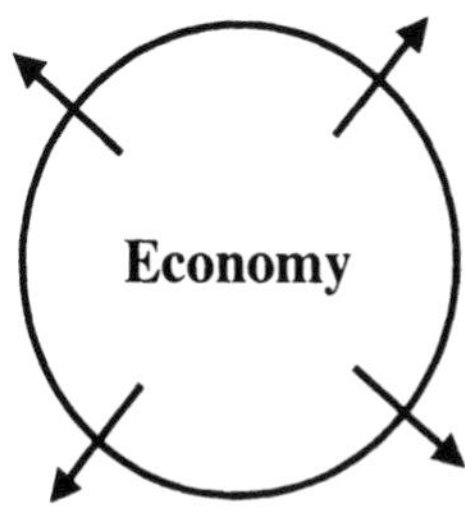

Figure 5.5 The centrifugal aspect of the economy (exchange)

Source: Leimgruber, 1999, p.199

After World War II, the boundaries in Europe maintained their role as barriers for several years, before regional transborder contacts and cooperation gradually started to evolve. After a timid start in the 1950s they have increased to flourish in a growing number of European boundaries, and they have received a singular stimulus by the fall of the Iron Curtain in 1989. The efforts to overcome traditional antagonisms in Europe have led to the creation of transborder regions all over the continent, often based on traditional regional contact patterns that had been interrupted by wars. In this way, formerly marginalized border regions have developed a life of their own, and a new form of relationship to the central authorities had to be established. Although formal transborder contacts are still an affair of foreign ministers, regional authorities can prepare such contacts in an informal way, and countries with a federalist system (Austria, Germany, and Switzerland) even allow their regional states some form of foreign policy.

The opening of the boundaries in western Europe and the generally good relations across the borders have led to a plethora of initiatives to value the potential of border areas. The modern society has changed the outlook on the boundary from centripetal or centrifugal to a combination of the political and the economic logic (Figure 5.6): exchange is to be promoted, but some form of containment is still desirable. Centres on the boundary received a new impetus, and formerly marginal regions could become zones of intense interaction. This process was favoured by an overall positive image the people hold of their neighbours and that can be regarded as a major element in the making of regional transborder networks (Leimgruber, 1981).

Figure 5.6 The double role of the political boundary in modern society

Source: Own elaboration

The new information and communication technology has added another dimension to cross-border relations. Radio waves diffuse independently of political boundaries, although the signals can be jammed and cables interrupted. Control over Internet connections is an instrument of power to shield populations against obnoxious ideas from abroad. At the same time, the new media offer considerable advantages for economy and scientific work; hence governments find themselves in an ambiguous situation about control and permission – mirroring the situation in Figure 5.6. Besides, telecommunication in itself is big business, and while the state wants to maintain a certain level of control, private corporations are sharing the market between themselves. Despite tensions on the US-Mexican border in the field of migration (see, e.g., House, 1982), there is cooperation in the field of information and telecommunication (BISN,1995).

In most cases, transborder cooperation originated from private associations, assembling politicians (as private citizens), representatives of the regional economy and of cultural life, and ordinary citizens. Examples can be found all around Europe, formally in the many Euroregions (Koter, 1994; Van Houtum, 1998), and less formally along many other boundaries (Klemencic and Bufon, 1991; Leimgruber, 1991; Bufon, 1994). The example of transborder cooperation in the region of Basle will be presented below.

Marginal border areas

The marginality of a border zone largely depends on the permeability of the boundary. A city on a closed border is more marginal than a rural area on an open boundary. Many factors interact to promote or impede intensive transborder collaboration. The advantage lies not only with cities as central places attracting customers from the two or three countries in the border situation; rural areas can also benefit from the border situation, particularly where the agricultural situation is asymmetrical (intensive farming on the one side of the boundary, abandoned land on the other). This is the case in some areas in the French and Italian border zone where Swiss farmers own and cultivate land abroad and in this way expand the basis of their enterprises. Bilateral agreements regulate the trade with the products from these fields and pastures. While farming issues were often at the roots of border conflicts throughout the Middle Ages, they are now promoters of contact and mutual understanding.

Despite such optimistic outlooks, most border regions are or feel neglected and are very marginal. The reasons for this vary. The persistence of the military function acts as a deterrent against settlements; settling marginal regions can be a strategic task, although this type of settlement enhances marginality rather than reduces it. In some cases, the natural environment is not very inviting for human occupation (high mountains, dense forests, alluvial flats etc.), in others, the major transportation routes bypass a region and contribute to the brain drain. This is the case, for example, with the entire Jura mountain region that is shared by France and Switzerland (Figure 5.7). A large part of it lies above 1000 metres a.s.l. and has a rather unfriendly climate. The major cities lie on either end (Basle and Geneva) or at its edge (Belfort, Besançon, Lausanne, Neuchâtel, Bienne). The towns inside the Jura are small and of limited importance, with the exception of La Chaux-de-Fonds, a regional centre with an emphasis on watch making. There are few transversal routes, and they are of local and regional significance only; the major axes are the A 36 in France and the A 1 in Switzerland that bypass the entire region. Access is thus a problem.

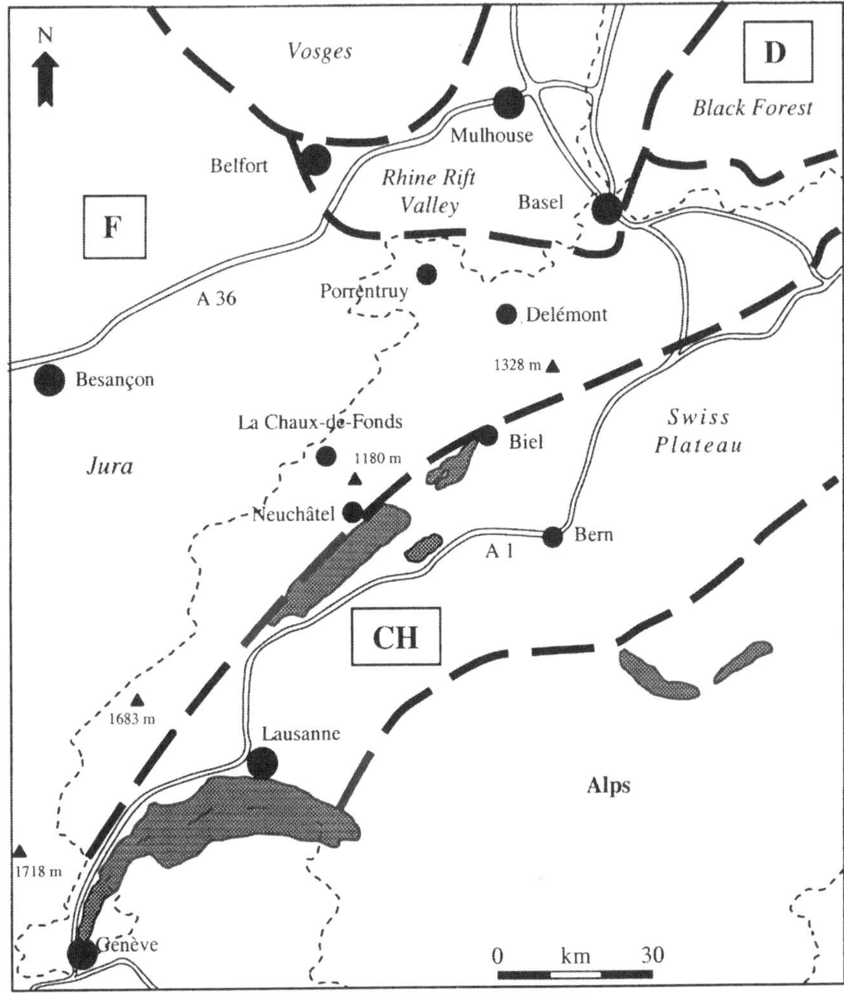

Figure 5.7 The Franco-Swiss Jura region

Apart from cattle farming, the economy of most of the Jura has been dominated by watch making, traditionally in small factories or at home as a subsidiary source of income on the farms. This industry went through a deep depression in the 1970s, and since it has only partly recovered: more than half the jobs in industry were lost (Gigon, 1991, p.514). The service sector has benefited from the crisis in the industry, but during this period, emigration deprived the region from part of its manpower. This structural weakness holds good for the French and the Swiss side.

While there are transborder relations along the entire French-Swiss border, they are significant in the urban regions of Basle and Geneva only; in between, they concern few people and only in selected areas (e.g. La Chaux-de-Fonds and Porrentruy). The railway network is weak, although the main lines from Berne and Lausanne to Paris cross the region, they do not have a particular impact. Even the TGV uses a track that is about one hundred years old and is far from meeting the expectations of fast rail transport. Lezzi describes the Jura region as a place 'where two peripheral areas meet' (Lezzi, 2000, p.92) and that 'is in danger of being detached from the European transport network (rail and road) and the surrounding economic centres such as Basle, Lausanne, Geneva, Dijon and Besançon' (ibid., pp.93 f.).

In order to improve the situation, the Community of Interests of the Jura (*Communauté de travail du Jura*, CTJ) was created in 1985. It has so far benefited from funding under the EU's Interreg programme. The initial accent was laid on cultural cooperation with the goal to improve tourism. The transborder museum passport, initiated in France in 1992 and extended to the Swiss part in 1997 (ibid., p.94) has been the first step to more intense cooperation that may eventually reach beyond culture and include economic issues of common concern. Populations who used to turn the back to each other are getting opportunities to meet.

The EU Interreg programme has been developed since 1991 in order to 'make the Single European Market (1.1.1993) work better' and 'to reduce the fears that enhanced free trade within the EU will deepen the economic cleavages between the economic centres and the periphery' (Lezzi, 2000, p.74). The programme had initially been conceived for three years (1991-1993), but then re-implemented for another five years (Interreg II, 1994-1999) and continued (Interreg III) from 2000 until 2006. The assistance offered by the programme covers a wide range of activities, from tourism to professional training of craftsmen, research in agriculture, study of air quality, mobility etc. With this programme, both a top-down initiative (from the EU, in order to ensure smooth working of its single market) and bottom-up ideas on transborder cooperation meet. Its effects are not spectacular but rather behind the scene, and the outcomes are, for example, regional tourist guides, a functioning regional consulting agency, an international museum or tourist pass, etc., results a visitor or an entrepreneur can experience without necessarily knowing much about their background.

Regional transborder tourism has benefited from Interreg in many ways. In the Regio TriRhena, culture has been one of the domains where this support is particularly welcome (Schäfer, 1996). Elsewhere, an interesting initiative has been supported around Lake Geneva (Figure 5.8), following an initiative of local and regional tourist offices.

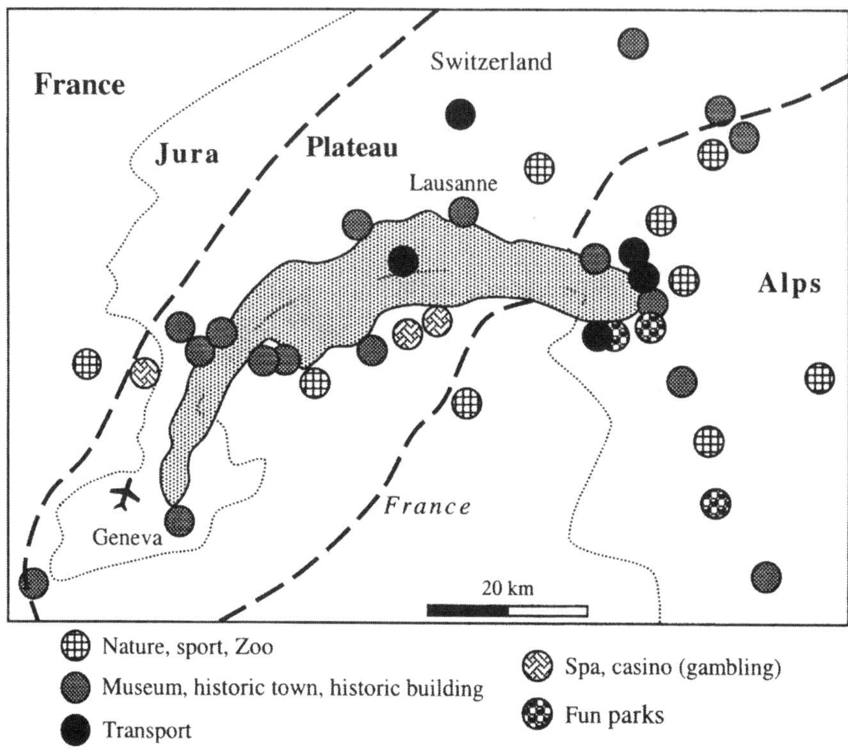

Figure 5.8 The *Léman sans frontière* transborder tourist region, situation in 2003

Source : Own elaboration after Léman sans frontière, 2003

Under the heading *Léman sans frontière* (Lake Geneva without boundary), a number of tourist organizations (12 at first, 16 by 1997; Leimgruber, 1998b, p.12) joined efforts of publicity and marketing in order to incite tourists to visit the region and profit of the variety of attractions and scenery that until now had not been appreciated very much. By 2003, the number of participating institutions had risen to 28. The region concerned is not a major tourist destination; rather many people pass through without seeing hidden treasures. Currently, these 28 tourist associations work together and market 37 attractions, ranging from mountain scenery and river gorges to steam railways, historic museums and fun parks (Léman sans frontière, 2003).

Further Interreg involvement has occurred in the Swiss-Italian border region where a transborder art exhibition in two villages on Lake Major gained support from the Interreg II programme (Leimgruber, 1998b, p.14). This kind of promotion is an excellent example of transborder cooperation where local and regional actors unite in a common effort and manage to ensure European support for their efforts. The bottom-up – top-down connection could not be illustrated in a better way.

Marginality has also been the result of the demarcation of the Finnish-Russian border in Karelia after Finland was granted independence in 1917. During the 19th century, when Finland was an autonomous Grand Duchy within the Russian Empire, this was a region of intense exchange, but in 1917 it became the state periphery (geometric marginality). Tourism suffered from the new political situation, as the border became an expression of state power, exemplified by border guards, and took over the role of a filter, exemplified by the precise location of crossing points (Paasi and Raivo, 1998, p.33). The centripetal paradigm came to dominate over the centrifugal outlook. When the ideological aspect of this boundary was reinforced during the Cold War, this fact attracted tourists from the West who were curious to see the ideological fracture in Europe and travelled to southeastern Finland for this simple reason. Contrary to the Iron Curtain in Germany and Berlin, there was no wall ; the absence of a fortification on such a 'hot' boundary seems to have been an attraction in itself. The inconvenience of this popularity was a lack of discipline: 'More and more people were arrested in the border zone, but the number of actual illegal crossings remained very small.' (ibid, p.38).

It would be too simple to consider all border regions outside major urban centres as marginal. Marginality need not be symmetrical. When the union between Sweden and Norway (1814) broke down in 1905, the border became a dividing line. The most strongly felt effect was the construction of railway lines parallel to it rather than across, thereby separating small centres in Norway from their rural hinterland in Sweden (Lundén, 1981, pp.130 f.). Two differing farming policies evolved, both influenced by the suitability of the land for agriculture and the farming tradition. Sweden favoured large holdings with high capital investment in mechanization, discouraged small farms and promoted the abandonment of less profitable land. Norway, on the other hand, promoted small farms and the expansion of agricultural land. 'These policies clash abruptly at the boundary' (ibid., p.133), giving rise to a peculiar border landscape, where the effect of marginalization on the Swedish side could clearly be detected: 'With the same kind of natural preconditions, Swedish fields are being converted into meadows or forests or just left to the weeds' (ibid.).

The same phenomenon could be documented by the present author on the Swiss-Italian boundary, where intensive agriculture and viticulture characterize the southernmost tip of the Ticino, whereas on the Italian side farming has been almost completely abandoned and not a single vineyard could be found (Leimgruber, 1987; 1991). Topography, climate and pedology are identical, but the social conditions are substantially different. The southernmost Ticino is one of the best wine growing regions in Switzerland, whereas in neighbouring Italy, the good wine growing areas lie much further south where the climatic conditions are distinctly better. The historical reason for the decline of local wine production in this area lies in the unification of Italy in the 19th century, which has created a single market; under these conditions wine could be traded inside Italy under new political and economic conditions, and the local subsistence production ceased.

When the former colonial boundaries in Africa were upgraded to state boundaries, they became elements of a centripetal outlook, destined to generate a sense of unity (nationality) in countries that were artificial creations with boundaries cutting across ethnic groups and farming and ranging areas (Harrison Church, 1981; Matznetter, 1981). A Eurocentric political view was imposed on the former colonies, including centripetal state boundaries and a politico-administrative apparatus. Legally, the border zones became state peripheries where the crossing was restricted to official crossing points, but it is almost impossible to enforce total control over the border; in fact the local populations maintained their old spatial practices and introduced as a new element the smuggling of goods in special demand in the neighbouring country. In their perception, the boundary did not really exist. 'Perhaps "unofficial" trade and these border markets are the truly African response to European-created boundaries' (Harrison Church, 1981, p.266). The border town of Kousseri in northernmost Cameroon (in fact a suburb of the Chadian capital Ndjamena) can be quoted as an example. Situated on the route from Nigeria to the Sudanese coast, its economy is essentially based on trade. Thirty per cent of the active population are traders, more than in agriculture and fishing (Kamdem, 1994, p.226 f.). It is essentially long-distance trade, but the local population also profits directly from the merchandise: 'goods ... are sometimes diverted from their final destination and sold in Kousseri' (ibid., p.231).

The transborder region around the international Basle conurbation

The opposite to marginal border areas are those regions where the neighbouring countries actively collaborate. This is the case in most Euroregions in areas of high population concentration. One such example is the 'Three Countries' Corner' in the Basle conurbation, where France, Germany, and Switzerland meet. Cooperation between these three countries (two of which are members of the European Union), has been going on for decades. Basle lies at the southern end of the Rhine Rift Valley, where river navigation on the Rhine comes to an end. The city has a long history of relations with neighbouring Alsatia (France) and Baden (Germany), but also with the entire Upper Rhine region as far as Frankfurt, due to its history as a trading town. The population of Basle looked traditionally towards the north; although Basle has been part of Switzerland since 1501, contacts towards the south remained limited. Even after the cementing of national boundaries, this old pattern of contacts was maintained, interrupted only during World Wars I and II.

The relations had been particularly severed from 1939-1945. While private contacts were possible soon after the end of the hostilities, formal relationships took more time to re-establish. The kick-off towards centrifugal thinking came from the academic world. In 1950, the Basle geographer Hans Annaheim, professor at the local university, published a study on the spatial organization of the Basle hinterland (Annaheim, 1950). In 1952 he wrote a school manual on the geography of the Basle region (Annaheim, 1952), and in 1959, the first volume of the journal of the Geographical and Ethnological Society of Basle was published, called *Regio Basiliensis* – again with Annaheim as one of the drivers. Gradually, the field for cooperation within this regional frame was prepared. The breakthrough came in 1963, when a private association, equally denominated *Regio Basiliensis,* was founded in Basle. A small group of 15 persons had the vision to improve transborder regional cooperation between Basle and her northern neighbours, the traditional hinterland (Speiser, 1995, p.30). This association served as a catalyst for further steps. In the following year, a French-German Community of Interests between Alsatia and German Brisgovia (CIMAB) was founded, and in 1965 the French *Régio du Haut Rhin* was created in Mulhouse. The early 1960s thus marked the beginning of intense collaboration, centred on the international Basle conurbation (c. 600,000 inhabitants) and with Mulhouse (in France) and Freiburg (in Germany) as partners in the two neighbouring countries. The formal recognition of these efforts arrived in 1971 when the *Conférence tripartite permanente de coordination régionale* (a permanent regional coordination conference of the three partners) was created, uniting representatives of the various regional authorities (ibid., pp.32 f.). This is remarkable because transborder issues are usually to be dealt with by the foreign offices of the nation-state. The federalist (decentralized) structure of Germany and Switzerland offer the regional authorities a certain margin of action, whereas centralist France may have more problems with such initiatives. However, this conference as well as other bodies created subsequently cannot take legally binding decisions. Their role is to improve the spirit of cooperation and to discuss issues that require a solution according to international law.

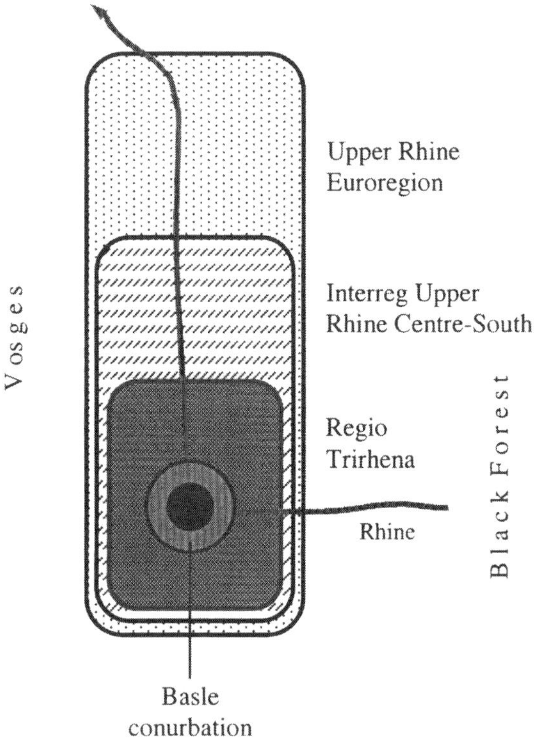

Figure 5.9 Transborder cooperation in the Upper Rhine region

Source: Own elaboration, after Regio Basiliensis annual reports

The region, nowadays known as *Regio TriRhena* (Figure 5.9), is inhabited by about two million people. In a way it acted as a pioneer for the subsequent regional initiatives. With the passing of the years, the spirit of cooperation extended further north, and nowadays, it is a part of the Upper Rhine Euroregion that comprises the entire Rhine Rift Valley between Basle and Frankfurt (Main) between the Vosges Mountains to the west and the Black Forest to the east. This Euroregion stretches for roughly 300 kilometres and is inhabited by about ten million people. It is an important transportation corridor with railways, motorways and river navigation. Regional contacts have overcome the political differentiation.

A particularly noteworthy example of transborder cooperation is the international airport Basle-Mulhouse (Walker, 1994; 1995). It has been built on French territory in 1954, jointly by the French and the Swiss. It comprises two separate accesses and check-in facilities, two restaurants (but only one kitchen!) and there is no possibility to cross the boundary other than on foot inside the airport. Access from Basle is via an international feeder road on French territory, which is completely fenced off. The EuroAirport is jointly managed and can be seen as the proof that transborder cooperation does work. It is a regional airport, with an emphasis on charter flights and air freight; scheduled flights connect it to other regional destinations as well as to the important European hubs of Paris, Frankfurt, and London for intercontinental connections. The economy in the 'Three Countries' Corner' is sufficiently strong to sustain it. With its chemical industries and the Bank for International Settlements, Basle boasts a certain weight in the regional and even national economy. The French (Alsatian) partners are equally interested, only for the German side, the benefit is at the moment marginal. The construction of a regional rail link would certainly improve the accessibility and increase the attractivity.

Basle has never been a marginal city. Throughout history, trade constituted the focus of economic activity, and even the cementing of state boundaries after the Vienna Congress in 1814/15 did not change very much. The economy has always kept the transborder spirit, and the chemical industry relied on the land and labour reserves in France and Germany. Cultural contacts are also very intense: Basle is the central place to a vast region also from this point of view. Formal cooperation in the field of culture was enhanced with the creation of a working group within the tri-national governments' conference (the former *Conférence tripartite*) in 1980 (Schäfer, 1996, p.28). To organize and manage cultural events is probably the easiest way of promoting transborder relations.

The case of Basle may be unique because three countries are involved and the conurbation is truly international, but there are many more examples of towns and regions that profit from the border situation. In Switzerland, Geneva and the southernmost Ticino display a similar situation to Basle; in the three cases, the populations profit from it: the Swiss economy in the border region offers jobs to transborder commuters at wages that are usually higher than in their own country. At the same time, many Swiss residents in the border zone do a considerable part of their shopping abroad, thus profiting from lower prices and a choice different from the one at home (Leimgruber, 1987; 1991). Similarly, in the Saarbrücken region (Saar-Lor-Lux), contacts across the border are very intense, and the implementation of the Schengen agreement in the EU has intensified the interrelations. Personal observations in 2000 and 2001 not only showed the existence of contacts that draw advantage of the border situation but also that the former buildings of police and customs control between Forbach (France) and Saarbrücken (Germany) had been dismantled. The German town of Aachen is a short distance from Belgium and the Netherlands and profits alike from the potential afforded by the border situation (Gramm, 1979).

Wars – the marginalization of people and environment

The problem of war is not frequently discussed in political geography. Is this simply because war has a negative connotation, or because it breaks out when politicians have failed to solve a conflict without resorting to violence, or because it has primarily a technical side? As concerns the latter aspect, the Gulf Wars fought in 1991 and 2003, and the war against Serbia in 1999 have seen the praise of sophisticated high-technology weaponry used with surgical precision (as long as they were programmed correctly). Missiles are fired far away from the target and cover considerable distance within a short time. Space becomes a negligible element in this new form of warfare. The hi-tech warriors ignore the territory between launch and impact and leave it to the ground forces to conquer. As to the second aspect, conflict resolution is a task that concerns the entire society; the politicians are usually the scapegoats if things fail, as they have been elected to marry opposing aims in the system of a society. Torn between various interests (of the economy, of their own prestige, of the population), they may opt for war as the simplest way out.

Since 1914, wars have taken on a new spatial dimension; they have entered the age of globalization (Holton, 1998, p.115), and weapons of mass destruction have been developed throughout the 20th century. The use of gas from 1915 to 1918 and the advances in the construction of airplanes gave the war a new dimension. World War I was the first global war with simultaneous belligerent actions around the globe. World War II was a repetition of this scenario, but in a much more violent way. In particular, by proclaiming a total war, Hitler chose to involve also the civilian population, as he demonstrated with the Blitz on London in 1940 and the subsequent use of V1 and V2 missiles on the same target. Needless to say that the Allies took up the challenge and began to do the same to Germany. By forcing children into the last battles, the Nazis showed their contempt for any form of human dignity. The construction and use of the atomic bomb marks a sort of climax in the race for military superiority and political hegemony.

The United Nations have banned wars, allowing them only in the case of self-defence and as the last resort once all other means of conflict solution by the international community (represented by the UN) have failed. This last motive is, of course, very elastic, as the discussions in the weeks before the 2003 Gulf War have shown.

War is an unpleasant phenomenon, resulting from the human incapacity to deal with problems in a peaceful way. The roots very often lie deep in a human mind, and very often wars are started because of an individual's particular obsession with an idea or perception of a given situation. The real explanation, whether biological (ethological, psychological) or social (liberal, socialist, nationalism or interest groups) may allow a classification of the causes of war (Frankel, 2002), but we contend that there are no objective causes: a war is always started by one or more human beings.

War and peace – the dual world-view has been the subject of many discussions. One of the most prominent peace researchers, Johan Galtung, has devoted his life and energy to promote non-violent conflict solutions. In a remarkable synthesis, he points to the four essential powers that are necessary for this process: military, economic, culture, and politics (Galtung, 1993, pp.170 f.). Cooperation and understanding are essential prerequisites, and so are diversity, the insight into the mutual benefits of solutions to all parties, i.e. a balanced situation of interests and needs that complete each other rather than compete with each other (ibid., p.173).

War is still a reality, despite thousands of years of experience about the suffering of people and the destruction of the foundations of living, and despite the countless efforts to solve conflicts through negotiations rather than violence. Witnessing the discussions about the war against Iraq, the author felt compelled to introduce this aspect into the present book. Besides, wars have a strong geographical component, as is demonstrated by the fact that the military were the drivers of cartography, and that in many if not most countries they still control the production of maps. Wars take place in space, and apart from causing human suffering; they also destroy landscapes and ecosystems. Geographers must not close their eyes to this unpleasant phenomenon.

There are many types of war, and a typology depends on the perspective chosen: the fields of operations (land, sea, air), the weapons used (conventional, biological, chemical, nuclear), the various motives that lead to a war (defence, conquest, liberation, prevention), the agents involved (nation-states, civil groups, *guerilleros,* terrorists, etc.), the dimension of a war (regionally limited action, total war), by technique (trench warfare, mobile war), etc. The existence of a typology, however, does not excuse the essentially negative side.

Wars are usually conducted by the military under the responsibility of the politicians – unless the two coincide or the military want to take control of politics themselves, as usually happens on the occasion of a putsch. This formal picture is, however, at times perturbed by the existence of paramilitary groups who occupy a shadow zone in this 'business'. Examples for such groups abound (Columbia, Democratic Republic of Congo, Indonesia, to name but a few). The politicians are primarily accountable for how a war is fought and for the consequences, although in the context of the field, they are unable to control events. The training of the soldiers includes not only the use of weapons and other items for self-defence and attack but also a psychological training, to avoid excesses towards civilians and other soldiers (their own colleagues and the enemies). Civilian victims are to be avoided, but this is almost impossible. There are the so-called 'collateral damages' caused by belligerent action: misguided bombs, stray bullets, and crumbling buildings can hit civilians (as they can also strike soldiers), but at times, civilian targets are also hit intentionally. What is certainly more difficult is to avoid damage to the environment; after all, wars are fought in the landscape, and the landscape has no defence mechanism against military action. Wars thus represent an extreme way of marginalizing the environment (see Chapter 7).

In the 1980s, the French biogeographer J.-P. Amat (1987; 1988) conducted research in forests in the Verdun area. This place had been a cornerstone of the French defence system during World War I, and the battle for Verdun was one of the bloodiest and most violent of that war. A visit to the area – indeed to the entire region along the former front-line in France – is still worthwhile more than 80 years after hostilities have ceased, because it reveals the extremely long-term consequences of war. We could identify at least three:

- the persistence of landscape changes, both in the micro- and in the macro-topography and in soil degradation;

- the long life of unexploded bombs, artillery shells, and grenades that have retained their explosive capacity for more than 80 years (in fact for an unknown period of time); and
- the presence of the war in the mind of the local inhabitants and its use as tourist attractivity.

The forest had become an essential element in the French defence strategy on the grounds that it offered shelter, rendered the soldiers almost invisible, prevented air reconnaissance, impeded the movement of people and reduced the effect of the newly developed shrapnel (Amat, 1987, p.222). The French army thus became one of the great protectors of the forests (ibid.). The forests also offered cover for the preparation of the decisive allied offensive in July 1918 (ibid., p.225). However, the forests also suffered from the war. The Argonne forest to the west of Verdun 'became a veritable laboratory for research in and experimentation of new weapons' (ibid.), such as the digging of tunnels to blow up the adversary's positions with powerful mines, or the use of mortars. The consequences of this kind of warfare have persisted until today, to say nothing of the trenches that still crisscross the forests and render efficient forestry extremely difficult and even dangerous. The Butte de Vauqois (Figure 5.10) is the most remarkable illustration of the consequences of mine-warfare. Both the French (to the SE) and the Germans (to the NW) had dug tunnels into the hill to undermine the adversary's positions. The top of the hill with the town-hall had been blown off by exploding powerful charges underneath, but the outcome of the battle was nil, either side managed to keep its position.

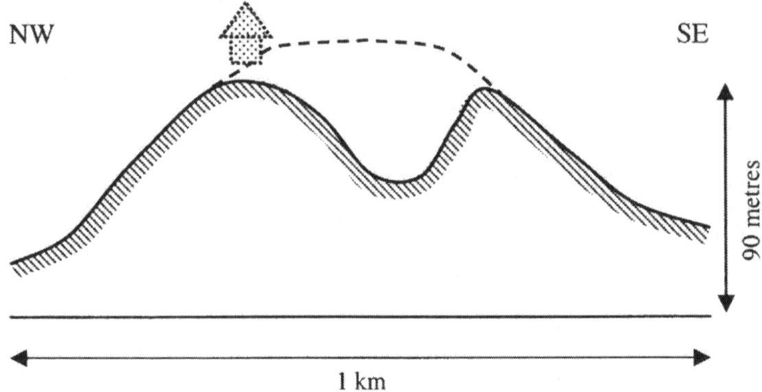

Figure 5.10 Butte de Vauquois, west of Verdun (not to scale)

During the battle, former fields were destroyed, the soils completely mineralised and the sequence of the horizons turned upside down. Deep funnels from the explosion of strong charges and complex patterns of trenches, sometimes containing metal debris from the war characterize vast surfaces (Amat, 1988, p.195 f.). These surfaces were no longer suitable for farming, but were afforested. 'Because of the absence of a structured soil, the surface corresponded well to the needs of tree species. It is loose, well aired, and with a good retention capacity for water, despite local tendencies for choking (as can be observed in certain funnels).' (ibid., p.196; transl. WL). The new forests have developed well, and it seems that from the ecological (especially edaphic) point of view the destructions of the war are judged as negligible, whereas economically they are considerable. 'In 1986, for example, an oak plantation replacing a former coniferous forest cost 7500 F/ha for the plantation plus an additional 6500 F/ha for the preparatory works of levelling.' (ibid., p.199, trans. WL). This corresponds to approximately 1,150 Euros for the planting and 990 Euros for levelling.

The problem of bullets, empty shells and non-exploded ammunition continues to haunt the populations of the former front-line in France. Until the present, amateurs are looking for the scrap-metal (copper, bronze, lead), and for many years, this has provided a supplementary source of income for locals, e.g. in the Somme region (Lavalard and Enbry, 1996, p.V). Collectors, however, run a considerable risk because of unexploded bombs and artillery shells. Between 1990 and 1994, at least four people were killed in the Péronne area (ibid.). These ammunitions are deeply buried in the soil, but they will work themselves gradually up; farmers tilling their fields with heavy equipment and deep reaching ploughs run the risk of touching them and provoking an explosion. Two soldiers from the Verdun garrison were killed in the Argonne forest in 1999 when they were collecting scrap metal and shells (information obtained from forestry workers in the Argonne forest, May 2000). The notion of 'time-bomb' is appropriate in this context and takes on a new connotation; looking at the quantity of landmines and cluster bombs that are scattered around the globe, one can appreciate the long-term danger emanating from relics of a war that has ended decades (or maybe centuries) ago.

The unexploded shells, the land mines and cluster bombs are direct material remains of military action. To them must be added further long-term consequences on people and the environment. By using Agent Orange to defoliate the trees in Vietnamese forests and destroy crops, the US disrupted entire local and regional ecosystems and caused considerable suffering both to the indigenous population and many soldiers fighting in the areas concerned. By setting fire to Kuwait oil fields, the Iraqis not only polluted the air for weeks and months but also contributed to the destruction of large tracts of land as a consequence of the fires and the many attempts to extinguish the flames. The 2003 Iraq war has left other disruptions of the ecosystem. The more than 2,500 tanks, personnel carriers and lorries that drove all the way from Kuwait to Baghdad have destroyed the hard sand surface, thus allowing the winds to blow the sand out to the detriment of the arable land along the Euphrates (Läubli, 2003). This will have dramatic long-term consequences for agricultural production and the provision of the population with food. Such damages to the environment are often considered secondary to the impact of, for example, depleted uranium that is used in bombs and shells to destroy enemy tanks and other objects. Both are, however, equally detrimental to the ecosystem and the society, and both are effective for many years to come. In particular, the use of depleted uranium bombs in the attacks in cities in Iraq will have serious long-term consequences for the civilian population.

It is the long-term perspective that is usually overlooked when a war is planned and started. Long-term means not only the material relics of a war (the unexploded bombs) and the devastation of the environment, but also the psychological and demographic damage to the people and the uncertainty about the future social organization. The first attempt to solve conflicts via the international dialogue, the League of Nations, has failed after less than 20 years of existence. Its successor, the United Nations, has survived 55 years until it was put to a severe test in the third Gulf War. To make international law work requires more than just smiling into the camera and opening Champagne bottles. It demands the readiness to negotiate in a way that enables all sides to support and accept a compromise. A war can also leave lasting problems in its aftermath, problems one never expected before and where a solution is difficult to imagine. On the European political map, the Balkans has been a particularly delicate region for a long time, and they continue to be a problem area. The war planners of NATO, who led the intervention against Serbia and in Kosovo, had clear ideas about the time after, but they could not know what would be the outcome of the war. 'The military had warned that the war could not be won with bombs alone, but they then "did their job"' (Gent, 2000, p.214; transl. WL). The politicians had seen the war as the only means to solve the problem, but they had never taken into account that their plans could founder. What nobody really wanted – an internationally administered protectorate in the Kosovo with no clear idea as to when and how to get out of this situation – has occurred because of bad planning and a certain naivety. This, at least, can be read from Gent's (2000) account of this conflict.

Let us close with a somewhat more encouraging note. While we tend to emphasize the destructive and disruptive aspects of war, we overlook the fact that it can also have positive effects. This may sound strange, but a community under the stress of war is faced with new challenges for survival, both physically and psychically. Many observers have noted that under such circumstances, people will usually change their attitudes and perceptions. Everybody is in the same critical situation, everybody has the same fears and hopes, and all of a sudden, feelings of solidarity will appear where indifference reigned before. Besides, new insights occur. 'The obvious destruction caused by a war therefore has a much deeper impact on humans than one that meets the eye. It reminds us daily of our mortality, and by destroying our cultural artefacts it reminds us that there is no way in which we can achieve permanence' (Maček, 2000, p.46). The same holds good for any unexpected event, such as the September 11 attacks on the World Trade Centre in New York. One develops a sense for danger even when everything seems to be quiet (ibid., p.52).One particular survival technique consisted in 'imitating the peace-life routines as much as possible' (ibid., p.71). New kinds of social solidarity 'came about under existential threat and in short and intense time' (ibid., pp.106 f.). While the life of the individuals and of the community is suffering from the war situation, it is driven by the desire for survival under difficult circumstances. A stress situation always frees new strength and creates a capacity for resistance – this is common knowledge, but nobody really wants to go through this particular stress.

The question of Human Rights: lip service or serious endeavours?

After World War II, the international community founded the United Nations Organisation with the intention to solve conflicts henceforth in a peaceful way. In 1948, the Universal Declaration of Human Rights was accepted. In this case the hope was that human dignity would receive world wide recognition and that cruelties would disappear. The failure of the League of Nations and the atrocities of the war had left an imprint on the societies and the politicians – the future looked bright.

Human Rights can be regarded as guidelines for political action. In fact they 'limit state power' (Weston, 2002). In the course of time, Human Rights have evolved as new domains were added. A helpful distinction has been made by dividing them into three generations: the first generation comprises civil liberty rights (right to life, equality before the law, non-discrimination, no distinction of race, colour, religion, etc., freedom of speech, etc.), the second generation covers socio-economic rights (right to work, of social security, to found trade unions, protection of the family, etc.), whereas the third generation includes the right to development and to a sound environment. This comprehensive list would ensure happiness to everybody on Earth ...

As is well known, during the last 55 years, humanity has not lived up to the expectations. Hot and Cold Wars have been fought, sometimes in self-defence, more often, however, out of a regression into the old conflict pattern (Figure 5.1). Similarly, Human Rights have not been respected everywhere – the entire system ably developed throughout the last decades of the 20th century by the international community is blocked by a grave deficit: the lack of a neutral enforcing agency that can effectively halt wars and outlaw violations of fundamental human rights and call offenders back. Every nation-state disposes of a judicial system that can enforce the law. International law is deprived of this mechanism; it can only function as long as every partner respects the international agreements. As long as the law of the jungle reigns, the range of action of the United Nations will thus be limited by the will of the member states to cooperate, to set solidarity before egoistic goals (see Chapter 3 on these issues).

It is not our aim to depart on an extensive discussion of Human Rights in this context. However, a few words may be appropriate, given the connection to the topic of war and the relevance they have for humanity. In particular, the notion of 'universal' has to be questioned. We must not forget that the idea of Human Rights is based on a Christian background, if we look at the original drivers of the idea, although they have no specific religious background. They are based on the belief in human dignity, respect and tolerance, i.e. of values anchored in every religion, but coupled with the Christian claim of universality. However, in practice Human Rights are central to the political discourse only, but rather marginal to everyday political practice.

Although about one third of the world population professes to be Christian, they are not necessarily better people than the adepts of other religions or persons who profess to belong to no religion at all. It is presumptuous to declare one's values as universal simply because one is in the majority, but this is a reality. Not surprisingly has the Islamic world developed its own declaration of Human Rights, and have the African people acted in a similar way, both in 1981. Cultural differences are as much a reality as are Human Rights, and they must be respected – this is also a Human Right. After all, there are elements in other religions that are missing in Christianity and might as well be included (e.g. the obligation of

Muslims to give alms, non-violence as a Buddhist principle, the respect of the family in Chinese culture).

The reality has rendered independent supervision of respect for or violations of Human Rights necessary. Out of this concern arose Amnesty International, a non-governmental organization, founded in 1961, that reports on violations and takes an active part in protests against them. In a way it is a shame that such a 'watch-dog' is required, but obviously mankind cannot survive without. One of its permanent tasks is to report on the practice of capital punishment, still practised widely around the globe, although it violates Article 3 of the Universal Declaration of Human Rights, Article 4 of the Banjul Charter of the African people, and Article 1 of the Declaration of Human Rights in the Islam (even if in this case the *šarî'a* may grant exceptions). The extent to which the death penalty is still common practice around the globe is an indicator for the respect governments pay to this elementary right of human beings.

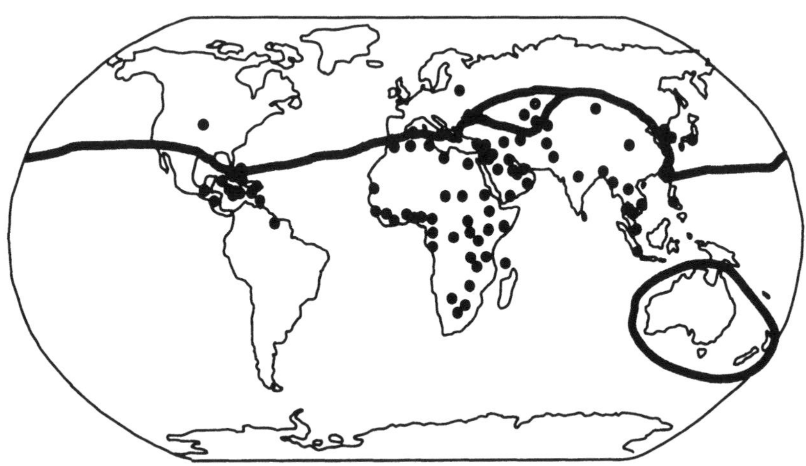

Figure 5.11 Countries practising capital punishment in the beginning of 2003, in relation to the contemporary *limes*

Source: AI, 2003a, p.7

The map (Figure 5.11) provides an approximate picture of the present-day situation of capital punishment around the globe. Whilst 112 countries have abolished it in law or practice, there still remain 83, i.e. a significant portion of the world where it is still practised (this number includes the Palestinian Authority). The abolitionist countries, however, fall into three categories: 76 have abolished capital punishment for all crimes, 15 for ordinary crimes only (i.e. wartime crimes can still be sanctioned by the death penalty), and 21 have not executed anyone for at least ten years or made a commitment not to use the death penalty any longer (AI, 2003b). It would be wrong to simply point to the 'black sheep' as being the backward part of the international community. It is our task to try and convince those countries that have not yet abolished it to do so, by pointing out the humanitarian and the spiritual reasons behind. But so far, the 'developed world' has failed to be the model to follow, as is shown by its position on the northern side of the contemporary limes. In this context it may be reminded that among the countries that abolished the death penalty for all crimes in the 1990s, we discover Switzerland (1992), Italy (1994), Spain (1995), Belgium (1996), and the UK (1998).

A global policy – or global deregulation?

Every empire had a global outlook that corresponded to the current knowledge of the world. The *mare nostrum* of the Romans can only be understood in its full significance if we take the vision into account that the Roman Empire was The World, and that any peoples living beyond the *limes* were the barbarians, non-civilized and a potential threat to The World. In a way this holds good still today, albeit under entirely different circumstances. If Rufin (1991) adapted this image to the modern world, he is fully aware of the fact that the North considers the South as a menace. In fact, the attitudes of northern states towards the (usually illegal) immigrants from Latin America, Africa, Asia and the former Soviet satellite states mirror the attitude of the Romans towards the barbarians. The North has something to lose – its affluence and privileges – which are those things the immigrants aspire to obtain as well. The new *limes* has already been built between the US and Mexico, and around the Spanish territories of Ceuta and Melilla on the Moroccan coast, and it is being constructed along the eastern border of the expanded European Union, towards Byelorussia and Ukraine (House, 1982; Leuthardt, 1999). This new *limes* is provoking substantial re-regulation, on the one side in favour of the free exchange of goods, capital and people, on the other towards increasing the obstacles for exchange. The global world is being divided up into two 'earths', one of the haves, and one of the have-nots.

The increase in global interrelations, in particular the creation of the United Nations as an association of all independent countries of the world (with the exception of the Vatican) has dramatically altered the global political map. International law has received a prominent position in the judicial world, and on all levels, it has come to dominate the relations between nation-states. In many instances it replaces national law or is at least the last instance towards which an accused can appeal. However, international law can only function if all members of the international community accept the rules, play the game, renounce individual privileges and specific claims – I repeat this statement intentionally. The UN has no police force that can arrest politicians for not complying with international standards. To obey the international rules is a *moral* obligation, founded on the *word* of acceptance and on global *solidarity*. Little of this can be observed at present. The law of the jungle prevails as soon as specific interests are at stake. Will it be possible to develop global politics under such circumstances? Will it be possible to arrive one day at a regulatory system that is based on mutual respect and solidarity? Will the global society be able to develop into a true community where – this sounds utopian – no rules will be necessary because all human beings live in harmony? Global de-regulation, a global community – this is a form of utopia that mankind under current conditions will not attain. Too different are the interests of the individuals, the various groups and entire societies. Maximum goals and minimum efforts are opposites and it is not possible to marry them. 'A true community needs no police. The sheer existence of an institution that has to enforce the law demonstrates that something does not work as it should' (Somé, 2000, p.75; transl. WL). What may hold good for local African societies is obviously too simple for the global society.

Chapter 6

Society, Culture and Marginality

Everybody has been confronted with the problem of marginality in social life – be it at school, at work, at a party, or even inside the family. Sometimes, a marginal situation lasted for a short time only (punishment in class), sometimes it can continue for a prolonged period in life (the position at the workplace), but it is hardly ever a permanent situation. Social marginality is tied both to specific situations and to one's place in life (in the group and in the society), and events can suddenly change it – everybody knows and experiences ups and downs.

What holds good for individuals is also valid for groups, communities and entire societies. In all cases, marginalization is an endogenous as well as an exogenous process, and it has a lot to do with perception and values. The way we perceive us (the auto-stereotype) and the others (the hetero-stereotype) influences our self-esteem and our attitude towards the others. History is full of examples; we shall present a few of them.

Every society is a system and as such finds itself in a dynamic equilibrium. Processes of marginalization and demarginalization are therefore a customary phenomenon. As an open system, a society requires permanent external inputs in order to survive in the long term. The exclusion of a new idea is a strategy to maintain a *status quo*, and secure power, but as a conservative attitude it will block change and the evolution of the respective social system. The same holds good for the long exclusion of women from politics and management. Women were not to take responsibility in public life or in the economy; as 'victims of global processes such as sex tourism and the transnational market in domestic servants' (Holton, 1998, p.193) they were servants to male lust and domination. Things have changed in politics and the economy, but the exploitation still continues. Stability (a sacred value) is not bad *per se*, but innovation (a secular value) must not be excluded; the two have to interact.

This chapter looks at marginality with regard to ethnicity, poverty, language, and religion. This selection is personal and takes a number of current social and cultural problems into account that have to do with values: discrimination as to origin and wealth, problems of multilingual communication, and divisions and conflicts according to faith. The chapter approaches the theme of culture through a number of topics at various scales. A personal definition of culture will be presented in the concluding remarks.

Ethnicity

The colonization of the world by the Europeans has been an example of systematic marginalization of entire peoples. Curiosity and trade motivated the first voyages of discovery, but soon territorial expansion and hegemonic aspirations became the real drivers of colonization. The confrontation with the native populations became inevitable, a confrontation that was not only guided by the intruders' objectives but also by the mutual perceptions of the other. In a report on the first encounter between a tribe of the Bapende people and the Portuguese on the North Angolan coast we read: 'White men came from the water; they spoke in a manner nobody could understand. Our ancestors were afraid, they said they were Vumbi: spirits that returned to Earth. They drove them back with swarms of arrows. But the Vombi spat fire and thunder. Many people were killed' (Bitterli, 1980, p.188; transl. WL). The interest indigenous people showed in their apparel astonished the Europeans for their part (ibid., p.189). Australian Aborigines sometimes found it difficult to distinguish between the clothes and the skin: 'It may not have been immediately apparent where the covering ended and the flesh began' (Reynolds, 1982, p.28). As with other objects, clothes were very much appreciated, they were items to be collected, though not necessarily to be worn (ibid., p.49).

A mixture of distrust and curiosity on both sides governed early relations. The ensuing epoch of systematic and frantic colonization, however, changed this almost idyllic picture: further European expansion was guided by a shift towards secular values. Driven by the greed for gold, furs, spices, slaves etc. and ideologically backed by a Christian missionary zeal, European lifestyles and legal systems were imposed on existing social systems that were discarded as being obsolete. The legal systems established by the colonial powers were usually inefficient in practical life and applied in a manner that resulted in a two-class society: the settlers on the one side, the original inhabitants on the other. South African apartheid has been the most brutal and longest lasting example of this segregationist policy. Examples abound; I shall illustrate the argument with reference to the Australian Aborigines. It is not necessary to look for further examples overseas. In Europe, the Roma have been and still are a marginal group, fighting to obtain minimal acceptance. The section concludes with a discussion of the question of refugees and asylum seekers; this issue concerns all scales, from international to regional, and the countries of the North are particularly concerned. It is as if they were harvesting what they sowed in the period of colonialism and with the diffusion of European and American lifestyles.

The Aborigines in Australia

The original inhabitants of Australia are among those who have suffered most under white colonial and postcolonial power and continue to suffer until today. 'The claim has always been that the Aborigines became subjects of the Crown from the first instance of settlement. But the facts speak for themselves. Despite coming under the protection of the common law, over 20,000 Aborigines were killed in the course of Australian settlement. They were not, in a legal sense, foreign enemies struck down in war although a few were shot down during periods of martial law. Most were murdered – nothing more nor less. Yet the law was powerless to staunch the flow of blood – and neither lawyers nor judges appear to have done much to bring the killing to an end. It is not an honourable record' (Reynolds, 1987b, pp.1 f.). The figures of those who died have been substantially revised since Reynolds wrote this: up to 600,000 aborigines were killed – a true genocide – according to an authoritative report published in 1987 (Pilger, 2002, p.192). The Australian example could be multiplied, and in every case we would arrive at the same conclusion: non-Europeans were regarded as non-humans ('They were part of the fauna'; Pilger, 2002, p.180), and they were systematically driven out of the newly established social system. As a side effect, they lost their land. This has happened to the natives in the Americas, in particular in North America; and in Africa, where the European attitude has been particularly vicious by making the continent a reservoir for slave labour.

Land questions have subsequently become core issues in the current debates between Australian Aborigines or Native Americans and their respective governments, but the road to satisfactory solutions is still long. Land has a particular significance to the Aborigines as well as to other indigenous peoples. For the Innu, the land had been given to them by the Creator, and they saw themselves as its custodians. It was an integral part of their life and could not be possessed – even today, their language does not have words to denote land ownership (Cordey, 2000, p.84). For the Aborigines, land and landscape were of mythical origin, specific to the individual tribes and regions (Ragaz, 1988, pp.16-26). Not surprisingly, rivers, springs and waterholes occupy a prominent part in these myths, and the latter are important elements in the groups' territorial organization (ibid., p.131). Often, the land was associated with the mother (Mother Earth), and the mother cannot be owned. As original inhabitants of the continent, the Australian Aborigines had developed a special form of land ownership that consisted of 'the gamut of forms of attachment to land that confer rights over that land' (Sutton and Rigsby, 1982, p.158). Such forms of attachment were in particular 'control over the stories, objects, and rituals associated with the mythological ancestors of The Dreaming at a particular place' (Myers, 1982, p.188). In other words, the myths referring to the land were a sort of mental maps (Moizo, 1989, p.165). Land possessed a sacred connotation and could not be private property; this attitude was strange to the European settlers who sincerely believed that they moved into uninhabited areas. They transferred their own concept of (private) ownership to Australia, not knowing that in this way they desecrated sacred sites and, by seizing the land, denied access to the former owners (Reynolds, 1982, p.67). This rational attitude also guided their use of the land that was, however, fundamentally different from the Aborigines' land use rationality. They based land utilization on certain principles, which had 'a moral foundation and may be expressed in religious idiom' (Williams, 1982, p.137). To settlers with a largely secularized world-view, the sacredness of the land was a strange notion. Land was declared Crown property; it was leased (not sold!) by the British colonial administrator, free from taxes for an initial period of five years, after that subject to an annual rent (see the Form of Land grant by Arthur Phillip, issued in 1791, quoted in Clark C.M.H., Selected documents in Australian history 1788-1850; translated and reprinted in Bitterli, 1981, p.266).

Native rights to land have thus systematically been denied. This attitude was based on erroneous assumptions by the first British discoverers, notably Joseph Banks, to whom the land looked totally uninhabited (Reynolds, 1987b, p.31). However, this was not correct. As hunters and gatherers, the Aborigines did not occupy fixed places but required vast territories for roaming. This did not prevent them from returning to the same place in the course of time. By all means, 'the land wasn't uninhabited. The Europeans clearly weren't the first occupants. The "savagery" and "heathenism" of the native gave no charter for expropriation' (ibid., p.40).

The right of indigenous peoples to their land was developed in the course of the 19th century. Native title as the 'legal right based on the fundamental principle of prior possession' came into existence in the US between 1810 and 1835 (ibid., p.46). It was recognized by Britain in the 1830s and led to 'the establishment of reserves, the recognition of rights of use and occupancy on Crown land and the provision for compensation to provide for education and welfare' (ibid., p.125). However, it was only as late as 1993 that the Native Title Act 'had removed from common law the fiction that Australia was uninhabited when Captain James Cook planted the Union flag in 1770' (Pilger, 2002, p.169). That there is still a deeply rooted racist prejudice in Australian society (in particular among the Conservatives) is shown by the 'Native Title Amendment Act of 1998, which waters down the 1993 laws, wipes out the universal principle of Native Title in all but name and takes away the common law rights that the judges said belonged to the Aborigines; nothing like it has been passed by a modern parliament anywhere' (ibid., p.176). The end of marginalization is not in sight.

The land question became an international affair when transnational mining companies started to exploit the mineral resources. In 1972, for example, Nabalco (now Alcan Gove) started bauxite extraction on Gove Peninsula (Arnhem Land, Northern Territories). The company had negotiated an agreement in 1968 with the Commonwealth of Australia, where details of land leases and royalties to be paid had been fixed (Nabalco, 2000). They also built the town of Nhulunbuy 'on a special purpose lease in 1971 to service its bauxite mine, despite opposition from the Yolngu. ... This led to the first major legal action over land rights and although the Yolngu case was lost, it led to the introduction of the 1976 Northern Territory Land Rights Act. Under this Act, Arnhem Land was returned to its Aboriginal owners along with other traditional lands in the Northern Territory' (EALTA, 2002). Native groups have to fight strongly for their land rights; in Australia, the situation seems to be more difficult than in the US (Reynolds, 1987a, pp.182 ff.).

Cooperation was also achieved in Weipa on the York Peninsula where relations between the Comalco bauxite company (part of the CRA Group, ex-Conzinc Rio Tinto Australia) and the local Aborigine community were at stake. Both sides had to change their attitudes, the company towards a more sustainable policy towards the original inhabitants of the area, the Aborigines towards a more peaceful image of the intruders. Howitt (1992) describes the two contrasting attitudes in the company's management: 'One group within the Comalco management team generally argues that the company's social obligations are limited to producing bauxite at the lowest cost and highest profit possible, and relegates indigenous rights issues to the status of optional public relations window dressing and support for state welfare programmes. [typically secular values; WL]. The other group accepts that mining does not occur in a social vacuum, and actively pursues the company's stated resource philosophy' (p.223 f.). This philosophy recognizes that bauxite is a finite resource and that a long-term strategy was deemed necessary (ibid., p.224) – it is much more oriented towards sacred values. However, the efforts undertaken by both communities 'have not transformed the underlying structure of Aboriginal economic and political marginalization, nor had a major effect on attitudes within the broader context of the CRA Group or the mining industry generally' (ibid., p.234). To influence attitudes is obviously a very demanding task.

In parts of Australia, efforts have been undertaken since the early 1970s to include Aborigine communities in planning processes, shifting from top-down to bottom-up oriented policies. Participation should 'empower local peoples to become architects of their own futures' (Rugendyke, 1998, p.257). Barbara Rugendyke reports a case of Arrunge (NW of Alice Springs), where the claims to traditional lands have been taken seriously. The public sector (the Department of Health and Community Services of the Northern Territories) has become engaged in planning in order to ensure acceptable living conditions and water supply for an Aborigine group living in an area, which had previously been used for pastoral purposes (Rugendyke, 1996; 1998). Despite many shortcomings and limitations to the planning process, the result has to be seen as positive. Nobody can expect a group which has for a long time been marginalized and excluded from participation, to work in the same way as a 'white' society which has gone through decades of formal training. Given the complexity of Aboriginal society, the real result of the exercise can be summed up in one phrase: 'The community planning process heightened awareness among members of the communities of the importance of the 'process' of planning rather than of the production of a formal, western style of plan' (Rugendyke, 1998, p.274). The people have been taken seriously, as human beings and not as outcasts; this change in perception and attitude is the real social and political challenge.

Such cases may offer a glimpse of hope, but they are probably rather isolated. The perception of the native Australians by the white majority is still very negative – apart from a few outstanding figures in the sports world, but even they have no guarantee that the respect and honour they obtain along with the medals is lasting (see Pilger, 2002, pp.180 ff. for examples).

Gove, Weipa and Arrunge are examples for the Australian mainland. In Tasmania, the situation is quite different. On this island, the last full-blood indigenous Aborigine died in 1876 (Scott, 2001, p.249) – according to Moizo (1989, p.157) they were massacred in 1888 – the remaining Aborigines are European-Tasmanian Aborigine mixtures, full-blood Australian Aborigines, and Torres Strait Islanders (Scott, 2001, p.251). Until the 1960s, Aborigines were officially non-existent or 'were regarded as outcasts in a white society' (ibid., p.250). Statistics on the number of Aborigines were therefore unreliable because many of them (especially the half-bloods) did not declare themselves as Aborigines. With the gradual change in attitude, the Aborigines dared to stick to their identity and census data began to look different (Figure 6.1); the number of Aborigines grew steadily through self-declaration, not because of immigration or a high birth-rate.

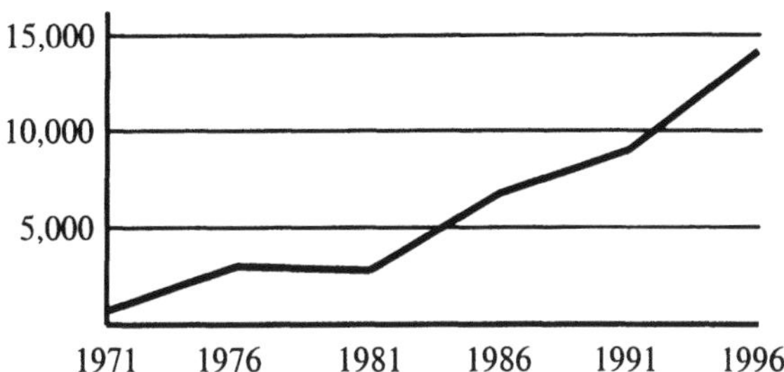

Figure 6.1 Aborigines in Tasmania (census results)

Source: Scott, 2001, p.250 f.

The self-declaration of Aborigines during the 1990s had consequences on the statistics: the socio-economic data of the 1991 and 1996 censuses can only be compared with caution (ibid., p.252). It seems that in particular persons with full employment 'among clerical, sales, and service workers' as well as paraprofessionals and professionals were among the newly registered Aborigines; the number of unemployed, on the other hand, was low (Scott, 2001, p.251). Globally speaking, the socio-economic situation of the Tasmanian Aborigines can therefore be considered as satisfactory (ibid., p.255). 'In sum, by comparison with Australian Aborigines and Torres Strait Islanders, Tasmanian Aboriginal communities ... have much less cultural involvement, are generally better educated, have a higher proportion gainfully employed, are better remunerated, less dependent on government payment, have most families owning their own home, less crime and delinquency, but poorer health.' (ibid., p.256). However, in their land claims the Tasmanians were less successful than their mainland counterparts, mainly because is has been difficult for them to prove unbroken links to the land. The government granted ownership (no native title) over a few areas only (ibid., p.261).

A particular situation exists in the Torres Strait Islands, the archipelago between Australia and Papua New Guinea. Here lives a population of about 8,000 people, 'originally of Melanesian race, but Polynesian, Asian and even European blood was introduced in recent centuries, producing an mixed ancestry and cultural milieu' (Ohshima, 1986, p.72). This community requires not only native land title over the more than 100 islands but also rights to the sea. This was claimed in the Torres Strait Development Plan of 1998 (TSRA, 1998). Island people generally have an intimate relationship not only to the land but in particular to the sea which furnishes a substantial part of their living. The regional authority has successfully supported the land claims, and it hopes to achieve recognition of the sea rights as well (TSRA, 2001a). The claim was lodged in November 2001 (TSRA, 2001c) and covers 'the sea, sea bed, subsoil, reefs, shoals, sand banks and waters'. It emphasizes the strong links the population of the Torres Strait islands have always had with the sea (TSRA, 2002a). This claim is a logical consequence of the attempts at greater regional autonomy, as has been formulated in 2001 (TSRA, 2001b) and will help to further promote public support for fishing (TSRA, 2002b) and other sea related activities.

Contrary to the Australian Aborigines, the North American Indians can at least draw on a popular image they still hold and that roots in frontier romanticism. However, this image is rather a perception by Hollywood than a mirror of past life. They have almost exactly the same problems as their Australian counterparts – indeed, aboriginal populations worldwide face similar problems of marginalization. Tsao (1997) reports that in Taiwan 'proportionally more child prostitutes come from aboriginal background than from any other ethnic groups.' (p.99). This is a direct consequence of the systematic marginalization of indigenous populations through 'the failure of education, and of cultural bankruptcy.' (ibid.). Once parents see no better future for themselves, they will even force their children into prostitution (ibid.). Land is an issue, but it is not the only one nor is it the most important. Dignity would be imperative.

The Roma in Europe

For a European, Australia and North America may look far away, and the current problems of indigenous groups can be regarded as simple relics of the past that will gradually be solved. True globality, however, demands that we feel with people even if we are not directly concerned (see Chapter 2). Ethnic marginalization (discrimination) no longer exists in Europe – one might think. The reality, of course, is different. Immigrants from other cultures face increasing difficulties, racism – a word that seemed to belong to the past – has become an issue again, and the ghost of apartheid is haunting us once more.

Among those people who are still victims of discrimination we find the representatives of the itinerant lifestyle, the Roma or Gypsies, the People of the Road. They are generally regarded with considerable distrust, and the attitude towards them is close to racist. Yet, they have a double image, somehow like the North American Indians, oscillating between romanticist admiration and fear of their otherness. Many people are fond of Gypsy music and dancing, and in his opera *Carmen*, the French composer Georges Bizet (1838-1875) has immortalized the Roma. On the other hand, whether settled or nomadic, they have their particular lifestyle and culture, which distinguish them from all other societies and do not please everybody.

Throughout the 20th century (to say nothing of the past) and into the present epoch they have been persecuted, and various attempts have been made to annihilate them. The worst persecutions occurred in the countries under the Nazi occupation when about half a million of Roma were sent to concentration camps where they finished in gas chambers. In Switzerland, Roma children and children of itinerant Swiss (not to be confused with the Roma) were taken from their families and placed into orphanages during the period 1926-1973; responsible for the 'Children-of-the-Road' programme was an organization called *Pro Juventute*! (Huonker and Ludi, 2000). This policy recalls Australian measures of separating Aborigine and Torres Strait Islander families from their children in the 1920s and 1930s (Pilger, 2002, pp.177 f.) with the aim of breeding out the colour. 'The policy was inspired by the fascist eugenics movement which was fashionable in the first two decades of the twentieth century and spread the fear that white women were not breeding fast enough and the "white race" would be "swamped"' (ibid.). Apart from regional details, it reads like a justification for racist Nazi policy or the Children-of the-Road policy in Switzerland. More recently, both sedentary and mobile gypsies went through a further ordeal after the transition in Central and Eastern Europe and during the wars in Bosnia and in Kosovo (Courthiade, 2001). In 1998, for example, the authorities of the Czech town of Kladno refused the young members of the Roma community to use the public sports grounds; following extensive protests, the government condemned the mayor for having promoted ethnic and racist hatred (Moutouh, 2000, p.42). In 1999, the authorities of the city of Usti and Labem in northern Bohemia started to build a two-metre high wall between a Roma area and the city, for the reason of noise, dirt and the different lifestyle. This case roused both national attention (President Havel in person demanded that the wall was torn down) and international criticism from the European Union. Clearly, Europe has seen enough of walls.

Roma live a life at the margin of society, 'the result of structural discrimination based on a history of systematic neglect by local and government authorities' (Groenewold and van Praet, 1997, p.69). However, they do not want to become part of the 'mainstream' because that would rob them of their identity. They find themselves in the difficult situation of opting for better conditions of living through improving their education and skills, thereby overcoming poverty and exclusion, and maintaining their culture and identity. The solution – integration into their respective national societies and respect for the particular values of either group – 'is a two-way street. It requires certain changes both from both majority populations and minority groups' (UNDP, 2002, p.5) – changes in attitudes and behaviour towards the Other.

The origin of the Roma lies in the Indian Punjab (Rishi, 1996, pp.1 ff., 41); they moved into Persia in the 10th century (Moutouh, 2000, p.20 ff) and arrived in Europe in the 14th century (1322 Crete, 1378 Zagreb in Croatia, 1407 Germany, 1417 Switzerland). By the 16th century they had reached Scandinavia (1515 Finland, 1544 Norway; ibid., p.28). Colonial powers like Spain, Portugal, France and Britain exiled them to their overseas territories (Angola, Cape Verde Islands, Brazil), other members of the community emigrated to North America (ibid., p.29). In this way they were distributed across the globe.

The Roma have their own language, called *chib romani* (Courthiade, 2001), belonging to the Indo-Aryan branch of the Indo-European languages (Rishi, 1996, p.85). They display a strong collective identity that helps them to distinguish themselves from other groups and societies (Courthiade, 2001). Their total number is estimated at about 12 million worldwide, the majority of which (about two thirds) live in Europe.

The Roma not only face social discrimination, they also encounter difficulties with securing their livelihood. A first problem is that the mobile Roma require adequate space for temporary shelter and the (legal) possibility to go about some professional activity to ensure survival. Authorities and populations are reluctant to offer satisfactory camping areas, although in France, for example, a law of 1990 obliges municipalities with over 5,000 inhabitants to set such areas aside (Moutouh, 2000, pp.91 f.). The situation is not much better elsewhere; in Switzerland, timid attempts are being undertaken on a cantonal basis towards regionally adapted solutions. The sedentary Roma are obviously not concerned by this difficulty.

A second problem lies with the evolution of industrial production and consumer habits. The traditional crafts the Roma used to practise as itinerant groups have almost entirely disappeared, partly because the objects themselves do no longer exist, partly because modern consumers shop in the supermarkets and replace broken utensils rather than have them repaired (plastic bucket are thrown away, whereas metal bucket could be mended, repairable reed baskets have been replaced by throw-away plastic bags). There are, of course, alternative professional opportunities, but they are usually more suitable for sedentary Roma than for the mobile groups (car mechanics, car traders). Even they, however, often find themselves in the informal sector, although this share very much depends on the level of education (UNDP, 2002, p.35). Flexibility and entrepreneurship are certainly the major qualities the Roma possess, and in the rapidly changing economic environment of the present this is vital for their survival (ibid., p.36 f.). We can only speculate about the long-term social and cultural consequences of such a development: how long will the family and clan solidarity continue if the gap between the educated and better-off Roma and those with low education and skills widens?

The UNDP/ILO study (UNDP, 2002) on the Roma situation in the five countries of Bulgaria, Czech Republic, Hungary, Romania, and Slovakia pointed to the major problem, which was unemployment (p.31). Looked at from the subjective interpretation (no steady or permanent job), more than two thirds of the persons interviewed (69 per cent) considered themselves as unemployed, with considerable differences between the individual countries (Czech Republic 46 per cent, Slovak Republic 85 per cent). This lies well above the ILO's broad unemployment rate (46 per cent for the five countries; UNDP, 2002, p.33) that does not take the steady (formal) job as the sole criterion of employment but includes income-generating activities in the informal sector. Among the causes for unemployment, education and lack of skills are frequently mentioned, but also ethnicity (discrimination) is advanced. The report suggests that this element must not be neglected, although it may be overemphasized in the subjective interpretation. Unemployment in the five above-mentioned countries is a fact that has to do with the transition from state-planned to market economy and is a structural phenomenon; it is, at the same time, related to the unsteady evolution of the continental and the global economy. Ethnicity may be just one additional element that makes access to the labour market more difficult for the Roma than for others. '… the skills of many Roma workers do not meet labour market requirements, whereas those Roma who possess marketable skills can still face barriers of prejudice …' (ibid., p.34).

The itinerant Roma are the only remaining nomads in Europe. The attitude towards this (minority) segment of the Gypsy population mirrors an interesting ambiguity of our western society. We have become sedentarized, used to a permanent residence and an administration that exercises a fair amount of control. Parallel to this, our society has become increasingly restless and is characterized by an extremely high mobility in many respects: professional, social, and spatial. The traffic jams at rush hour and during the holiday season suggest that we are longing for something we have lost in the course of the evolution of society. It is as if we had re-discovered the 'nomadic virus' in ourselves, yet at the same time we are opposed to true nomadism as a lifestyle. The true reason behind this opposition (that is not limited to the Roma in Europe but lies behind the sedentaization of nomads in general) is the question of power and control over everybody and everything, but also specific (i.e. well controlled and politically admitted) ideas about freedom.

Immigrants, refugees and asylum seekers

Social marginalization also accompanies people who flee their home country for some reason and try to find refuge elsewhere. Contrary to those who emigrate on the basis of a labour contract, they are 'unauthorized migrants' (Papademetriou and Hamilton, 1995, p.7) who leave their homes for highly varied motives, ranging from political persecution to natural disasters. Whatever the reason behind departure, they perceive their personal situation such that they deem it imperative to look for a new home, at least temporarily if not for good.

Refugees have been with mankind since very early times in history. They are a burden to the receiving societies and are therefore prone to marginalization; usually, they will be placed (concentrated) in refugee camps. They add to the population to be supplied with basic goods (food, clothing, articles for personal hygiene, etc.), but they do not necessarily see themselves as permanent dwellers in their host country. To some, a camp provides temporary shelter after a natural disaster or during a civil war, to others it is a necessary passage on the way to a final destination, generally in Europe or North America. Most of the refugees and displaced persons live in countries of the South, where they occupy a veritable archipelago of camps (Figure 6.2), separated from the resident population and often with little to nothing for their livelihood. Marginalized in their own country, they flee to escape persecution, but they continue to lead a marginal existence, hoping to return to their homes sooner or later. The Palestinian tragedy epitomizes this situation, but it is just one case among many. Nobody knows the exact number of refugees worldwide; Rekacewicz (2001) provides estimates between 20 and 50 millions. This figure includes people who leave their country and those who displace themselves inside their own country. At any rate, the 19.8 million people under the protection of the UN High Commissariat for Refugees early in 2002 (UNHCR, 2003) are far below the real number of refugees.

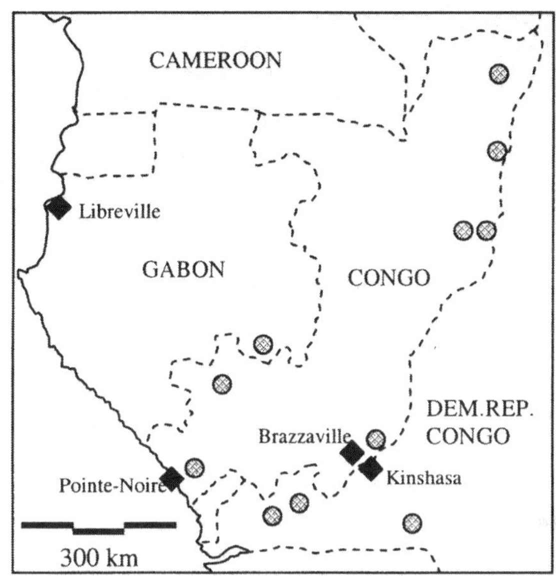

Figure 6.2 The archipelago of refugee camps in the Congo Republic and neighbouring border areas

Source: UNHCR, 2003, Population statistics, Table 11

The countries of the North have developed a particular attraction for refugees, based on affluence and the image of a paradise. At the same time, their migration policy is dictated by national (economic) interests, with humanitarian aspects ranging behind: refugees are in fact unwanted immigrants (Papademetriou and Hamilton, 1995, p.7), and refugees according to the UN definition mingle with 'economic' migrants who are simply looking for a decent life with a minimum income and some prospects of survival. This new tendency has 'transformed the migration issue into an economic dilemma' (ibid.), and anti-immigration movements have gained considerable momentum in most countries of the North (Birindelli and Bonifazi, 1993; Thomas-Hope, 1994; Frey and Mammey, 1996). The United States, who have been leading a restrictive immigration policy since the late 19th century, is not the only country where the popular attitude continues to be largely ambiguous. 'On the one hand, Americans proudly claim to be a "nation of immigrants." On the other hand, concern has often been expressed about the impact of too many immigrants' (Weller et al., 1994, p.110). On the whole, the US is no longer very open to to immigration (including refugees). The same ambiguity can be discovered in Canadian immigration policy (Basavarajappa et al., pp.7 f.).

Immigration policies in the North have become increasingly selective. Canada's history as an open immigration country ended with the Great Depression of the 1930s (Ghosh and Pyrce, 1999, p.233), and since then, immigration (including accepting refugees) has been regulated by the needs of the labour market, skilled labour force fluent in English and French receiving priority (ibid.). In the 1990s, strict planning was adopted with annual quotas and percentages for specific source countries (ibid., p.236). The total arrivals per year oscillated between 215 and 230 thousand. In 1998, for example, the total amount of immigrants were between 200,000 and 225,000, of which between 10 and 15 per cent only could be refugees (ibid., p.237). Compared to the numbers in war-stricken Sudan (four millions according to Rekacewicz, 2001), for example, this is ridiculous.

European countries have experienced intermittent labour immigration throughout the 20th century, but this was migration inside the continent, and many of these migrants have become integrated and even assimilated into the host societies, as has been the case in Switzerland (Leimgruber, 1992a). Even the occasional flows of refugees (Table 6.1) were well accepted and adapted into the host society. This has to be seen from two perspectives: on the one hand there is what we call the humanitarian tradition of Switzerland, on the other hand there is a political factor. Refugees from Hungary, Czechoslovakia, Tibet, Vietnam, and Poland fled Communist regimes. Those from Europe were culturally close; their main problem was to overcome the linguistic barrier. Until about 1980, marginalization of these populations was not an issue because their number was limited and the Cold War helped to maintain anti-communist feelings.

Table 6.1 Refugees into Switzerland after World War II

Year	Region or country of origin
1956	Hungary
1959	Tibet
1969 f	Czechoslovakia
1973 ff	Chile
1975 ff	Vietnam, Cambodia, Laos
1979 ff	Africa (Ethiopia, Erythrea, Zaire etc.)
1981 f	Poland
1982 ff	Turkey (incl. Kurds)
1983 ff	Sri Lanka
1989 ff	(Ex-)Yugoslavia

The situation changed when asylum seekers arrived in growing numbers and from countries with a different cultural background. Anxiety and hostility began to spread. The Swiss are not particularly prone to violence, and physical attacks were limited (aggression and arson, however, did occur). They try to solve problems through negotiations; as a consequence, the question of immigrants (more correctly: of foreign residents) has been on the political agenda since the 1960s (Leimgruber, 1992a, p.63 ff.). Right wing parties formed around the issue of *Überfremdung* (overpopulation by foreigners), and they tried to restrict the number of foreigners in Switzerland via the Constitution (as to the procedure, see the Prologue). Fortunately, they failed on all occasions, and the issue was resolved through the implementation of laws.

There would be a simple way out of the problem of *Überfremdung*, the easing of naturalization. There is no legal claim that a foreigner can be naturalized. To obtain Swiss citizenship is the object of a personal request, and it can only be undertaken after 12 years' residence in the country. The complex political system (see Chapter 5) does not facilitate the process. The request has to be addressed to the municipality of residence. Depending on cantonal legislation, the decision lies either with the municipal executive, or the legislative, or the assembly of the citizens (direct vote or vote on the poll). In the latter case, the result may be very much influenced by stereotypes and personal prejudices (as to everyday behaviour, country of origin, or religion).

Swiss identity is complex and related to the political three-tier system (municipality, canton, country). Due to the our history, a Swiss is first and foremost a citizen of his home municipality, as a consequence of this of the respective canton, and as a consequence of this, of Switzerland. There is a Swiss passport, but it has to be ordered at the municipal level.

Figures on foreign residents in Switzerland are thus misleading. It has been estimated that about a third of the 1.4 million inhabitants without a Swiss passport (December 2000) are second or third generation foreigners, i.e. they were born in Switzerland, have frequented schools and obtained professional training or went to universities in Switzerland, they are ordinary taxpayers and may participate in local and regional cultural and scientific activities – but for some reason or another Swiss citizenship has been denied to them or they have not dared asking for it because of prejudices. A further obstacle to many residents originating from countries of the European Union is that for them a Swiss passport is not pressing because of free mobility around the EU, although since 1992, when double citizenship has been accepted in Swiss legislation, an important barrier to naturalization has been removed.

To be a refugee is awkward, to be an asylum-seeker in Switzerland (or in other northern countries) is no picnic. As an unwanted person he or she is first and foremost regarded with suspicion, as he or she might be a false refugee. This is a problem of definition. Politically, the accent is laid on the UN definition of a refugee: a person seeking asylum and wishing to qualify as a refugee has to prove that he or she is persecuted. The reason for the flight must be a direct menace to life due to political, religious, cultural or ideological persecution. Indirect motives (hardship because of the political system) are not considered, and they may be even more difficult to prove than direct threats. Economic causes (poverty due to the general economic situation or natural catastrophes such as crop failure due to drought) are not accepted, although they might be the consequence of political repression and persecution.

The road to asylum and the status of a recognized refugee is long and painstaking. It begins at one of the four reception centres on the Swiss border (the archipelago) where asylum seekers first are directed to in order to channel the administrative procedures. This, of course, is a first marginalization in Switzerland. Those who will eventually be granted asylum may take up temporary residence, but they are directed to asylum centres in the various cantons with few possibilities to influence this allocation. Their marginalization thus continues in another way. In the regional centres, they are not always welcome – the cultural difference is a strong obstacle to even partial integration. Once they are allowed to find a job, this will usually be in the lowest segment of the work hierarchy, performing what neither the Swiss nor the older immigrants will do, at low salaries. They are not allowed to compete with the native workforce – another form of marginalization. To add yet another element, ten per cent of their income is then held back to cover the cost of procedure and assistance (Efionayi and Piguet, 2000, p.121). Having finally made it to the 'Paradise' is no guarantee that one will eventually feel there.

The purpose of the entire procedure is to make the country unattractive and thus slow down the flow of asylum-seekers. This has to some extent succeeded, as Figure 6.3 seems to suggest – but the drop in European asylum-seekers is in fact the result of return migration of Bosnian and Kosovo Albanian asylum seekers: their number dropped from almost 60,000 in 1999 to 26,000 in the year 2000 (FOS, 2002, p.79). In summer 2003, these countries were declared safe countries; this means that requests for asylum from people of Balkan origin will no longer be accepted.

It is true that Switzerland had a foreign population of about 19.3 per cent in 2000, second only to Luxemburg (37.3 per cent) in the European part of the OECD (OECD, 2002), but the 71,957 asylum-seekers in 2000 made up for five per cent of the foreign population only. This does not constitute a severe problem, in particular as they are in a very fragile situation. However, in the case of extreme events (the flaring up of a civil war or the 2003 Iraq war), asylum-seekers from the countries in question will not be sent back, even if their request has been declined. This, of course, does not ease their situation, but provides at least temporary relief.

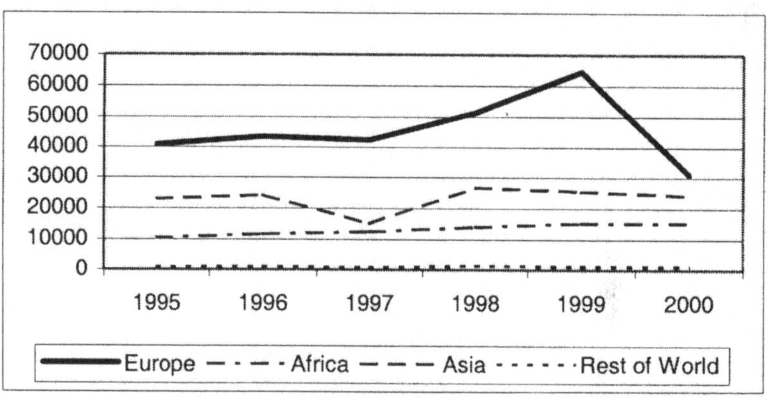

Figure 6.3 Asylum-seekers (provisionally accepted) in Switzerland according to continent of origin, 1995-2000

Source: FOS (2002), p.79

Poverty

Generalities

Marginalized groups often survive at or even below the threshold of deprivation and poverty. Social and spatial exclusion prevent them from striving for more than elementary education, to say nothing of training in their desired profession. They find themselves reduced to the status of unskilled and often temporary labourers, situated at the bottom of the income scale. To obtain access to higher education requires a special effort, and even if the school is free, the extra costs may be prohibitive. The attitude of the parents towards training is crucial but cannot replace lack of funds.

Poverty is an old problem and used to be considered an inevitable phenomenon in every society. There are many kinds of poverty and the reasons behind are diverse and by no means simply related to an insufficient production of food and other basic goods to satisfy everybody.

This renders the definition of the term and practical measures to fight poverty difficult. Generally, a person can be considered as poor once he or she does not dispose of sufficient means to cover his or her basic needs. What, then, are basic needs? We can define them as those material goods that are indispensable for survival under decent human conditions, i.e. enough food and drinking water plus those material goods a given society requires to secure health (clothes, furniture, a home). Non-material components may be added (see p.271). Since basic needs vary in time and space, poverty has to be put into the socio-cultural and socio-economic context of a society. The basic needs of a rural Swiss in 2003 are different from those in 1950 and 1900, and they vary from those of a rural inhabitant of Malawi in 2003. An infant has different basic needs from an adult and a retired person. The concept as such is clear but extremely flexible, and this flexibility makes it difficult to arrive at an objective definition.

To satisfy the material basic needs is one task, but to these we have to add what we may term 'social basic needs', required to participate in community life. To own such goods or to have access to them may be an important part of the quality of life. The kind of such goods depends on what is more or less universal in a society or group (Sen, 1999, p.89). Ownership of such goods (cellular phones, television, a car etc.) offers participation in mainstream social life (Andersson, 1994, quoted in Jargowsky, 1996, p.107), something that is quite normal and can prevent people from becoming socially marginalized. The choice between satisfying first the material or the social basic needs may not be easy when the income is at the lower limit of existence, and it often goes into the 'wrong' direction and exceeds the financial capacity of a household: 'Indeed, the paradoxical phenomenon of hunger in rich countries – even in the United States – has something to do with the competing demands of these expenses.' (ibid., p.90). The allocation of the various expenses in the household budget is a key to poverty.

Measuring poverty encounters difficulties precisely because of the vagueness of this notion. The measures adopted are based on average statistical definitions, without looking at concrete situations. Income support seems to be the obvious way of alleviating poverty. But even so, a political undertone cannot be overheard. 'For statistical purposes, the Census Bureau defines a person as poor if their total family income falls below the federally defined poverty level, which varies by family size and is adjusted each year for inflation.' (Jargowsky, 1996, p.9). As Van den Bosch (1999) demonstrated, however, there are different opinions about what indicator to use: is it disposable income or should household expenditure be considered? Furthermore, which unit should be chosen, the individual, the household, or the family? The choice of variables influences the comparability of data. And finally, where exactly is the limit between poor and not poor? Is this a simple statistical value (e.g. 50 per cent of the average income), or is it rather the poverty gap, i.e. 'the aggregate income shortfall of all poor households with respect to the poverty line.' (ibid., p.104)? When discussing poverty we have to bear these difficulties in mind (see also Rusanen et al., 2000).

As a substitute for a vague concept, a monetary definition has been adapted. This approach is dictated by the 'hegemony of the economy over social life' that has come to dominate our thinking since the end of the Middle Ages with the emergence of productivist rationality (Gaboriau and Gouguet, 2001). Money is considered a neutral measure, but it is not objective either. The value of a pecuniary threshold depends on the unit chosen (individual, household, family), on the spatial resolution and on the general economic prospects of this unit. The measures range from the gross national product (GNP) to the personal income of households or individuals, according to scale and available data. The most widely publicized international 'standard' poverty threshold is an income of one or two US dollars per day, propagated by the World Bank. This value is about as realistic as is the use of the GNP to describe the living conditions of the people in the South. It may serve as a global measure for comparison, but the internal reality of the people in the 190 countries of the world is 190 times different. With one US dollar a day, a person can buy a meal in India, whereas a US citizen might give this same sum as a tip to the waiter who has just served his meal. In Nepal, for example, a person paid between 50 and 90 Rupees for one kilogram of rice, one kilogram of bananas and one litre of mineral water (June 2003 prices in Kathmandu); at the corresponding exchange rate this amounted to between 0.70 and 1.26 US$ (personal communication P. Pradhan). The minimum income for survival varies internationally and even inside a country, and it also depends on age and lifestyle (Leu, 1999). On a global scale, poverty as 'monetary marginality' surfaces even among academics: often, colleagues from certain countries in the South can hardly attend an international conference because of the exchange rate. They will lead a normal life as a professor in their own country with a living standard comparable to any western professor, but they will be utterly poor when it comes to paying for the airfare in hard currency. For various reasons, the material poverty threshold is clearly insufficient to define poverty. It is based on one simple exchange unit (money, the US dollar) that happens to be accepted as a universal clearing unit. Because of its global outlook, it lacks the differentiation of the regional perspective. Besides it does not take the lack of goods and services, and the deficit in education and information into account.

Depending on scale, regional differences can be striking. Poverty varies not only according to the definition applied, but also according to the spatial units considered. Data, however, are rarely available for the same region on different scales to allow for comparisons. An exception is furnished by a study on Michigan (Mehretu et al., 2003), which details poverty on four different scales, macro-, meso-, micro-, and *in situ* scale, down from state level to census tracts. The authors can in this way show the difference between contingent marginality on the macro-scale and systemic marginality, dominant on the other levels of resolution. Their work allows poverty to be put into a spatial perspective, to illustrate the *babushka* principle.

The case of Switzerland (Figure 6.4) demonstrates that a very coarse spatial resolution of data is misleading, but as no more regional details are available, we have to be content with that information. By including it here, we want to illustrate the risks of interpretation that are inherent in small-scale representations. Nobody would expect large cities like Berne, Lausanne and Geneva or the region of Lake Geneva to be hot spots of poverty, despite the fact that in cities affluence and deprivation meet (see below). The aggregation of data has been necessary because the sample size of the national poverty research project did not permit a presentation by canton (two of the regions – Zurich and Ticino – are in fact cantons). The map must therefore be read with great care. However, it shows an East-West gradient that exists also when we look at unemployment figures (see Chapter 4). Poverty islands exist in the affluent (or less poor) regions just as wealth islands occur in the 'poor' part of Switzerland. The use of the federal tax data is in this respect far more illustrative (see Chapter 4).

There is more to poverty than just money. Among the non-economic connotations, we find intellectual capacity, i.e. the faculty to adapt to circumstances and make the best of a given situation, education and skill, physical strength, and health. These elements can be more important indicators of poverty than an internationally defined minimum salary. It is true, however, that income and the possibility to develop one's intellectual capacities, to promote skills and to ensure good health are related, albeit not causally. Insufficient income 'can be a principal reason for a person's capability deprivation' (Sen, 1999, p.87), but money alone does not make people intelligent. It is also the differentiated allocation of money within a family, for example, that can influence capacity building (preference for boys over girls, of the elder over the younger) and hence decide upon affluence or poverty and marginalization (ibid., p.88 f.).

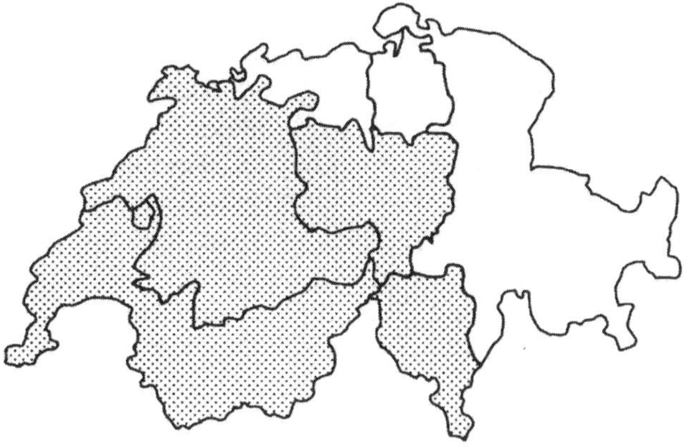

Figure 6.4 Poverty in Switzerland, 1992

The map highlights those regions where poverty lies above the Swiss average (5.6 per cent) after ordinary social service remittances had been paid.

Source: Suter and Mathey, 2002, p.35

There is a further aspect of poverty that is never taken into account. People who are rich by money standards may be spiritually poor, which manifests itself in an endless greed for more money, more power, more material goods etc. Similar to marginality, poverty is to some extent a matter of the mind, in particular if the basic needs can be (totally or almost entirely) satisfied. Finally, poverty is often a combination of factors internal and external to the individual and the group: the faculty to see a chance and seize it (or not) may combine with the general economic situation at a given moment in time, the attitude of the socio-political environment towards disadvantaged groups, attitudes towards schooling, personal difficulties at a given moment etc.

These statements demonstrate that poverty depends on specific contexts and a variety of factors combining objective with subjective elements. Based on such considerations, Hemmer and Wilhelm (2000) have enlarged the notion of poverty by taking various dimensions into account and putting the material aspects into a more general perspective (Table 6.2).

Table 6.2 Duality of the poverty concepts

1	absolute poverty	relative poverty
2	material poverty	immaterial poverty
3	objective poverty	subjective poverty
4	primary poverty	secondary poverty
5	voluntary poverty	involuntary poverty
6	temporary poverty	permanent poverty

Source: Hemmer and Wilhelm, 2002, p.3

The two authors demonstrate in this way the inadequacy of the simplistic World Bank value. The conventional view holds that poverty is a state of destitution, a situation that cannot be changed. To distribute money in order to alleviate material poverty is one possible method, but the faculty to deal with it in a sustainable way does not automatically accompany the money.

Briefly, the six types of poverty can be characterized as follows (Hemmer and Wilhelm, 2000, pp.4-9): Type 1) is based on the satisfaction of people's basic needs, in an absolute way and by comparison to other individuals, the distribution of means within a group. Type 2) relates to goods and services (material side) and the lack of 'meta-economic values, i.e. ethical, religious and cultural factors' (ibid., p.6). Type 3) is a matter of perspective (who assigns poverty to whom?). Type 4) is interesting because it refers to the availability of resources and the way these resources are used or allocated. Voluntary poverty (type 5) may occur when people spend money on immaterial purposes and will thus lack the resources to cover their basic material needs. Type 6), finally, points to particular events that may render people poor for a certain time because of climatic fluctuations (droughts), natural catastrophes (floods, landslides etc.), economic fluctuations, illness, etc. Permanence may exist when individuals or household lack the material and immaterial sources to quit what may be a vicious cycle.

Poverty also has a political component. The public awareness of the problem may lead to policy decisions and the creation of corresponding instruments to alleviate poverty – it all depends on the political will to distribute the resources and the wealth or to reserve them to a few. Such a policy, however, requires long-term thinking. Governments are asked to govern (i.e. to look ahead, beyond the next elections), not to administer the current tasks like any day-to-day business. Unfortunately, 'the contemporary political culture is rushed and tends to favour the superficial and the spectacular' (Baudot, 2000, p.64). This is one facet of the *Zeitgeist*. A second aspect 'is the lack of interest for issues of equity and *a fortiori* equality between individuals, social groups, communities, countries and regions' (ibid., p.65). In this way, the poor will remain useless because they are seen as unproductive members of society or of the world community.

Poverty in cities

Although cities can be described as a sort of centres of wealth, they are in reality mirrors of the entire society. Affluence meets poverty, sometimes within very short distances, both in the spatial organization of a town (rich and poor areas) and in daily life (the beggar on the High Street watching the luxury limousines pass by). Segregation dominates; there is hardly any exchange between these two worlds. This kind of marginality is found at a very large spatial scale, it has consequently been termed microperipherality (Blom, 1998), or micromarginality (Mehretu et al., 1996).

Social segregation produces the spatial coexistence of affluent and poor regions. For the past 30 years, the phenomenon of gated communities has diffused from the US in particular to Latin America, creating, as Coy and Pöhler (2002, p.269) put it, 'islands of the rich in the oceans of the poor'. To the visible walls and fences erected to separate the privileged from the less fortunate comes the invisible wall between these two groups, a wall constructed of a mutual fear: the inhabitants of gated communities fear for their security, those outside fear the uncertainty of an unfamiliar area.

Researchers of this phenomenon have controversial opinions as to its origin. To some the security issue dominates, and they see people's wish to lead a life free from criminality and harassment as the primordial drive for deciding to live in such a 'ghetto'. This reasoning is supported by studies in a number of Latin American cities (Coy and Pöhler, 2002; Kanitscheider, 2002; Kohler, 2002). Others see this aspect as merely one of several reasons for the emergence of this form of urban fragmentation. Fischer and Parnreiter (2002) do not exclude the two related aspects of criminality and poverty, but they also point to tendencies of the liberalization of global capital flows and the gradual withdrawal of the state whose regulations are not always taken seriously (Kohler, 2002, p.279). The existence of 'superfluous capital' (Fischer and Parnreiter, 2002, p.247) is an important motor for investment in luxury apartments that will inevitably be reserved for the upper class. Janoschka (2002) is even more radical. To him, referring to criminality as the sole motive for socio-spatial segregation is simplifying a complex process. It is obviously a useful argument for the builders of such complexes (p.296), who play with the fear of the inhabitants. Personal security (for one's life and material goods) is indeed a major problem in our secularized society, and the public hysteria is cleverly used by many economic agents, and gated communities are the fruit of this fear. However, Janoschka (2002) points to other factors such as the public neglect of urban infrastructure, the wish for life in a socially more or less homogeneous environment, changes in lifestyle, and the general dynamics of the cities. Whilst violence is not to be neglected, he sees it as one factor among many, a point where he meets Fischer and Parnreiter (2002). Even Kohler (2002) in her study on Quito demonstrates that the quality of life is an essential element in the segregation process, and that the well-off obviously have a larger choice than the poor.

An interesting phenomenon is likely to accompany the gated communities. In many cases the inhabitants have to look entirely for themselves, organizing their own essential infrastructure, building and maintaining the roads within their condominium, perhaps even private schools. To some extent they live separate from the social reality, in their own well-organized world. In such a context, a feeling of solidarity and common responsibility must needs arise, otherwise the community could not function (Coy and Pöhler, 2002, p.275). It is not just paying money; it is also taking part in a decision-making process that usually takes place within the public administration and with the politicians. Thus, while on the one hand the gated community favours secular values (individualism) it also promotes a return to sacred values (common responsibility). However, this solidarity is a within-group phenomenon and does not extend to the other social groups that remain excluded.

To measure poverty on the basis of indicators, we can use, for example, the number of persons or households addressing themselves to social welfare institutions in order to receive income support. However, this goes along with immaterial aspects of poverty mentioned above. Studies in the United States have shown that poverty is based on factors like education, employment and ethnic structure (Mehretu et al., 1996; Blom, 2000). Poverty breeds apathy and negligence, as has been shown by Jargowsky (1996, p.11); he describes areas with a high (statistical) poverty rate as characterized by 'dilapidated housing, vacant units with broken or boarded-up windows, abandoned and burned-out cars, and men 'hanging out' on street corners.' Similar observations have been made in Britain (Coleman, 1985; Smith, 1989). In high-poverty areas, unemployment is therefore a major problem. In neighbourhoods with (statistically) less poverty, the outer appearance was generally more pleasant, and the majority of the inhabitants appeared to belong to the working class or the lower middle class (Jargowsky, 1996, p.11). From Blom's study of New York emerges that in Manhattan districts with a large proportion of white population, the percentage of people receiving income support is relatively low (Blom, 2000, p.189). A further example of a city centre containing poverty islands is London where we encounter 'the concentration in the inner city of what is frequently termed an 'underclass' population. These are the poor and the disadvantaged who find it difficult to participate in the changing urban economy.' (Diamond, 1991, p.219). The same phenomenon can be observed almost everywhere, and it is usually a combination of ethnic, income and educational segregation which characterize the poverty islands: the inhabitants are essentially immigrants who work in low-pay jobs for unskilled manpower (ibid., p.222). Blom (2000, p.181) adds another social dimension: 'early retirees, single-person households and single mothers' display a particularly high percentage among the residents in inner cities in Sweden – we could probably add elderly people in general. These persons belong to groups who usually have to cope with limited financial resources and can thus be classed among the poor or near-poor population. This is the material side of poverty.

An important element in poverty (and marginality) in the urban setting in general is the lack of interpersonal contact, the anonymity of the society: 'despite the fact that there are a large number of people in a limited space, few know each other personally.' (ibid., p.182). This is the immaterial side, less easy to seize but maybe even more important than the material aspects. The quality of life is not only made up of money and physical objects but comprises social and psychological elements and spirituality. Immaterial poverty and marginalization is in fact independent of the affluence of the persons; wealth and an expensive apartment cannot guarantee true friendship and happiness. Economic affluence and psychic or spiritual poverty meet in the same microcosm (family, individual).

Rural to urban migration flows account for large poverty areas in cities of the South. Their roots lie in the colonial period when cities were founded as centres for the colonial administration and trade with the motherland. Expectations in Zimbabwe as elsewhere in Africa were that the city would be reserved for the Europeans and those indigenous people who had found employment, whereas 'the unemployed would always return to their Tribal Trust Lands (renamed Communal Lands) in the rural areas set aside for them' (Mutambirwa and Mehretu, 2003, p.215). This, of course, did not happen, and the demand for residential areas for natives increased steadily. The city of Harare (Salisbury prior to independence) was divided between white and black neighbourhoods, the former lying closer to the CBD, the latter further and further away. There is only one African township close to the centre (Mbare), built in the early years of Harare, which, at that time (1907), was relatively small. Racial segregation came to an end with independence (1980), but little changed for the inhabitants, and 'the majority of African residents in Harare continue to live in varied conditions of racially based contingent and systemic margins of the city.' (ibid., p.220). In concrete terms this means that people suffer under competitive inequality in the market economy (contingent marginality) and a predominantly top-down policy (systemic marginality). Marginality is primarily of an 'economic' nature; the inhabitants in the high-density suburbs lack the means to change their lifestyles (ibid.). They are practically confined to their residential areas as the cost in time and money to travel to the centre is prohibitive. Poverty thus resides at the outskirts of Harare. Even Mbare that lies fairly close to the centre has not escaped this fate. As the oldest African township 'it has been a residential area of first choice for most low-income families and a first stopover for most new African immigrants into Harare.' (ibid., p.221).

The coexistence of rich and poor in urban settings is characteristic for all countries, but not everywhere can solutions be found to marry the interests of both groups as they have been reported from Bangkok (Bähr and Mertins, 2000, p.21). The marginal residential quarters in the Thai capital are concentrated in flood-prone area on private land. The poor inhabitants occupy the land plots on the basis of a lease. When the landowner decides to develop his property, he will proceed to a sort of 'land sharing': supported by public subsidies, he will use a part of the land to build blocks of flats for the poor who can thus obtain a simple owner-occupied flat, and use the remainder of his plot for commercial purposes. Such agreements are based on the leasehold contracts between the landowners and the poor dweller, but also on 'religious-moral motives' (ibid.).

Despite the many efforts to eradicate poverty worldwide, we shall have to live with it unless the 'rich' change their attitude towards the 'poor'. The process of ex-urbanization has even aggravated the situation of the cities. This can be observed even in a rich country like Switzerland, where the population in large cities displays a relatively low average income compared to the well-to-do suburbs. The figures of the mean tax income of the direct federal tax, presented in Chapter 4 and exemplified by the Zurich conurbation are eloquent in this respect. Certainly, Zürich is not a poor city, but compared to the suburbs she looks it. This is also manifest in the local taxes which – though variable from one year to another – are considerably higher in the city than in municipalities on the 'Gold Coast' (factor 1.6 in 2002).

Rural poverty – a forgotten problem

In an urbanized world, problems in cities are usually taken very seriously because they touch regions with the highest concentration and a great variety of population. Residential segregation concerns many people (in absolute and relative terms), and poverty issues can also be seen from the angle of a town's image.

The countryside, for some reason or another, is simply forgotten. Few people live there; they are a *quantité négligeable* in a world where quantity counts more than quality. The rural world is often seen from the idyllic perspective of the farmer who as an independent person produces sufficient food for himself and is cunning enough to make a comfortable living at the expense of the rest of a population. The reality is, of course, different. Farmers are in a certain way a marginalized group, even if many of them manage to live comfortably thanks to standardized mass production and state subsidies and price guarantees. They have to produce the food we eat, but the price must be as low as possible (see Figure 6.5), they often have to submit to specific regulations of an agricultural policy that restricts their independence; they depend on the weather and thus have no guaranteed regular income. These factors may explain why farmers in Switzerland are the socio-economic group that spends the highest percentage of its income on food (see below). The opening of the global markets has not been to the advantage of everyone. Swiss farmers emigrated in the 1880s and 1890s because their production was no longer competitive when cheap grain began to be imported from North America. Reports from contemporary Russia and other countries of the former USSR show that the lack of a coherent farming strategy and the image of the city as paradise leaves entire stretches of the countryside in poverty. The young move out, the old stay behind, supplied with an inadequate pension and an infrastructure that is far from what the people used to know in the old days of communism.

In their study on Finland, Rusanen et al. (2002) demonstrate the convergence of poverty in areas of very high and of very low population densities, urban centres and the rural periphery. These two extremes not only had the lowest disposable incomes but also the lowest state taxable incomes. Hence they profited from the highest income transfers (p.52 f.). Apart from this rather coarse differentiation, the authors also recognized a particular spatial pattern of poverty: very high in the urban centres, it declined rapidly in the immediate urban periphery, and gradually increased the further away one moved from the centre. Between the centre-and-periphery, the poverty gradient develops in an almost linear manner, although there are some irregularities that still await explanation (ibid., p.57).

High birth rates and a fast growing population in rural areas of the South are often at the roots of poverty. Scholz (2000) and Wälty (2000) describe the difficulties rural populations in Java and Bali (Indonesia) faced and have in part still to live with. The colonial economy and the Japanese occupation during World War II have left traces in the rural economy that could only slowly recover. In either case, subsistence agriculture had to be replaced by surplus production and market orientation. Rural Java is the chief rice producer of Indonesia, the basic food of the entire population. New brands of rice, developed by the International Rice Research Institute IRRI, were introduced from 1968 onwards (Scholz, 2000, p.15) and permitted three harvests per year instead of two. Together with technical improvements in the fields of irrigation and use of fertilizer the production could be more than doubled (from 2.6 to 5.4 tons per hectare). Commercialization and mechanization were the consequences (ibid., p.15 f.). In remote rural Bali, on the other hand, various cash crops came to dominate the rural economy of the northern part of the island. Coffee, cloves, oranges, tangerines and vanilla are certainly a good mix of cultures, but their price is not determined locally but in a global context (Wälty, 2000, p.29). Thus the well-being of the population largely depends on external factors, i.e. the world-market demand for coffee, cloves, vanilla etc. and on the competition with other producers in the South.

Poverty is thus a field where urban centres and the rural periphery converge. The difference may be that with increasing urbanization, rural areas are upgraded in people's perception and may profit of the exurbanization process, whereas in inner cities poverty will continue to increase. 'The continuing trend towards metropolitan deconcentration also contributes to neighborhood poverty' (Jargowsky, 1996, p.211).

Working poor: a new face of poverty

Poverty is not restricted to persons who cannot find work or who live on a small pension only. Neither is it limited to countries of the South. In recent years, a new form of poverty has emerged in the rich industrialized countries, a phenomenon that sounds astounding but has become a reality with which we have to live. The representatives of this 'new poverty' are the 'working poor', people who have a job (often full-time) but whose salary is insufficient to cover the basic requirements of their household and have to demand financial assistance from the public welfare system. Victims are single-parent families, families with small children who cannot afford the costs of a baby-sitter, a day-mother, or a nursery, but also retired people whose pension does not keep pace with inflation. Their financial means may suffice to cover the material basic needs but not the social ones. Here they will arrive at the limits of the household's financial capacity. While in this particular case financial poverty is to some extent related to a person's lifestyle and his or her ability to manage the personal finances, it is nevertheless an indicator that something is wrong in society.

Basic requirements in the industrialized countries of the North have changed radically since the middle of the 20th century. The period of affluence has brought about an unprecedented increase of material goods that were deemed essential for a modern life. This movement has not to the least been influenced by the application of modern psychological knowledge in marketing, which resulted in the growth of the publicity and advertisements sector in the media (see p.80). Not to follow current trends in material equipment and lifestyle meant to be excluded from the mainstream and to be marginalized. The consumption pattern has changed accordingly, as can be seen from the changes in household expenditure in Switzerland (Figure 6.5). The mean figures are, of course, not representative of the reality of low-income families. They demonstrate, however, that simply in the 15 years between 1985 and 2000, the cost of food has declined markedly, while insurances and taxes are on a steady increase.

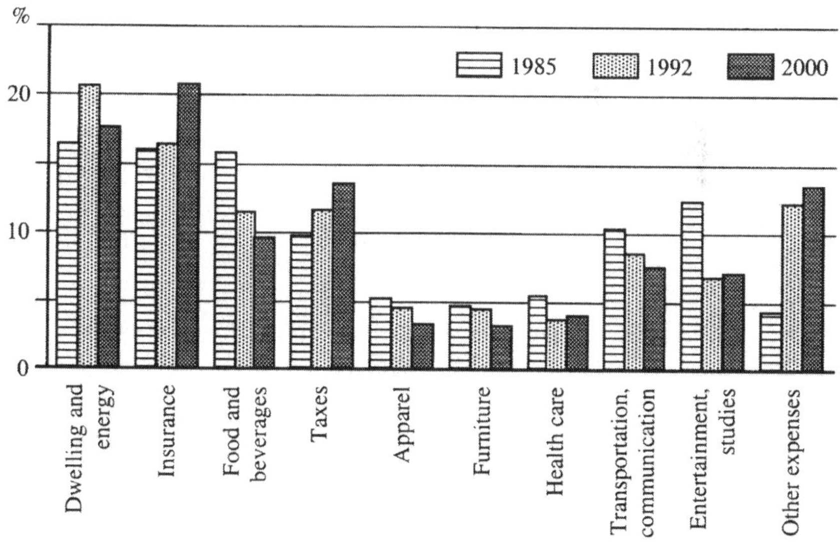

Figure 6.5 Mean household expenditure in Switzerland, 1985-2000 (in per cent)

Source: Statistical Yearbook of Switzerland, 1986, 2000, 2002

Household expenditure surveys in Switzerland reach back to 1912. Before 1975, the statistics distinguished between 'workers' households' and 'employees' households'; this differentiation permits us to discover that the workers were generally less well-off (not really poor) and had to use more of their income simply for food (Table 6.3). Since 1975, this distinction is no longer made and we have to limit ourselves to average figures.

Table 6.3 Food expenditure in Switzerland as per cent of total household expenditure during the 20th century

	Workers	Employees
1912	44.1	36.5
1920	44.9	36.2
1936/7	32.9	24.8
1945	37.6	30.6
1955	32.2	25.6
1965	25.0	20.9
1973	18.7	15.6

Source: Statistical Yearbooks of Switzerland

Looking at the difference in price of food, an essential part of life, one can see that it has lost quite a bit of its former appreciation. Food has become a simple commodity, but people with a low income still need a substantial part of their salary to survive in the physical sense, and even more if they look for high quality food from organic production. The two socio-economic groups who spent most of their income on food in 2000 were retired people (8.5 per cent of a mean monthly income of 5,761 Swiss francs) and, interestingly enough, the farmers (15.9 per cent of 6,352 francs). In general, they are not poor people, but the average values hide the reality of the elderly widow or the small farmer who live at the limit of existence. The real new poor, the working poor, do not show up in the statistics because they are few in number (about 250,000 in 1999, 190,000 of whom are full-time workers) and hardly influence the calculation of mean values.

Working poor are essentially foreign workers (12.2 per cent in 1999), single-parent (especially single-mother) families (29.2 per cent), and families with three and more children (18.0 per cent). The statistics (which reach back to 1992 only) are silent about details of the qualifications of the working poor, but they indicate that about one third has no further training beyond compulsory school (FOS, 2000, p.527) and are therefore working in low-pay jobs. Temporary workers and people with contracts of a limited duration are particularly vulnerable, and women are usually more concerned than men. From a regional perspective, the French speaking part of Switzerland has more working poor than the German or the Italian speaking regions.

There is an official poverty line, fixed according to household size and taking regional variations into account. For a single-person household, it varies between 1,055 and 1,165 Swiss francs per month, for a four-person household it oscillates between 2,260 and 2,495 francs. There is no minimum-salary regulation in Switzerland, and the working poor have thus no legal possibility to demand higher salaries. Even fixing a minimum salary of 35,000 francs would not greatly reduce the share of working poor because of the considerable social and regional differences (ibid.,). The problem has to be tackled from various angles; the regional factor cannot be eliminated, but education and an increase of stable (permanent) jobs could offer a solution. Farming, retail trade and restaurants are occupations where there is a high risk to be a working poor (ibid., p.528).

Poor is not poor, there are at least three distinct groups: those who have somehow lost the contact to reality and end up as vagrants, those who are marginalized for ethnic reasons, and those who work regularly but whose salary does not suffice to ensure decent survival without social service. The former are visible in the streets, but the latter disappear in the mass of the population who is usually not aware of them.

Language

The growing influence of the English language has been at the origin of a controversy in Switzerland on the first foreign language to be taught at school. Currently, French comes first in the German-speaking part of the country, and German in the French-speaking region. The Italian speakers are obviously in the difficult situation to choose between the two or to learn both at the same time. French is easier from a linguistic point of view, German is more important for political and economic reasons. The losers in the 'game' are the Romansh speakers, the small minority in the Grisons whose language is in fact Latinized Celtic. They are practically all bilingual (usually with German, some also with Italian). The controversy started when the Canton of Zurich wanted to introduce English on primary school level and thus replace French, one of our three official languages. The law was refused by a referendum in 2002, but in 2003 it was decided to proceed with 'early English' as planned.

The discussions in Switzerland about teaching and using English mirror the status this language has achieved as a global communication vehicle since World War II, replacing French that had held an important position until then. After 1945, the world was systematically anglicized as a consequence of the Allied victory. The Marshall Plan and the Bretton Woods institutions (World Bank, International Monetary Fund) were and are run from the United States and contributed to the diffusion of the language and terminology – and they were readily adopted around the globe. In a way, this is a better solution than resorting to one of the many artificial languages (such as Esperanto) that have been developed but lack the vitality and cultural background of a living language. However, national languages have to some extent become marginalized, at least in certain contexts. Internal scientific cooperation in Switzerland may at times well have to resort to English because not all participants in a group understand the language used by his or her partners – speaking is not necessarily required. The French have to some extent resisted this trend by deliberately translating at least part of the terminology into their own idiom (*l'ordinateur* for the computer, *le logiciel* for the programme), and so have the Finns who managed to differentiate between the (political) satellite (*satelliti*) and the man made satellite that orbits the earth (*tekokuu*, i.e. made moon). By simply adopting a terminology a language looses its vitality, and the linguistic diversity is menaced.

German, French and Italian are the three official languages in Switzerland. All legal texts must be published in the three idioms, and all three versions are qualified as originals. The fourth language, Romansh, has the constitutional status of a national language, adopted in a referendum in 1938 under the menace of territorial expansion by Fascist Italy. It has maintained this position in the new Constitution of 1999. However, certain legal texts are nowadays translated into the new high-language. Romansh is split into five distinct dialects that have developed in isolation. After various attempts to create a common high-language since the early 19th century (Gloor et. al, 1996, p.59), success has been reached in 1985, when a grammar and a dictionary in *Rumantsch Grischun* were presented (ibid., p.62). Acceptance was not unanimous in the beginning, but the steady use of the 'new language' in the media has helped to overcome major obstacles. However, it still leads a marginal existence. The hope lies now with the young generation that may be more open to it.

In Switzerland, German is the uncontested majority language, followed by French. Since the end of World War II, however, there has been an interesting shift in the distribution of languages (Figure 6.6). Immigration has resulted in an increase of both Italian speakers and other languages, among which we find Spanish, Portuguese, Slav languages, Turkish, to name but a few European tongues. Serbian and Croatian in fact ranked fourth in the census of 2000. The decline of German and Romansh is relative rather than absolute, but the trend continues. While the Italian speakers had little difficulties in becoming integrated, people of the first generation immigrants who spoke more remote languages were facing social isolation. Second and third generations, on the other hand, could link up well with the Swiss society and culture.

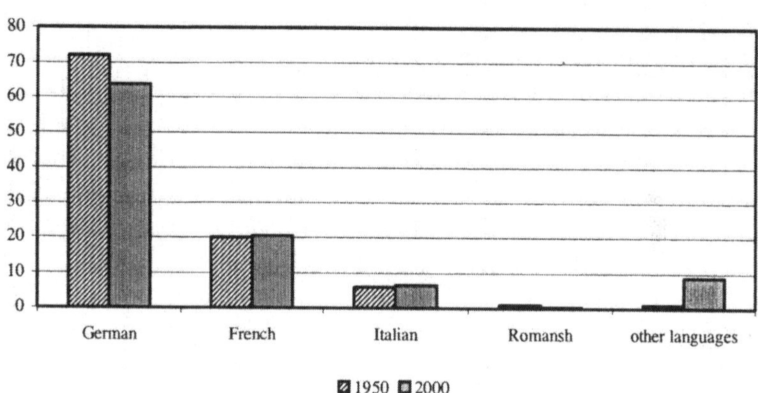

Figure 6.6 Languages in Switzerland, 1950 and 2000 (per cent of resident population)

Source: Statistical Yearbooks of Switzerland

Language is an expression of thinking and is one of the many means to declare one's identity. It is also a means of communication and serves the organization of daily life and the transmission of information and social rules (Raffestin, 1980, p.87). This is a plausible (political) motive to restrict the number of official languages spoken and written in a country, and the globalization process since World War II has favoured changes in language use and other fields of culture (Inoue, 1997a, p.14). Linguistic diversity, however, is an indicator of the cultural richness of the world, of a continent and even a country. Europe has had a plurilingual tradition for centuries, although it was mainly confined to the intellectuals (scholars, writers, composers; Pfeiffer 1997, p.190). This tradition continues in smaller countries surrounded by large neighbours, in particular 'Luxemburg, Belgium, Switzerland, and to a lesser extent Denmark and the Netherlands.' (ibid., p.190 f.). It is even a declared policy of the European Union to maintain this rich tradition, despite the technical difficulties it implies and that will inevitably increase with the admission of new members. However, with currently 11 official languages in the European Union, one can say linguistic diversity is taken seriously (and offers jobs for translators).

The marginalization of languages takes place in a different context. It is indigenous languages that are most often threatened by extinction, idioms spoken by minorities who in this way manage to uphold some sort of cultural particularity. Tsao (1997) points out that 'the erosion of indigenous languages and cultures' is due to two main reasons: the propagation of one national language (to promote national unity), and the worldwide expansion of languages of wider communication (LWC), especially English (p.102). There is little that can be done to oppose the latter, but the former tendency can (theoretically) be averted. The role of a language is intimately related to the power structure of its society (ibid., p.89). 'To rule a monolingual country is certainly easier than to be confronted with a variety of languages' (Leimgruber, 2003b, p.245). The idea of 'one nation, one language' underlies the state philosophy of many countries and is often driven by some form of nationalist fanaticism: the Spanish dictator Franco prohibited the use of the Catalan language, Turkish is the only national language permitted in Turkey, and the Taiwanese government promoted Mandarin as the sole official language of the island. In the latter case, we have to bear in mind that the nationalist government who retreated to the island in 1949 gave priority to the task of introducing Mandarin in order to unite the entire population around one common idiom. Only at a later stage did it start to promote the feeling of 'ethnic authenticity among its ethnolinguistic populations' with the goal of 'creating a sense of nationhood' (Young, 1988, p.337).

France is another case in point. From the 16th century onwards, the French raised their language above all other idioms spoken on the territory of the hexagon. In 1539, King Francis I decreed the Paris idiom as the official language of his realm (Edict of Villers-Cotterêts); in 1635, Cardinal Richelieu established the *Académie Française* and thus demonstrated 'the political weight he attached to a standard language' (Brücher, 1992, p.47). The Revolution of 1789 reinforced the position of French; all other languages (as far as they had survived) were considered inferior dialects which had to be annihilated (Raffestin, 1980, p.98). Language diversity was undesired and socially unacceptable, and it has remained so until France embarked on decentralization in the 1980s. Through this process, attitudes gradually began to change and a slow revival of regional languages can be discerned since (Leimgruber, 2003b, p.245). However, the situation continues to be complicated. 'The 1992 Constitution declares French the official language of the Republic, including the overseas departments and territories (Sibille, 2000, p.195). As a consequence, the proposal to declare Tahitian a second official language in French Polynesia was rejected by the Constitutional Council in 1996 on the grounds that it was anticonstitutional.' (Leimgruber, 2003b, p.247). As a member state of the European Union, France found herself forced to take position on the European Charter on Regional or Minority Languages. The government had signed it in May 1999, and Parliament was about to ratify it, but the Constitutional Council declared this charter anticonstitutional and prevented the ratification. There is a mismatch between the French Constitution and European Law, maybe a minor problem but it is a thorn in the flesh.

The case of languages illustrates the loss of cultural diversity, as Trudgill (1991, p.62) noted: 'In these last years of the 20th century, languages are dying out at an increasingly catastrophic rate, without being replaced.' Although this has been a standard process throughout history, the current situation is alarming because it 'is the speed with which languages are dying out and the extreme improbability of their being replaced in the traditional Stammbaum manner because of modern demographic and communications conditions.' (ibid.). This process also concerns varieties within languages. Many islands in the Atlantic and Pacific were at one time settled by British whalers or served as refuge for shipwrecked crews or mutineers (e.g. the Bounty mutineers on Pitcairn); they lived in isolation and maintained a specific state of the English language. Nowadays the young generation is emigrating from such remote places, and these specific languages are slowly disappearing ('English as an endangered language', Chapter 14 in Trudgill, 2002, pp.150-158). This process is of course part of the long-term evolution of every cultural phenomenon. Contrary to physical artefacts (houses, furniture, paintings and sculptures), a language cannot be preserved as such. Written witnesses may survive, but spoken recordings can only survive if the technology of sound reproduction in the future will allow the playing of the medium that was used for sound recording in the past. The largest record or tape collection is worthless without the correct record or tape player.

Language is not only an objective reality; it is also a social construct. The marginalization or the disappearance of minority languages can be interpreted as a threat to democracy. Linguistic uniformity suppresses ways of thinking and expression, it is 'a clear obstacle to a democratic control of the choices made by a handful of scientists. It discards the public national spheres from any participation in the scientific enterprise, ...' (Pfeiffer, 1997, p.198). To some extent this can be witnessed in the relations between science and the public. Science has developed a language of its own, incomprehensible to those persons who have not proceeded beyond compulsory education. But also within the scientific community, numerous specialized jargons have evolved that resemble secret languages. Understanding a colleague inside the same faculty may still be possible, but across faculty limits it may become impossible. At times, specialization inside disciplines can result in mutual incomprehension because of the different jargons in use. Under such circumstances, the 'ordinary citizen' (who is also the taxpayer and in this way helps to finance the universities) will be completely marginalized.

Religion

A first introduction to the topic of religion has been provided in Chapter 3. Religion is about a person's relation to a Supreme Being or a Supreme Force, God, the Creator, the supernatural or spiritual world. Religion is not the same as spirituality, although the two are related. Spirituality is limited to the immaterial sphere and has to do with mysticism, attitudes and the personal conscience, whereas religion is connected to material manifestations such as a (hierarchical) organization, rules, rites, social networks, buildings, doctrine and dogma, etc. The two are interrelated, but not mutually exchangeable. Simply said: no religion can exist without spirituality, but spirituality lies outside religion.

The author has been brought up in the Christian tradition, and his thinking has been deeply influenced by it. Belief in what the institution behind a religion declares has been taught to him as to every adept of a religion. However, the contact to other religions has permitted to detect similarities and differences. The similarities in the spirituality of all religions are striking, and so is the influence exercised by the human representatives of a religion on their respective followers. The priests (to choose a general term) used to be viewed as representatives of God, and their word posses considerable weight. They influenced social and political life, and in many regions of the world they still do.

As a consequence of Enlightenment thinking, Christianity has radically discarded spirituality from daily life, but retained and intensely cultivated its institutional sides. The Enlightenment revolution has replaced the belief in mysticism by faith in the rational, in science and technology (see Figure 3.1); the personified God was gradually superseded by the human race.

Following the process of rationalization, the Church institutions (both Catholic and Protestant) lost a large part of their influence on most members of society. In a hedonistic world, words of moral count little, and a priest who does not live up to what he preaches will lose his 'sheep'.

People no longer know the significance of certain events that play an important role in our religion. Pentecost, for example, suggests nothing to most people 'aside from the fact that it means a long weekend off' (Voyé, 1997, p.159). Writing from her Belgian background, Voyé deplores the fact that most people no longer remember the role of the Church in the building of her country, of the development of the school and the hospital system, and they have forgotten its 'unique mode of conflict management, in virtue of which it is often consulted by various countries needing to solve cultural, i.e., ethnic and religious problems analogous to those solved by Belgium without bloodshed and without thus far being torn apart' (ibid.). The Church is obviously perceived as intruding into one's private sphere, the mystic message is no longer understood, but maybe it is also badly conveyed. The worldly role with a spiritual background (conflict management) played by the Church is rather seen as a project to re-conquer the worldly hegemony it has lost than as an effort to re-introduce spirituality. After all, the hierarchy within the Church is a mirror of the political and the military hierarchy (as is demonstrated by the organization of the Salvation Army); contrary to the state and army institutions, however, the Church is a voluntary organization where the superior is not necessarily taken seriously by the public. By emphasizing the external manifestations of the dogma, the Church has itself contributed to the secularization from the Age of Enlightenment onwards. The way it handled the claims by Martin Luther, the Augustine monk who eventually started the Reformation, is a clear expression of the conviction that its behaviour was beyond criticism. Luther protested against the commercialization of absolution, an abuse that could not be tolerated because it was directly opposed to the spiritual side of the confession and the absolution of sins. But instead of examining its own errors, the Church played the power card and banned Luther. This way of acting is a simple way out of self-interrogation, and it can be observed right into the 21st century! Reformers are people who point to errors in a system; for this they are marginalized. Jesus was a reformer, a rebel, an innovator, misunderstood by the masses but well understood by the ruling class – and for this he was crucified.

The Church lives in an ambiguous situation: it sees itself as a moral authority, but it also wants to assume the function of disciplinary authority (ibid., p.173) and continue to exercise its role of censor over the dogma This seems to be unrealistic in an epoch when personal freedom is advocated. It is true that we have been given the free will, but once we use it, we also have to bear the consequences – something most people do not contemplate. The Church has lost its role of a spiritual guide but has developed a power structure with claims to hegemony over the human conscience. The marginalization of religion in large parts of the Christian world finds a ready explanation in this ambiguity.

However, spirituality has not entirely disappeared, but it has been separated from the traditional Christian religious institutions. In Europe, in particular, we witness a resurgence of all kinds of spiritual movements, some linked to Christianity, some seeking support in the Eastern religions (Buddhism, Hinduism), and all intending to fill in the vacuum left by the traditional Churches. Needless to say that there are also groups with purely materialist goals and a power structure of their own, which, however, usually remains covert and discrete.

The process of globalization has also interested religion. Knowledge about other religions has been diffused over the new channels of electronic communication (the 'Online Bible', for example), and proselytising has become very easy. The new information and communication technology can also be used in a negative sense. Inoue (1997b, p.94 f.) recalls the case of the Aum Shinrikyō (the sect that launched the Sarin gas attack in Tokyo on March 20, 1995). 'The doctrines of Aum Shinrikyō were a hodgepodge adopted from a variety of existing religions' (pp.94 f.). They recruited adherents from around Japan and as far away as Russia, and they mixed elements of Japanese with Christian and Buddhist religion. They proselytized in particular among young people who were experts in natural science (ibid., p.95), combining a 'heavenly message' with deadly objectives.

Inoue points to an important process that is occurring in religious life as a consequence of globalization. 'As the traditional religions lose their overall legitimacy and authority, the social integrating role of religion will be weakened' (ibid.). Through the new information and communication media, knowledge about other religions and their rituals are diffused worldwide. The younger generation readily captures these new ideas because they offer a change from traditional routine. 'For example, traditional religious observances [in Japan; WL] celebrating New Year are decreasing, while those centred on Christmas or other religious holidays of foreign origin are becoming more popular. One sees increasing unconcern with the Buddhist seasonal holidays of *setsubun* or *higan*, and more with St. Valentine's Day and Halloween. Needless to say, the latter are not being accepted as 'religious' rituals, but only as a kind of non-everyday event or secular festival' (ibid., p.95 f.). Traditional religious celebrations are therefore no longer imbued with mysticism but are simply ordinary festivities that have lost their meanings – just as Pentecost in a large part of Christianity.

Concluding remarks

The present chapter has attempted to highlight a number of social and cultural elements and their relations to marginalization and globalization. It can be read as a critique of modernity, although the processes of ethnic segregation, pauperisation, dynamics of language and opposition to religion (or rather to religious leaders) are not new at all. Globalization has, however, had a number of specific consequences. In particular, and this is a positive point, it has permitted ideas and information to spread around the globe within almost no time. To witness what is happening around the world, how people live and cope with new situations may be beneficial to us in our understanding of the world. The globalization process has intensified the confrontation with and created the potential to understand the other. However, and here comes a first warning, it can lead to a 'cultural mix' that may be neither fish nor fowl. The result may be a global culture that 'is here and now and everywhere,' a sort of huge 'cosmopolitan patchwork' (Smith, 1990, p.177). It is timeless, not linked to a specific space or place, and it uses 'its folk motifs in a spirit of detached playfulness' (ibid.). Such a diffuse global culture would lack the role as the creator of a specific regional and social identity; it would stand for mankind as such or for nothing at all. Viewed from this angle, a global culture is a 'stateless culture' (Inoue, 1997a, p.13). 'The globalization process results in the constant formation of new, but yet-unknown cultural alliances, alliances that differ from those which formerly were established by the nation or specific ethnic groups. As a result, it produces not only convenience but also a vague anxiety' (ibid., p.12). Another warning has been pronounced by Warner (1999, p.68): 'The universality of science is a Trojan horse that allows the western culture to enter the traditional societies by force' (transl. WL). From this statement one can understand why in Burkina Faso the traditional leadership must still be respected (at least *pro forma*) when the members of a NAAM group want to introduce innovations in agriculture (see Chapter 9). These critical remarks can be understood from the role culture has for its bearers. Culture is conservative, tending to look back in order to offer the individuals and the groups an anchor in known values. Globalization, on the other hand, is forward oriented, a child of the economy, business and technology, all of which are constantly striving for innovation. No wonder that globalization produces the fear of innovation.

We have not attempted to define and discuss the notion of culture at the outset of this chapter, but we want to conclude it by clarifying what culture means to us. We consider it the sum of values and customs inherited from the past, successively modified and completed and likely to be carried on into the future. It includes the world-view or the spiritual adherence (religion), ways of transmitting knowledge (language, education), social order (political and non-political), attitudes, taboos, techniques and material objects related to the way of living. Culture is specificity (Friedman, 1990, p.312), it is a plurality of ways of life (Ryu, 2000, p.26). It is a group-related concept although every person will have his and her own culture, depending on his and her experience in life. After all, it is the variety that makes the culture concept at the same time a difficult and a fascinating field of study.

Chapter 7

The Marginalization of the Environment

The topic of this chapter may appear out of place, but there is good reason to include it and dwell on it. Whilst marginality has always been looked at from the economic and social perspectives, the role of the environment in human thinking has hardly ever been considered. It has been accepted as something that is simply present, that offers the foundation of life, our resources, space, and that accepts our waste – but nobody has paid particular attention to the way humans really perceive it. It had been taken for granted and marginalized in our minds.

The chapter wants to address this problem by trying to answer a number of questions: to what extent has humanity excluded the ecosystem from its thinking and acting? Where are we now? What can we do and what are we doing to remedy the situation? We shall first look at changes in perception, afterwards discuss landscape transformations, nature protection, and the role of agriculture in this process, and conclude with a few thoughts on the problem of resources. This chapter intends to lift the marginality discourse on a less material level.

Shifting perceptions of the environment

In Chapter 3, we referred to Charles Dickens and George Orwell who described environmental degradation and pollution and thereby criticized the inconsiderate attitude of their respective society towards the environment. Long before politicians decided upon concrete measures, enlightened people complained about what has become a major issue since the industrial revolution and the source of heated debates in national parliaments and during international conferences. All of a sudden, the environment has obtained a central place in many people's thinking, while before it was somewhere at the margin (if at all). Awareness in the scientific community can be traced back to the 1950s when William L. Thomas organized the symposium 'Man's Role in Changing the Face of the Earth', chaired by C.O. Sauer (Williams, 1987, p.220). 'One of the first and most influential examples of a holistic, integrative interpretation of the past, present, and future' (ibid., p.218), its significance was not felt at the time (ibid., p.229), although the resulting book (Thomas, 1956) has later exercised a considerable influence on the geographical community. The central message of the book – the interplay between humans and the ecosystem – did not become a really geographical issue until the 1970s. The 'Man and Biosphere' programme, launched by UNESCO in 1970 and preceding the Stockholm conference on Man and the Environment (1972) can be seen as a starting point.

Since then, global environmental change has become a research field of its own, pioneered by geographers but multidisciplinary in character. Global in this context is seen as a spatial notion, implying '(i) the impact of events or processes that are global in their effects, such as volcanic eruptions; (ii) the impact of processes that are themselves global in extent, such as deforestation; and (iii) the impacts of events which occur at the global scale, such as climatic changes' (Thornes, 1995, p.358). To this we could add the thematic dimension: global environmental change affects not only single elements (such as the forests) but the ecosystem as well as the social system. Thornes' statement can be completed by another classification of the processes that transform our environment:

- natural processes beyond human control (earthquakes, volcanic eruptions, long-term climate change, hurricanes);
- natural processes as unintended consequences of human actions (soil erosion following forest clearing, desertification due to overgrazing, climate change due to the emission of greenhouse gases);
- intentional transformations (flood control, dams, afforestation, urbanization).

These processes act on different scales, from local via regional to global, and they will affect both the ecosystem and the inhabitants of villages, towns, valleys and entire continents. By tying them to the particular scale of the entire globe, one renders them more abstract, removes the threats from people's perceptions. Sometimes, the consequences are quite rapidly visible (a landslide or a flood), sometimes they can only be discerned after years of measurement (the steady retreat of alpine glaciers). To see behind these processes and their physical manifestations requires not only technical knowledge of the determinist natural processes but also insight into the irrational human being.

Focusing on marginality and the natural environment demands a shift in thinking. Marginality is not a topic within the ecosystem where every element has its place and role to play, whatever its position from the point of view of human interpretation. In the domains dealt with in Chapters 4 to 6, human decisions and actions were at the centre of reflection, but now the extra-human world will be the core of attention. We shall call this world 'nature', 'environment', 'ecosystem', and 'landscape', using these terms consciously as synonyms, although they in fact their meanings differ. While nature can be understood as including humans (as physical beings) as part of the animal realm (Cencini, 1999, p.279 f.), environment and ecosystem are terms used in science, emphasizing relationships rather than elements, 'relationships which, in most cases, are not captured by our senses' (Zerbi, 1999, p.269). Landscape, on the other hand, belongs to the domain of experience and of emotion (ibid.). By not differentiating between the four notions, we want to stress that behind them are deterministic processes, and their evolution is to a large extent spontaneous.

The concept of ecological marginality roots in the place the ecosystem occupies in the human mind. The role of values and world-views as fundamental drivers for our attitudes and perceptions has been discussed in Chapter 3 – they are the key-factors to understand the way humanity deals with the natural world around us. As long as the world-views were dominated by sacred values, the environment was imbued with holy spirits that inhabited both the a-biotic and the biotic environment. These spirits had a religious meaning, they were worshipped or at least respected – and they were a sort of guarantee that no excessive harm was done to the waters, forests or mountains concerned. The idea of the holy mountain was not just superstition but an expression of respect and hence a form of environmental law against abuse. Nature protection was thus guaranteed through shared values.

The perceptions of the environment and the attitudes towards it have varied over time and in different cultures. Native people ('uncivilized peoples', as they used to be called) usually display a high respect for nature, manifested in the attribute of sacredness. This does not preclude the use of the land for human needs, but it required a respectful utilization: 'The land was not untouched, but it was unspoiled' (Hughes, 1966, p.132). After all, the place of humans is inside the environment, not outside, but the growth of technology has led to the belief that mankind can survive (almost) without. Nature and environment have become marginalized in our thinking and acting (Leimgruber, 2000), despite overt social and political manifestations to the contrary (such as environmental associations, international environmental agencies, national legislations on the protection of the environment, etc.). The shift in the world-view and in values (Chapter 3) makes itself felt in a dramatic way.

The demystification of nature can be traced back to the Age of Enlightenment, however difficult it is to define this important period in history. With 'observation, classification and comparison' as the new scientific methods (Livingstone and Withers, 1999, p.1), the secrets of nature were gradually unveiled, and in the long run, paradise as 'nature bountiful and uninscribed' (Withers, 1999, p.87) has been replaced by the ecosystem. From this time onwards, humans have changed their attitude and examined nature increasingly from a utilitarian perspective – economic thinking took priority. The environment had since been taken for granted, perceived as the inexhaustible source of an unlimited number of resources, and as a useful bottomless sink for all sorts of waste, degradable or not. It was only in the late 20th century that there was a slow turnaround in thinking, epitomized by book titles that suggested changes in paradigms (Maude, 1975; Bendixson, 1977; Capra, 1982; Wilber, 1982; Abt, 1988). They were all published after the first report to the Club of Rome (Meadows et al., 1972), a book that had aroused controversial discussions by showing that humanity was approaching critical limits of the ecosystem and menacing its own existence. Investigations conducted 20 years later, confirmed the trend (Meadows et al., 1992).

This shift in perception has demonstrated that the intrinsic value of nature has been (re)discovered. This value 'is independent of mankind and requires a specific ethical attitude which allows each species and living community its right to exist' (Grosjean, 1984, p.74). Embedded in the intrinsic values is the aesthetic value, dominated by visual perception, although 'the total perception of a landscape cannot be separated from the other senses' (ibid., p.75). The perception of landscape is an important factor in the development of tourism; landscape is in fact the capital of tourism, 'a "raw material", the basis of the existence and the economic driving force of tourism' (Krippendorf, 1984, p.429).

The evolution of such attitudes over time is difficult to trace, given the lack of sources. In Europe, mountain areas have undergone a gradual re-evaluation over time, depending on the various actors and their social surroundings (Figure 7.1).

Fear has thus given way to admiration, respect, and then to curiosity. Once the mountains had lost their mythical attribute, their resource value was discovered and they were increasingly exploited. Only in recent decades, respect has returned. The various initiatives in the Alps show that the populations have become aware of the real threat to the alpine ecosystems. The Alpine Convention of 1991 (signed by the seven countries that share this mountain arc) and the referendum in Switzerland in 1994 on the protection of the Alps from excessive transit traffic (see Prologue) are indicators of a new awareness. It may be very difficult to transmit this feeling to everybody, and the fact that the Alpine Convention has not yet been completely ratified by the Swiss parliament demonstrates the plurality of interests that are at stake: regional transportation, future economic growth, installations for tourism. The discussions on road and rail transit traffic across Europe (through the Alps) are an excellent illustration of the dilemma between economy and ecology. As long as Europe views commodity transportation solely from the point of view of cost, there will be no reduction of mobility. Under such conditions, the environment is still marginalized, as the price of fuel is too low. The railway is not competitive yet, and such conditions do not promote a modal change (Tiefenthaler, 1990, p.158).

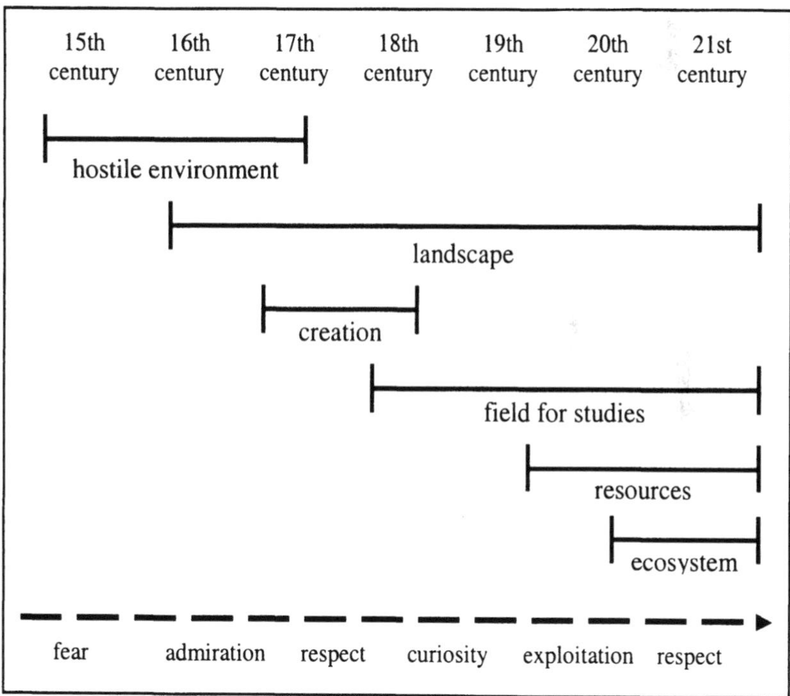

Figure 7.1 The changing perception of and attitudes towards mountains

Source: Leimgruber, 1992b, p.123, modified

Varying interpretations of the environment

Apart from the temporal shift in perception, the environment is also interpreted in a different way by the various actors of our time. From whatever perspective we study the relations between the environment and society, it becomes clear that nature is under increasing pressure. This pressure concerns the exploitation of resources, the encroachment of human activities (due to the growth of the population, the increasing space demand and the changing nature of activities), and the pollution of air, soil and water. Basically, certain pressures are inevitable given the dynamics of the society, on the other hand the environment still occupies a marginal position in many actors' minds. Politicians, planners, developers, conservationists, and the (anonymous) general public have different ideas about what is nature and to what it is good for. They may agree on one fundamental point, the fact that the environment is the basis of human existence, but beyond that they will seek different answers as to how much nature is required and the degree of spontaneity natural processes are to be granted. This latter point can be illustrated by one element in the rhetoric used to promote the construction of an entertainment and commercial complex at the outskirts of London. The project, Universal City, was to transform a site much frequented by birds and home to a wetland ecosystem into a pleasure 'island'. One of the various arguments included the idea 'that nature could be engineered and created by skilled technocrats.' (Harrison and Burgess, 1994, p.301). This mechanistic idea included the transplantation of species and the creation of 'the very precise ecological circumstances that rare species require if they are to support viable populations.' (ibid.). After what we might call the systems revolution in thinking one is surprised to discover a reductionist approach that is completely out of date. Yet, according to this economic (profit-oriented) logic, the end justifies the means.

One question that obviously nobody asked is whether such an entertainment and commercial complex was really necessary, whether it responded to a demand, or if this demand was to be created in the course of construction. In a hedonistic society, such a question is obviously heresy as it challenges utilitarian thinking and pleads in favour of unused sites that exist for themselves (i.e. for the ecosystem) and have no material but 'only' an idealistic purpose.

Rainham Marshes is but one example; due to its size and location, it may be outstanding, but there are many more sites that have been and will be subject to the same processes. In old gravel pits and quarries (or in unworked section of pits and quarries in general), for example, specific small ecosystems have developed. Rehabilitating such 'wounds' in the landscape will inevitably lead to the destruction of such small habitats. They are too small to justify conservation, and a gravel pit or a quarry may offer areas that can be used profitably. Conservation in this case in judged unnecessary. Nature, the ecosystem, the environment – they are all social constructs, useful at certain times, but uncomfortable at others.

Landscape dynamics as an expression of changing values

Landscape has been a key notion in geography until the 1960s when it was temporarily banned from discussions; from the 1980s onwards, however, it began to make its return. What has not changed since is the uncertainty about the definition and understanding of the term. One way of putting it into perspective is to compare it to *(geographical) space*, a term sometimes used as a substitute. In a very simplified way, *space* can be seen as something formal, abstract, and rational, whereas *landscape* would be affectionate, concrete, and emotional, filled with meaning. This becomes obvious in the confrontation of painting with cartography: there are landscape paintings, whereas maps represent space. From another point of view, space can be considered the physical support of landscape (Grandgirard & Schaller 1995, p.30), situated on one side of the perceptual filter interposed between space and landscape. Thus space belongs to the objective, landscape to the subjective realm. In this way, landscape is defined as 'a portion of space perceived by an observer at a given moment in time, from a certain place on the earth's surface' (ibid.; transl. WL) . Perception produces images related to feelings and values. Hence, landscape is very much dependent on the value system of a society, and the notion of landscape conservation reaches beyond nature conservation that is often motivated by species protection).

When discussing values in Chapter 3, their dynamic aspect was particularly emphasized. Humanity does not follow one extreme set of values for a long time, but it is bound to adapt to changes in society and environment. As a social construct, landscape is subject to constant changes that must be seen in the interrelationship between individuals, the society, and the environment itself (see Figure 2.6). The cultural landscape is landscape consciously shaped and transformed by human decisions and actions, and as a consequence it is endowed with 'multiple layers of meaning' (Ryu, 2000, p.39), conferred by a variety of human agents, from dominant as well as marginal social groups (ibid., p.34 f.). While the social elite usually succeeds in leaving its imprint on the cultural landscape, the socially marginal groups tend to reject the values of the dominant class and manage to design their own cultural landscape (ibid.). However, this differentiation varies from one society to another: in a highly polarized society, traditionally cultural landscapes may persist for a long time, whereas in a relatively levelled society the values of the elite will lead to a constant adaptation of the cultural landscape to new ideas (modernization). Landscape is a palimpsest and mirrors the depth and complexity of history (Piveteau, 1995, p.163).

Ecologically marginal regions are areas characterized by substantial human impacts and disturbances of the natural cycles, whereas an ecologically central region will be pure nature with operational natural cycles. This point deserves particular emphasis because the ecosystem is the basis for human existence. Humanity can only survive if it takes care of and respects the needs of this basis. However, in order to survive and pursue its activities, human impacts on nature are unavoidable. The major problem lies with the degree of intervention and where the limit between 'too much and still tolerable' is drawn. Just as we always shift between the extremes of sacred and secular values, there is also a swinging between natural and cultural landscape. This polarization can be visualized by the landscape continuum (Figure 7.2). The two terms 'natural' and 'cultural' landscape are defined according to the degree of human interference or appropriation. They are value-free and in no way linked to any ideological background. Man is an agent who transforms the environment consciously and intentionally, whatever principles lie behind his actions, and the unintended effects do not depend on an ideology either. These transformations are guided first and foremost by human needs and wants, and by the perception of and the attitude towards the environment: the environment is 'a set of possibilities' and 'a set of affordances' (Ingold, 1986, p.2 f.).

The similarity to the value continuum (Figure 3.3) is not accidental but intentional. Processes always take place between extreme positions or claims, with the two extremes constituting ideal (or pure) situations. Such ideal conditions do not exist in the real world, neither in the present nor in human history – with one exception. The very first humans were like animals, harvesting (i.e. collecting and hunting) the fruits of nature as their survival strategy, but as soon as they started to satisfy needs which they defined themselves and which went beyond purely physiological requirements, they began to transform the landscape according to decisions related to such needs and wishes, impacting the environment to a varying degree. Deterministic processes were thus gradually sided or even replaced by probabilistic ones, intentional actions began to dominate over spontaneous reactions.

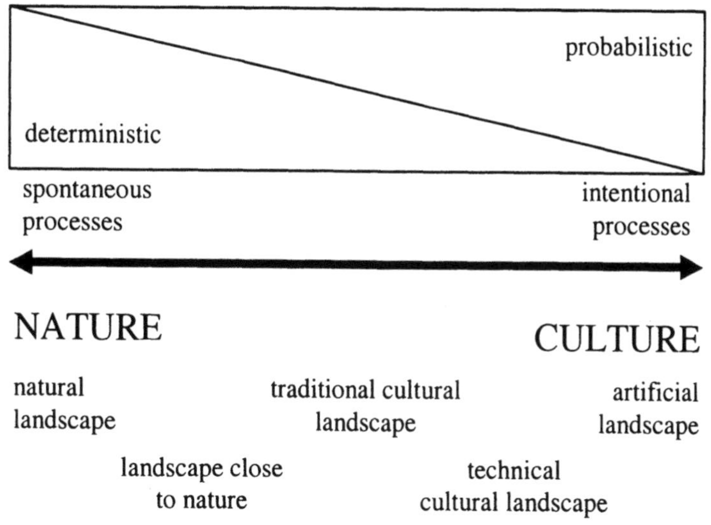

Figure 7.2 The nature-culture continuum

Source: Leimgruber and Hammer, 2002, based on Ewald, 1978

The outcome has been an increasing modification of the environment through pollution and a growing artificialization of the landscape. In a study on Switzerland, von Felten (1998) describes the present-day results of this process. By combining two data banks (plant associations and human constructions) on a kilometre grid, he was able to distinguish fourteen different degrees (or classes) of artificialization (hemerrobia). He described about ten per cent of the country as being 'non-artificialized' landscapes, but it would be wrong to conclude from this figure that human impacts are entirely absent there. The author could, for example, not include air and water pollution in his analysis nor the problem of noise. Skiing off the signalled tracks and heli-skiing (where skiers are transported by helicopter to a starting point) have considerable consequences for the natural environment. The forests suffer because skiers damage trees that grow under difficult conditions and will not recover from wounds as quickly as they might in low-lying regions (Hormann, 1997, p.27). Animals are scared by the skiers who run through their roaming areas, and nesting birds are disturbed by the noise of the helicopter (ibid., pp.28 ff.). Seemingly natural alpine regions are artificialized, even if the hemerrobia-values may be low. The important findings from this pioneering research can be summed up in the simple statement that 'Switzerland still possesses important virgin spaces or areas which are used in a very soft way' (Felten, 1998, p.38). In fact, the four classes 0-3 (least human impact) comprise 66 per cent of the national territory, whereas the highest degrees of artificiality (classes 9-13) concern only 1.45 per cent (ibid.). In the present terminology this can be interpreted in two ways: first, environmental marginalization appears to be a very rare phenomenon in the Swiss landscape, confined to the major urban areas, and second, from an ecological point of view, most of the landscape occupies still a strongly central position. The regions concerned lies essentially in the Alps and Prealps as well as in the Jura, areas that are very thinly settled. More precisely, it is the high parts of the mountain regions that are least influenced by man (on the scale studied by von Felten), whereas the main valleys (which carry important communication routes) excel by high degrees of hemerrobia. It has to be repeated, however, that important human imprints on the environment could not be taken into account. In addition, it has not been possible to include the horizontal visual dimension into the analysis (landscape aesthetics). Looking from a skyscraper into the open land or looking from a hill towards an urban zone leave two different impressions on the observer, hence the judgement on the artificiality of a landscape will vary accordingly.

Landscape changes are nothing new. As has been pointed out above, throughout history, there has been a constant flux between exploitative (secular) and economical or sparing (sacred) consumption of nature's gifts, according to the various societies' organization, technological level, and patterns of needs (Figure 3.4). Since the ecosystem has its own rhythm of regeneration, many results of human interventions that have happened centuries ago, are still visible in our time, in particular the negative ones. The naked mountains of the Mediterranean, which are due to the overexploitation of the forests, are a striking example. In his play *Crizia*, Plato describes the devastations in Attica that had taken place as early as the fifth century BC. The city of Athens, numbering more than 400,000 inhabitants, required enormous quantities of firewood and charcoal for the needs of her population (Lévy, 1995), and the city-state needed timber for the construction of the fleet. Plato deplores the loss of the forests and the runoff of the water on the naked soil (Krebs, 1997, pp.49 f.). Like Dickens and Orwell, Plato was well aware of the risks this destruction of the landscape entailed, and be it only the high cost of firewood and timber that had to be transported over increasing distances. The Weald in Southeast England was at a time – as its name suggests – a large extent of woodland, but being one of the many iron-working areas in England, it lost its forest cover in the 16th and 17th centuries (Hoskins,

1969, p.109). On a shorter time-scale, the impacts of wars have been pointed out in Chapter 5.

The protection of nature and environment

Nobody would seriously contest the fact that a well functioning environment is essential to human survival, indeed to the survival of all living beings (Ingold, 1986, p.2). Since its inception in the 1970s, the ecology debate has finally entered the stage in the 1990s where even persons high up in the world political hierarchy dare to remind our society of this simple fact. During the World Management Forum in Davos in January 1999, the United Nations' General Secretary, Mr. Kofi Annan, addressed the managers of the world with the following words: 'I call on you, individually through your firms, and collectively through your business associations, to embrace support and enact a set of core values in the areas of human rights, living standards and environmental practices.' (quoted in Gschwend et al., 1999, p.1/1). As representative of the United Nations, Mr. Annan has no material power to enforce such core values, but his appeal may be understood as an alarm bell for our society. By integrating society and environment in one single plea, he emphasized that our social and ecological problems are closely linked, that they mirror the current state of our value system, which focuses exclusively on materialistic goals. His statement carries at least a moral weight, and he has since reiterated his position in his address to the Association of American Geographers in 2001 (IGU Newsletter, 2001). His repeated appeals demonstrate that awareness has been transmitted from the scientific community and the environmentalist movement to the political platform. The fact that a number of influential countries of the North ignore this message and snob international conventions on environmental issues (such as the Kyoto protocol of 1997) reminds us of the diverging interests and the lack of solidarity with the global community. Neoliberal thinking (the law of the jungle) dominates, but it works neither in society nor in our relations with the environment.

The following discussion on nature protection starts with a brief glance at its history. A society's attitude to nature is in fact mirrored in the rules and norms it establishes in order to regulate its use by humans. We shall see that the interests behind protection were not necessarily ecological but related to the power structure of a society. The leaders, who in former times were not the population but a ruling class or an individual monarch, define the society's norms, and they may include the protection of the environment. By safeguarding their hunting grounds, the upper class involuntarily edicted nature protection, at least in limited areas.

Forests, parks and gardens

The first references to nature protection concern the forests. Germanic tribal laws as far back as the 6th century include regulations promoting the protection of certain elements of nature. They combine late Antique, Roman and Christian with indigenous Germanic ideas (Jäger, 1994, p.17). Hunting was the privilege of the ruling class; it was a sovereign right, hence illegal hunting (poaching) was punished. These rules demonstrate that the protection of nature was not an ecological but a social objective (ibid.). The same holds good for forest laws, which can be traced back to Charlemagne, although in those cases the interests of the public were also taken into account (regulations on the feeding of acorns to pigs, on poaching, the collecting of timber and firewood etc.; ibid., p.18). Although the laws were made in the interest of humans, not of the environment, one can discover a concern for the sustainable use of forests as far back as the Middle Ages.

The practice of protecting forests for hunting ultimately led to the creation of *parks*. The nobility required sufficient open space for leisure and pleasure, with an adequate stock of deer and other animals to be hunted. References to the creation of such parks in England take us back to the time before the Conquest, possibly to the reign of King Alfred the Great (849-901): 'Woodstock Park (now Blenheim) ... was fenced around and separated from the surrounding forest of Wychwood, as a game-preserve for the Anglo-Saxon kings, before the year 1000' (Hoskins, 1969, p.130). Further evidence dates from the 13th and 14th centuries (ibid., p.75; p.129 ff.), and the aim is always the same: setting aside forests and open land for hunting and for embellishment.

The question poses itself if this socially motivated nature protection really led to an ecologically positive result, i.e. if nature as such was really protected. The answer, however, tends to be rather negative: in fact, this kind of human intervention resulted in a gradual reduction of the woodland within the enclosure 'as grazing animals thinned out the trees and prevented their natural regeneration by consuming the seedlings' (Hoskins, 1969, p.130). By reducing the roaming range of the deer, man had in fact – unwillingly – concentrated their impact on a limited zone. The philosophy of this kind of protection was that nature was to serve man and had no purpose in itself hence it has been marginalized.

These hunting parks can be interpreted as a sort of predecessor of the English landscape garden, but the story of this 'nature imitated' is more complex and has to be set within the European tradition of park and garden architecture. The origin of this movement in England may have been the tales told by travellers to Greece and Italy, and the paintings by Lorraine, Poussin and others, depicting wild landscapes with temple ruins and broken classicist columns (Darby, 1976, p.45 f.). The first parks in the new fashion, however, were built in 17th century France. Based on the geometrical outline of the Italian Renaissance garden, the first French garden *sensu strictu* was outlined by Louis le Vau in Vaux-le Vicomte, 1656-1661, and served as a model for Versailles (Grimal and Lévy, 2002). This fashion of straight alleys, sometimes also labyrinths, and trees neatly cut into different shapes also made their way into England. In the 18th century, a proper English garden style developed, imitating nature but entirely artificial. Apart from seemingly spontaneous vegetation (endemic and exotic species), there were Greek and Gothic temples, Chinese pavilions, romantic grottos, and other unusual features. In either case, nature was at heart of the protagonists of this garden architecture, but at the same time they dominated it according to their own perceptions. Architects such as William Kent (1685-1748), Lancelot 'Capability' Brown (1715-83) and Humphrey Repton (1752-1818) were the drivers behind the new look at parks and gardens. Their new approach was at the same time a reaction to the static formalism of the French parks, which forced nature into a rigid geometric framework (ibid.). The result was 'a new and peculiar character to the general face of the country' (Darby, 1976, p.48). The English garden architecture is to some extent indebted to the Chinese and the Japanese gardens that were places of contemplation and meditation, in relation with religion and philosophy (Grimal and Lévy, 2002). Nature in this case sustains the human spirit in its way to perfection by offering the quietness of an artificial garden, robbed of every object that might disturb the process of meditation.

Garden architecture therefore subjects nature to human will. In any case, human intention is confined to shapes and materials, but it continues to use natural materials (shrubs, flowers, grass, sand and gravel, rocks), and the result mirrors aesthetic values of a particular period and culture. Nature has only partially been subdued to Man.

However, the same question as above has to be asked, whether the creation of parks and gardens really benefits the ecosystem, and the answer is not entirely positive. The English landscape garden is an artificial human creation, the natural spontaneity is intentional, and the planting of exogenous species put the local ecosystem to a test. The idea behind the landscape garden was to give the visitor (not the general public at that time) the illusion of walking through a natural landscape, but in fact nature and spontaneity have once more been marginalized. The gardens served the upper class as a sort of romantic and exotic nature refuge.

The situation changed towards the end of the 19th century, when the negative consequences of the industrial revolution became visible to a growing degree. Nature protection for the benefit of the entire society gradually became a public issue. In an essay, published in 1880, the German musician Ernst Rudorff (1840-1916) discussed the relationship between modern life and nature (Andersen, 1987, p.143), complaining the fact that industrialization was destroying nature. His plea roots in Romanticism and presents an idealized view of the past. It was a nationalist bourgeoise critique that opposed him to the visions of the internationally minded working class (ibid., p.144). His ideal was a crafts and farming society, which, however, was an unrealistic goal, given the evolution of the capitalist industrial society. Rudorff demanded the protection of 'untouchable sanctuaries of nature and history', which were to become a sort of reserves where any transformation should be banned. However, no discussion on the methods of industrial production took place, 'the possibility of an ecological reorientation of the production was never taken into account' (ibid., p.145; transl. WL).

Nature protection received support from another side. Increasingly, air and water pollution due to mining and heavy industries were seen as detrimental to the output of agriculture and to the renewal of forest resources, but also as a burden to the quality of life. Before the idea of nature protection began to appeal to a wider public, it was the foresters who criticized that forests could not regenerate adequately as a consequence of air pollution – their argument was economic, not ecological (Andersen, 1987, p.146). The movement of *Heimatschutz* (protection of the countryside with its natural and historic heritage) can therefore be said to root in both a romantic and a utilitaristic attitude. The campaigns grew in importance towards the end of the 19th and in the beginning of the 20th centuries: the first association for the protection of nature and landscapes dates from 1901 (France), Britain and Germany followed in 1904, Switzerland in 1905 (Walter, 1996, p.83 ff.).

Similar movements could be observed in North America during the 19th century, albeit under different social circumstances. In densely settled Europe, the impact of industrialization and urbanization put pressure on the remaining quasi-natural areas; in North America, the fast expansion of the white settlement areas threatened what was perceived as the wilderness. The romantic side was not missing: 'artists, poets and writers as William Cullen Bryant, Thomas Cole, James Fenimore Cooper, and George Catlin slowly developed the theme of "wilderness"' (Nelson and Butler, 1974, p.293). Catlin, well known for his paintings of the North American Indians, in particular cultivated this kind of Romanticism, emphasizing the role of the Prairie and the Indians who roamed around on horseback – he saw the native inhabitants as part of wilderness (ibid., pp.293 f.). All these advocates developed the same philosophy, rooted in the worry about disappearing natural landscapes and the growing imprint of man on nature: to set aside nature and wildlife reserves, to create national parks and wilderness areas for the society as a whole. The protection of nature was thus considered a public task. This movement can be interpreted as a reaction to the industrial progress, as the creation of an urbanized society wanting to save nature (or what people thought to be nature) from destruction and from disappearing as long as there was still time. At the same time, they mirror people's curiosity for 'exotic' natural features, and a growing scientific interest. Finally, they also reflect the increasing concern of the state (the public sector) in landscape and environment, which are seen as essential bases for life.

The nature protection initiatives have not only been influenced by romantic ideas but also by legislation in favour of woodland. This is exemplified by the protection of Yosemite Valley with its giant sequoias *(Sequoiadendron gigantea)* in 1864 and the subsequent creation of the world's first nature reserve in Yellowstone Park in 1872 (Blume, 1978, p.86).

Created and managed by man, national parks and nature reserves are areas where human interference is restricted. Wilderness areas, national parks and nature reserves are not spontaneous nature but controlled nature-like zones. Human interventions have become necessary because the natural equilibrium has been disturbed. Dilsaver (1992) recalls the example of the Yosemite Valley, which, prior to European immigration, had been used in a fairly sustainable way by the native inhabitants. The arrival of European settlers and, later, of tourists with their specific behaviour has uprooted this balance. Recreation demand, request for preservation, and tradition compete with each other; camping, concession housing and the general congestion of visitors were prohibitive to the (traditional, i.e. native) use of fires as a means of ecological regulation. When fires were systematically suppressed, trees began to encroach on former meadows, an undesired process which required felling in order to preserve the former open character of the valley. Whether this open appearance was really the natural (i.e. spontaneous) landscape prior to the arrival of the Europeans in 1851, or if it represented the result of the valley's use by the natives, is a question that cannot be answered by 'yes' or 'no'. Moderate as they had been, native activities had nevertheless included the use of fires and therefore resulted in the transformation of the natural to a (traditional) cultural landscape.

Perspectives on nature have changed since 1851, as have the attitudes and expectations of the society. 'Recreation, preservation policy, and tradition have become so entangled that a coherent vegetation management policy may never be fully accepted.' (ibid., p.135). Indeed, even very large parks, 'managed under the principle of natural regulation', cannot be sheltered entirely from human influences – they may be simply too small 'to even come close to sustaining themselves, and even the biggest ones cannot be isolated from the harmful effects caused by activities in nearby areas' (Miller, 1996, p.629). Holistic thinking in the case of one particular ecosystem must be completed with the notion of distance to and interrelations with neighbouring ecosystems.

Despite such reservations, parks are important for the future of our environment, even if they are 'tamed nature'. They are in the interest of biodiversity, vital to the survival of the human species (Leimgruber, forthcoming). Unless a new way of thinking leads to a new appreciation of nature and a more rational coexistence of man and the ecosystem, the separation of near-natural and cultural landscapes may be the only answer to the threat to the environment from human activities. What is required is a way to ensure that man can live in nature, use it and still respect and protect it. This ideal appears to be possible in biosphere reserves.

Biosphere reserves

The discussions on environmental conservation and sustainable development are at the roots of the latest 'child' of protected regions, the *biosphere reserve.* Contrary to the protected areas discussed above, however, this concept is characterized by a highly differentiated approach. National parks and wilderness areas are to be kept free from human activities; they are a sort of museum and can be visited according to specified regulations. Biosphere reserves, on the other hand, are not museums or untouchable regions; on the contrary, they are settled areas comprising economic activities. More than any other type of protected area described above, biosphere reserves are based on the idea of long-term sustainability, where spontaneous natural and intentional human processes operate side by side.

The biosphere reserve idea originated at the 1968 UNESCO Conference on the 'Conservation and Rational Use of the Biosphere'. Two years afterwards, UNESCO launched the Man and Biosphere (MAB) research programme on a worldwide scale. The biosphere reserve concept was formulated in 1971, with the objective of striving towards 'a balance between conserving biodiversity, encouraging economic and social development, and preserving cultural values and variety' (Leimgruber and Hammer, 2002, p.131). In the following years, this idea was steadily developed and put into action. The biosphere reserve philosophy is to respect the interaction of the natural and the cultural systems; it admits that nature protection alone cannot be an aim in a world whose population is growing at an accelerated pace. Humans are recognized as being a part of the biosphere: they are not only social beings but also part of nature – a fact that seemed to have been forgotten under the impact of Carthesian thought.

In February 2003, 425 biosphere reserves existed in 95 countries on all continents ((Table 7.1; UNESCO, 2003b), and their number is steadily growing (in March 2001, their number was 391; Hammer, 2001, p.281). Many of these reserves are situated in what we would call marginal regions, and they often originated in national parks and nature reserves. The designation of a reserve will be made following a national request that is based on the criteria a biosphere reserve should fulfil, e.g. 'be representative of a major biogeographic region, contain landscapes, ecosystems or animal and plant species, or varieties which need to be conserved, be of an appropriate size, and have an appropriate zoning system, with a legally constituted core area or areas, devoted to long-term protection' (UNESCO, 2003a). The legal side is an important component, as it will ensure public support and collaboration with the authorities and the reserve monitoring organization.

Biosphere reserves are based on the idea that protected areas and human activity zones should exist side by side in the same region, but without being subject to conflict. By defining three-zones (Figure 7.3), the authors of the model wanted to create a buffer between total protection and sustainable (moderate) use of the landscape and avoid a sudden break between an entirely protected area and surrounding belts of land use. The core need not be a contiguous zone but can be divided into a number of isolated protected areas. The buffer zones around them and the fact that human activities must be environmentally friendly ensure that these individual protected zones will not gradually be degraded. The concept is flexible and allows to take local circumstances into account. This is very important for popular support and acceptance, in particular where the population concerned is politically active and displays a strong sense of identity with the region.

Table 7.1 The evolution of the biosphere reserve idea

1968	UNESCO conference on the 'Conservation and rational use of the Biosphere'
1970	Foundation of the UNESCO research programme 'Man and Biosphere' (MAB)
1971	Biosphere Reserves created as a MAB-policy objective
1972	United Nations Conference on the Human Environment in Stockholm
1974	Development of the Biosphere Reserve concept
1976	Initiation of the Biosphere Reserve global network (approval of 53 Biosphere reserves in eight countries)
1983	Minsk conference (UNESCO & UNEP) in biosphere reserves
1984	Action plan for biosphere reserves
1995	Sevilla strategy: reformulation of goals, establishment of consistent criteria
2002	20 new biosphere in 14 countries reserves approved, bringing the total to 425

Source: Leimgruber and Hammer, 2002, p.132, modified

The three zones are defined as follows (ibid.):

- The *core area* is a strictly protected territory with little to no human influence. It can consist of various areas in separate locations, but they all enjoy the same degree of protection. Research and monitoring natural changes in representative ecosystems are permitted, but its main purpose is the conservation of biodiversity;
- The *buffer zone* is the region surrounding and protecting the core. Only low impact activities are permitted (research, environmental education, recreation, limited human settlements) that guarantee a soft transition to the third zone;
- The *transition area* is the outer zone where sustainable use of resources by local communities is encouraged. 'It is here that the local communities, conservation agencies, scientists, civil associations, cultural groups, private enterprises and other stakeholders must agree to work together to manage and sustainably develop the area's resources for the benefit of the people who live there.' (ibid.)

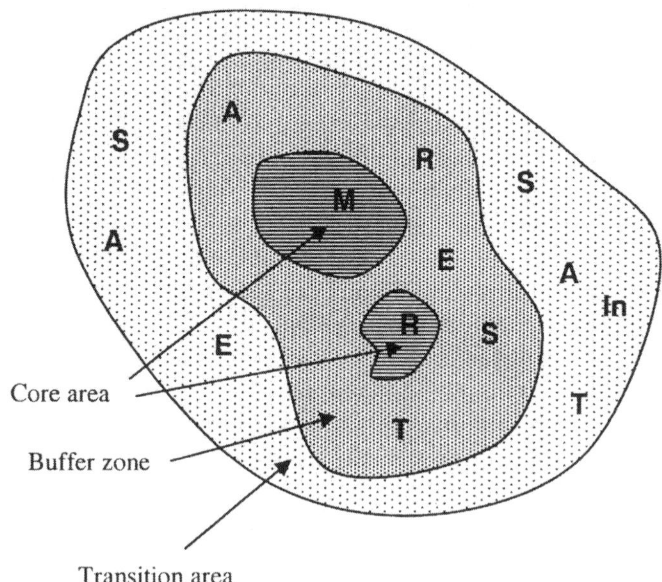

Figure 7.3 Conceptual model of the UNESCO biosphere reserve

A = Agriculture; E = Education; In = Industry; M = Monitoring; R = Research; S = Settlements; T = Tourism

In the core, biodiversity should obtain a chance to be preserved and develop according to its own rhythm; therefore strong protection is crucial. Humans have to be kept out. Settlements and certain human activities (such as environmental education, ecological observation, soft tourism, extensive traditional and/or intensive sustainable farming) can be tolerated in the buffer and transition zones. In the latter area, manufacturing industry drawing on local resources will be permitted, as long as the exploitation remains sustainable. Intensive cultivation and other activities (e.g. industry based on external resources), on the other hand, have to be located outside the biosphere reserve. Economic development in a biosphere reserve has to be guided by the idea of sustainability, i.e. a long-term vision towards equilibrium. To achieve these ambitious goals, the biosphere reserve will be scientifically monitored in order to guarantee the long-term survival of the area.

Biosphere reserves belong to those large surface protected areas that have received increasing attention in the past decades. They have to be seen not only from a conservationist perspective but also in the general framework of regional planning and development. (Hammer, 2001, p.280) identifies three elements that have influenced the discussion about regional protected areas: the ecological dimension, systems thinking, and 'the cultural landscape as the expression of ecological and anthropogenic processes' (transl. WL). The region has been discovered as a scale that is gaining in importance between the local and the national level (ibid.). These technical aspects are completed by the important shift in thinking among conservationists who have understood that segregation (i.e. the separation of nature from the society) is not the only solution to nature protection (ibid.). Rather does it make sense to see it in a wider context and include the ordinary citizen into the process by improving his and her awareness for the positive sides of conservation.

Biosphere reserves have been allocated three basic functions (UNESCO 1996, p.4), resulting from the discussions leading to the Sevilla strategy (1995):

- a *conservation* function: to contribute to the conservation of landscapes, ecosystems, species and genetic variation;
- a *development* function: to foster economic and human development which is socially and ecologically sustainable;
- a *logistic* function: to provide support for research, monitoring, education and information exchange related to local, national and global issues of conservation and development.

The Sevilla strategy's first goal is to use the biosphere reserves for the conservation of natural and cultural diversity. By doing so, it establishes a link between the conservation of the environment and the requirements of regional development. Although the text of the strategy emphasizes the biosphere (i.e. the physical environment), it connects it to human efforts in land management, research, and training with the long-term goal of sustainable development. This integrative approach represents an innovative way of thinking that will gradually be brought home to the population – and, we should add, also to entrepreneurs and company managers. In this way, conservation is transformed from a technique to an attitude that has to be shared by the entire society. Sustainable development means that no element of the ecosystem and the social system can be neglected or marginalized.

Another important element in the Sevilla strategy of 1995 is the collaboration among local communities, government institutions, NGOs and private enterprises. Recommendations are given for various levels:

international, national and regional (the individual reserves). By emphasizing such a wide range of cooperation, responsibility for our ecosystem has been allocated to all political and economic actors and has visibly influenced Kofi Annan's address to the World Management Forum in Davos in January 1999, quoted above. The biosphere reserves, in particular after the Sevilla strategy, represent an approach that combines tradition with modernity. This philosophy allows marrying customary knowledge with the achievements of modern technology. Hammer (2002b, p.195 f.) quotes examples from West Africa where traditional societies are encouraged to keep traditional values and complement them with new information on working techniques.

The biosphere reserve concept as such looks very promising; and it is likely to improve the integration of cultural and natural landscape, of human activity, planning, and the protection of the environment. In this way, spontaneous processes may supplement purely intentional ones, and the landscape may at least partly become re-naturalized (or de-artificialized). Progress will certainly be slow, but even small actions do show effects. Not so long ago, small streams and hedges have been considered obstacles to rational agriculture. Streams were canalized or even put underground, and hedges were uprooted. However, such 'disturbing' landscape elements play a significant ecological role as habitats of small ecosystems of plants and animals; birds nesting in hedges feed on all kinds of insects that modern farmers kill with pesticides, and the dense network of leaves and branches offers shelter for their breeding activities. As a result of the new awareness, the process has been reversed. In various parts of Switzerland, streams have resurfaced and have been re-naturalized, and hedges have been planted again – all at a considerable cost, money that could have been put to better use if planners and farmers had always reflected in ecological terms instead of production increase.

Contrary to the 'reductionist' concepts of national parks, landscape gardens and nature protection areas, biosphere reserves are based on organic thinking (Wood, 1987). The biosphere concept favours cooperation of the various approaches (ecological, economic, social and cultural) instead of competition, integration instead of segregation. The ecological world-view, characteristic for the biosphere reserve, may avoid what has happened with other types of reserves, viz. dissatisfaction of certain members of the community because the protection was too absolute.

A biosphere reserve is the result of a political process. To persuade a population of the value and profit in the long run is an arduous task. It requires convincing arguments as to the value of an intact environment and the positive side effects of the decision to support the idea. The principle of long-term thinking has to be taught to people whose time-horizon is strongly influenced by human life expectancy and the fragility of our existence. The question of identity may play a key role as well: can the inhabitants still feel at home even if their region is one among many on the long list of biosphere reserves around the globe, if researchers from around the world may congregate to discuss issues and consequences, if restrictions have to be accepted for the sake of the common good? The first Swiss biosphere reserve in the Entlebuch (Canton of Lucerne) has gone through this long process of fact-finding, scientific discussion, information of the public, and finally the referendum on whether to support the project or refuse it. The campaign had lasted from 1997 to 2000, and in the referendum of autumn 2001, the population of the eight municipalities involved accepted to grant financial support to the project for ten years with a 90 per cent majority (Hammer, 2002a, p.111). This result demonstrates that a plan can be put into operation if the promoters ensure careful preparation, conduct an open information policy, and practise transparency as to legal, physical and monetary consequences. It also shows that such a process need not take an excessive amount of time if the project is open to public debate at any time. And finally, the multiple aspects put forward by the Sevilla strategy – protection, education, research and monitoring, conservation of biological and cultural diversity, model function for planning, regional development, land use – are certainly very convincing arguments to pursue the idea once it has been launched in a region (Hammer, 2001).

The biosphere reserve is a novelty and acts as a stimulator for innovative ideas. Although commercial success is not the prime goal, the population, in particular the farmers, usually hopes for positive feedbacks for the regional economy. To this effect, a regional label with the denomination 'Entlebuch Biosphere Reserve' has been created in order to indicate the origin of the products. Such regional labels have become very popular wherever regional production chains are to be privileged, both for regional consumers and tourists (e.g. the LaNaTour project [Landscape-Nature-Tourism] in the Canton of Valais; Leimgruber and Imhof, 1998, p.392). They have both economic and identity significance for the region in question.

Agriculture – still part of nature?

Despite being the economic sector where employment has diminished considerably over time, agriculture is still the only economic activity humanity cannot survive without. We depend on the surplus food produced by farmers, basically no matter how they do it, and their products have to be put onto the market. Parallel to the decline of the farming population, we have experienced a process of rationalization through the use of machinery, consolidation of land holdings, and a growth in capital input. Important increase in production and productivity was achieved in this way, assisted by the growing input of synthetic fertilizers, pesticides, fungicides and herbicides that enhanced soil fertility and killed unwanted pests and weeds. Total world agricultural production in 2000 was 21.6 per cent higher than ten years before, but only 5.8 per cent higher per inhabitant (Berié et al., 2002, section Wirtschaft). The highest growth had taken place in China (+ 65.4 per cent total production, + 49.6 per cent per inhabitant; ibid.). Such figures look very promising, in particular if we take into account that the land surface devoted to farming has been slowly declining over the past 30 years (Aubert, 1996, pp.8 and 32). The spectacular rise in wheat output from 23.6 million tons in 1957 to 106.4 million tons represents an increase from 0.9 to 3.5 tons per hectare – almost a four-fold growth (ibid., p.22). However, it is precisely China that had been confronted with the negative effects of increasing fertilizer input. Such a stunning advance in agricultural production could not have been achieved without massive chemical fertilization. The input has in fact grown parallel to the grain output, but the chemical fertilizer efficiency has diminished in the period investigated. In 1969, 30 kilograms of grain could be obtained per kilogram fertilizer, in 1974 even 50 kilograms, but in 1992 and 1993, the figure had dropped to about two kilograms (ibid., pp.10 f., 34). The law of diminishing returns seems to be effective, although Aubert thinks that the trend might be reversed 'for some period through better use resulting from more favourable policies. Its effect may therefore be partially offset on the long term' (ibid., p.11). This is a cautious statement that depends on the rather fragile statistical base the study had to be built upon, and it remains an open question how far this reversal will go (if at all).

This example reveals the risk modern conventional agriculture runs if it goes on to ignore certain basic laws and experiences in both economy and ecology. One of the major ecological problems are the deposits of toxic substances in the soil, originating from pesticides and herbicides, substances that can penetrate into the groundwater and pollute wells and rivers. Excess application of fertilizer results in soil degradation and requires more input every season. Pesticides for their part are only temporarily efficient due to mutations that make pests resistant.

A shift in thinking has gradually taken place among the farming population (not necessarily among the very large landowners), first in countries of the North, but increasingly also in the South. Ecologically friendly farmers had been idealistic individuals in the 1960s, were ridiculed (marginalized) in the 1970s, and suddenly found themselves in the forefront of a movement that has been gaining momentum in the 1990s. They were pioneers, having sensed the prerequisite for a new approach to agriculture: the awareness of the long-term risks of conventional farming methods, and a vision of what farming should be like.

Indicators of the public recognition of environmentally friendly farming came from the consumers on the one hand, through reforms of agricultural policies on the other, and also through the statistics and the promotion of research. Consumers started to demand food produced in a less harmful way, agricultural policy began to emphasize the importance of ecological methods, and statistical offices started to collect information on organic farms. Since 1990, we dispose of relatively good data on organic farming, but next to nothing is known for the time before. Research institutions began to emphasize the importance of environmentally friendly farming methods – something that had been at the margin of interest suddenly exercised a mighty influence on various sectors of the society.

Taking care of the environment is not limited to the actual production of vegetable and animal food but comprises upstream and downstream processes, such as the kind of seeds a producer uses or the disposal of excess manure. In the Swiss canton of Lucerne, for example, farmers have been blamed for the eutrophication of small lakes. For years, various measures to increase the oxygen content of the waters have been tried, but to little effect. Farmers have now started to build retention ponds where the excess liquid manure is allowed to settle (Stähelin, 2003). This measure will become effective in the long run only, but it represents a change in attitude and increasing awareness among the farmers involved in the project. Besides, these retention ponds develop into special biotopes that are of botanical interest.

There are several ways in which a farmer can operate in an ecologically friendly manner. Traditional European farming methods, until the mid-19th century, were respectful of the environment. When the German chemist Justus Liebig demonstrated in1840 that plants are able to transform anorganic substances to organic ones, he prepared the field for the use of mineral (chemical) fertilizers. However, his outlook was holistic, and he repeatedly warned against the exclusive use of mineral fertilizers (Fischer, 1980, p.12). In many countries in the South, farmers do not have the means to purchase chemical fertilizer and pesticides; their production continues to follow the traditional methods that were adapted and harmless to the environment. The output depends on the local soil situation, their personal effort, and the little means they might have to enhance soil fertility. Theirs is a subsistence production that hardly produced any surplus beyond the family needs, and these figures never turn up in statistical tables. In advanced agriculture, it is possible to keep to a strict minimum of synthetic input and complete this with organic fertilizer. The farmers in this way try to integrate traditional and modern techniques and will certainly obtain higher yields than without fertilizer – in Switzerland, Integrated Production (IP) has been adopted as the official name. The impact on the ecosystem is thus greatly reduced, but it is hoped that at least a number of producers will go a step further at a later stage and turn to organic production.

A more radical shift occurs when farmers renounce chemical input completely and turn exclusively to natural methods, applying organic matter and using techniques that will minimize the risks from pests and weeds. This method, organic farming, has become very popular since the 1990s. Around 1990, it had in fact begun to develop considerable dynamism, and since that time it has entered agricultural statistics and can be studied with some precision. The movement is, of course, much older. What has happened in the late 20th century is rather 'the rediscovery of old knowledge and its application under new social, economic and ecological conditions' (Leimgruber et al., 1997, p.161) than a true innovation.

Organic farming represents not only a way of agricultural production but is an entire world-view, dominated not by financial profit but by an emotional attitude towards the land and its products. It combines economic with ethical and ecological considerations (ibid., p.162). A farmer who converts his enterprise from conventional to organic farming makes a deliberate choice and adjusts his or her value system from the secular towards sacred side. Ecocentric conviction and financial incentives (higher prices for organic products, ecological compensation payments) combine with social responsibility to motivate a farmer towards the conversion from conventional to organic production (Afangbedji, 2000, pp.69 ff.). His or her outlook corresponds to the 'small-is-beautiful' philosophy, quality is more important than quantity (Table 7.2).

As with many innovations, organic farming was pioneered by an outsider, the founder of anthroposophy Rudolf Steiner (Leimgruber, 2001, p.21). He pushed the idea of biological-dynamic farming very far: not only was he opposed to the use of synthetic fertilizers and pesticides, he also propagated the observation of the cosmic cycles of the moon and the planets, and their influence on sowing and harvesting. Later pioneers concentrated on the natural cycles, fertilizer input and the use of harmless methods to fight pests and weeds.

Table 7.2 Options of agriculture

organic farming	conventional farming
nature-oriented	civilisation-oriented
labour input	capital input
sustainability	profit
traditional	modern
existentialist	productivist
extensive	intensive
small plots	large surfaces
long-term	short-term
quality	quantity

Source: Leimgruber et al., 1997, p.162

Since 1995, organic farming has increased by about 25 per cent every year (Berié et al., 2002; section *Umwelt*). On a worldwide scale, it is nowadays present in more than 120 countries and practiced on about 10.2 million hectares (ibid.). This impressive surface represents, however, only 0.2 per cent of the global agricultural land area (4,974 million hectares), certainly a negligible surface but with an overall positive trend. Data for individual countries are more reliable. In Germany, for example, three per cent of farms (with 3.2 per cent of the agricultural land) have turned over to organic farming. In the European Union, it is Italy that dominates the statistics: of the 3.7 million hectares organically worked surfaces, over one million lie on Italian territory. Within the EU, 1.9 per cent of the farms and 2.9 per cent of the agricultural land are organic. In Switzerland, almost seven per cent of the farms are organic, working 7.7 per cent of the farming area (Statistical Yearbook of Switzerland, 2002). More than two thirds of these farms are situated in the mountain regions, and 90 percent of the surface is therefore grassland (ibid.). Organic milk and milk products are the chief products on the market from these areas. Cereals, on the other hand, account for only nine per cent of total cereal production. In synthesis, environmental friendly practices are marginalizing conventional farming (Figure 7.4) with integrated production (IP) taking the lead.

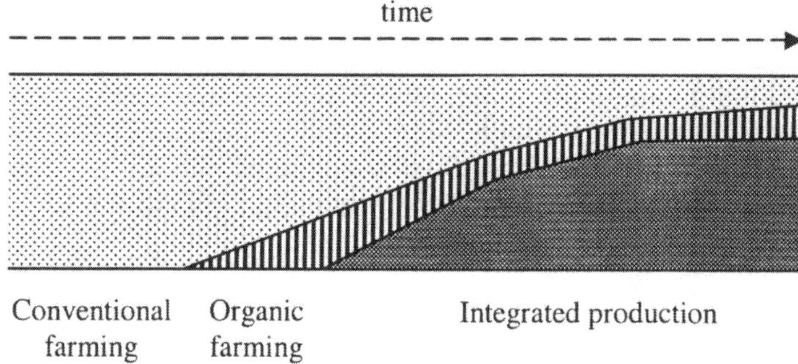

Figure 7.4 Shift in farming practices in Switzerland in the late 20th century

Source: Own elaboration

Elsewhere, organic farming has made its way both in North and Latin America, and to a lesser extent in Asia. Since April 2002, Japan has a legal basis for it (Berié et al., 2002, section *Umwelt*), and there seems to be a considerable demand as well. Indeed, the market for organically produced food has increased, at times even faster than the farmers could produce. The shift in attitude has taken place not only among the producers but also with the consumers. This is a remarkable phenomenon because prices for organic food are higher than for conventional products. Consumers have to make an extra effort in order to satisfy their demand, whereas the farmers are often motivated by public transfer payments in addition to a change in attitude. It is certainly difficult to find out to what extent the change in practice was motivated by insight and not by money. Whatever the reason, the essential result is that the environment profits as well.

Organic farming is one of the many elements that characterize post-productivist agriculture (Wilson, 2001, p.85), although it is conspicuously absent from the long list of dimensions Wilson furnishes in his paper (ibid., pp.80 f.). The post-productivist paradigm includes the new consumer awareness that manifests itself in the demand for high-quality (healthy) food. This request passes through consumer associations that have mushroomed in the North and find their equivalents in the South (see Ekins, 1992, pp.130 ff.). Consumer associations receive complaints from their members, engage in political and administrative action, and arrange for comparative tests of a variety of products. Their task is to inform the consumers and thereby exercise pressure on the producers. Post-productivist thinking follows the existentialist way of thinking (Cunha,

1988) that privileges a return towards sacred values. By now it has spread to a large part of the (northern) society, manifesting itself in 'concern for the environment; adoption of environmentally-friendly farming practices; acceptance of new forms of policy regulation; changing perceptions of role of farmers and agriculture; acknowledgement of multiple actor spaces in the countryside' (Wilson, 2001, p.87). Agriculture is one element in it; it is not the only activity in the countryside but an essential component of the complex activity pattern.

Resources and the future of mankind

One of the issues related to sustainability is the renewal of the natural resources for which the ecosystem is the one and only source. The resource issue concerns human beings who currently have a mean life expectancy of up to 80 years. To put it into a time perspective: which resources can in fact regenerate within this period? To answer this question, we propose to look at a number of materials we need in our daily life.

- Obviously, the renewal of mineral resources need not be discussed at length. Although certain minerals are constantly produced through volcanic activity, particularly in the oceans, such deposits are not and will not be the primary stock our society is using. The question of the poly-metallic nodules on the deep-sea plains illustrates this point. While potentially rich in valuable minerals (iron, manganese, nickel, copper, cobalt, lead, zinc, titanium; Kent, 1980, p.29 f.), the cost of recovery is exorbitant, and the techniques are still being developed. It would be very unjust (to say the least) to leave this arduous task and cost to further generations simply because our society exploits all terrestrial sources, but it would be equally wrong to exploit them now alongside the terrestrial resources and leave nothing for future generations. At current levels of consumption, there is still iron ore for another 300 years, for example (Berié et al., 2002, section *Bergbau, Rohstoffgewinnung und –versorgung*).
- The same holds good for all types of fossil fuel that had the same origin and took an extremely long time to develop. Oil has become a much sought after resource since the advent of the combustion engine, and it continues to haunt politicians and corporations. The real quantity of oil reserves on earth is not known; the estimates have been increasing every year, and so has the consumption; this has fostered the belief that there might be no end. Current calculations point to reserves

for 45 years (120 years if oil slates are included; ibid.). However, even if more reserves will be discovered in the future, they are likely to be exhausted within about two generations. If we cut down on consumption and developed alternative energy sources, the span might extend to three or four generations. Natural gas is certainly a better alternative as concerns environmental impacts, but again, stocks are limited (reserves of about 75 years; ibid.) and not renewable in human terms. Coal seems to be still widely available (reserves: 200 years, and 230 years for lignite), but burning it has an even more negative impact on the atmosphere than oil. The relation between time frame and quality is telling in this case: high quality coal comes from the carboniferous period, some 300 million years ago. Lignite, mainly formed during the Tertiary (less than 60 million years ago), slate-coal found in Swiss moraines from the Pleistocene (less than one million years ago,) and peat, dating back to the post-glacial period, are all of low to very low quality. – The Methanehydrates discovered in 1996 in substantial quantities on the ocean floors (Tecflux, 1999) will not be a solution, although one cubic metre may contain up to 170 cubic metres of gas. They are a fossil fuel, limited in quantity (not renewable in human time perspectives), and when burnt will contribute to global warming just as any other fuel. While their discovery has been hailed as the solution to the fuel problem, one has to be very cautious about their real impact on the global ecosystem – the problems will not be solved but simply displaced into the future.

- Forests, hailed as a renewable resource, definitely have a shorter turnover time, but in most cases it still exceeds human life expectancy (or at least the professional life of a forester from training to retirement). Growth speed and ageing/maturing of forest trees depend on light conditions, to say nothing of soil, altitude and latitude. Finnish forests, for example, grow relatively rapidly in the southern part of the country (mean annual growth of more than 4 cubic metres per hectare), but slowly in the northeast (less than 1 cubic metre per hectare; Häkkilä, 1974, p.19). Plantations of fast growing stock may have a shorter turnover time, but they still have to obey natural laws of growth and the annual temperature and light cycles. Such 'tree-farms' may be a renewable energy source, but they are monocultures and no forests, plantations and not ecosystems. Their ecological value is therefore almost zero, they respond to the productivist paradigm.

Time is one critical element when discussing resources. While we are worried about the limited time span of many non-renewable resources, we must also take care about renewable resources, as has been pointed out above with reference to trees and forests. The *conditions of survival* are as important for the preservation of resources as is the time-factor. Every resource is part of a system and in close interrelationship with other elements. Due to inertia, systems and their elements possess some 'elasticity', a certain amount of scope within which they can vary without taking harm. Once a process overshoots this range, a system will loose its equilibrium and gradually develop a new balance. The felling of the trees in the Mediterranean, for example, did not make room for new trees and forests; on the contrary, as a consequence of the radical forest clearing, soil was rapidly eroded and the mountains remained bare. The destruction of the tropical rainforests around the globe offers the most recent example of the marginalization of the environment. A Rainforest Conference, organized by the Geographical Society of New South Wales, had addressed the problems of recovery in 1981 already. Australia has some experience with the elimination of rainforests. According to one estimate, 'over 500–1,000 years would be required for the natural recovery of cleared rainforest to its original biological composition and structure' (Webb, 1981, p.53). In other words: natural ecosystems can regenerate themselves, but they do it according to their own rhythm, which usually lies beyond human time spans (the above figures would mean something between 20 and 40 generations!). In the case of the regeneration of soils after soil erosion, the time dimension becomes even more dramatic.

Even resources with a shorter turnover face a critical future, as can be demonstrated by the depletion of fishing stock in the North Atlantic and elsewhere. Despite ample information and numerous appeals, strict policy measures (restrictions and prohibitions) were required to try to save at least part of the fish swimming in the oceans. It is not only the excessive quantity of fish caught that causes concern; it is also the industrial fishing techniques that are wasteful. In the waters off Madagascar, for example, it is estimated that 'for one kilogram of shrimps caught, five to ten kilograms of fish were lost in this way' (Chaussade, 1997, p.73). Traditional fishermen used to separate edible from inedible fish immediately after the catch; the latter and also young fish were immediately thrown back into the sea where they survived and ensured the continuity of the stock. Industrial fishing separates 'useful' from 'useless' fish far too late; when the 'useless' fish are thrown back, they are already dead. Ultimately, it is a problem of time: it may be unlikely that the last fish will disappear from the North Atlantic, but the reconstruction of a sufficiently substantial stock for human consumption will require many years. Fish farming is no alternative because it is a monoculture with all the risks involved. It can be considered a complementary source of food but not a substitute.

Concluding remarks

In environmental terms, the problem of marginality is located in people's minds: it is the way they look at and use (exploit) nature that causes environmental problems. A region that is marginal from this perspective is the civilized technical landscape where nature has either completely disappeared or where it is strictly confined to spaces that are artificial, controlled and bare of any spontaneous processes.

The loss of spirituality is doubtlessly a major factor in this process of marginalization of the environment. Traditionally, religion evoked feelings of solidarity, but by abandoning it, many people have followed the route of individualism. The recent shift from exploitation to respect (Figure 7.1) can therefore be interpreted as a sort of counter movement to the secularisation process, which has characterized western thinking since the Age of Enlightenment. Land has been sacred to our forefathers, and it still is to many people in the world. Africans retain a 'mythic-religious relationship with the land (some places were either holy or bewitched)' (Bidima, 1995, p.67), and these cosmological relations have been considered by the Europeans as a sign of underdevelopment as they gave rise to a fatalistic attitude (ibid., p.56). Unfortunately, the 'civilized' western society has developed a utilitarian attitude and reached out its hands to exploit the world's resources according to its own values, pretending that they are the only ones valid for the benefit of mankind. If a return to more spirituality is occurring (as it seems in the early 21st century), it may still be appropriate to view the references to religion and the Holy with some attention. Sacredness can also be used as a pretext in political conflicts: most wars are fought in the name of God, whether He likes it or not. The conflict between Israeli and Palestinians has similar roots; both Judaism and Islam found their claims to the land in Palestine on 'religious attachment', on the 'holiness ascribed to the land', on 'eschatological associations with land in general' (Cohen, 1993, pp.6, 27, 38 f.). Sacred values are in this case used for secular purposes.

One of the ways out of the environmental dilemma would be to develop more respect for the ecosystem, to demarginalize it in our minds, to change our attitude, and to present actions instead of words. As Schaller (1998) has demonstrated, the gap between talking and doing is still wide, among the population as well as with politicians and entrepreneurs. Even a rapid change, however, would only yield visible results in the medium to long run, not immediately. It is first and foremost a cultural task, not a technical one, which lies ahead. In a message to a landscape conference held in Capri, the local community council explicitly demanded more respect for all aspects of the island (the rocks, all forms of life such as forests and their animals, rare species, rural art and architecture). The council members used very strong words when demanding that the latter are to be protected 'from the ignorance and the concupiscence of the new rich – the shameful rich – …' (D'Aponte, 1999, pp.256). This may sound harsh, but a straightforward language may be necessary under specific circumstances.

The environment is a domain that requires global action. While local and regional problems can be dealt with at the appropriate levels, they may reach beyond this restricted scope and touch the national and international scale. Humanity is increasingly menaced by man-made risks (Beck, 1992) that ignore political boundaries. Pollution does not respect borders or state sovereignty. The case of the Chernobyl radioactive cloud (Gould, 1990, p.39) is the most striking example Europe has witnessed so far – even if national legislators reacted differently to the challenge (e.g. fishing was forbidden in the Swiss parts of the Lakes on the Swiss-Italian border, but not on the Italian side). Sometimes, industrial or energy plants are deliberately located close to a state boundary, in places where there is little risk to one's own country, but where problems will be shed on the neighbour. Such decisions as well as the construction of factories that use hazardous materials in countries of the South demonstrate to what extent not only the environment but also entire societies have become marginalized – the Bhopal disaster in 1984 with more than 2,000 dead is but one example. The internationalization of environmental issues is linked to the globalization of economic activities. The fact that transnational companies often act in complicity with local elites (Holton, 1998, p.124) does not make things better, in particular if these elites will deposit their monetary gain on banks in the North.

Such considerations were doubtlessly behind corporate initiatives that are popping up here and there, usually in cooperation with an NGO with environmental objectives. The partnership agreement between WWF and Lafarge (a globally active producer of building materials) from 1998 (Lafarge, n.d.) can be mentioned in this context as well as the engagement of the cement industry within the World Business Council for Sustainable Development (WBCSD, 2002). In either case, the respect for the environment and the need to protect it as far as possible to ensure continuing economic activity has been the driving force; the insight of a long-term conservation strategy seems to dominate the philosophy of short-term profit. The ecological restoration of quarries is one goal among several in the first of these two agreements, whereas the cement industry action plan is wider in scope and includes also the domains of health and security for the workforce.

Public awareness has reached the highest levels of politics already in the 1970s, but as political processes are usually very slow, the results take a long time to show up. Setbacks are inevitable, and the attitude towards the populations within its own country and towards the neighbours determines a government's reaction. The slow reaction of the Spanish government when the tanker Prestige broke apart off the Galician coast (Northwest Spain) in November 2002 and polluted the sea and the Galician beaches can be seen as characteristic for the attitude towards a region that Reynaud (1981, p.186) regards as a dominated periphery. Every international agreement is only as good as the will of the partners to ratify it and comply with its regulations. The refusal by the United States to sign and ratify the Kyoto Protocol on atmospheric pollution signifies that a major polluting society in the world dismisses the concerns of the majority of the people as non-applicable to the special case of the US. The same attitude holds good for the United Nations Convention on the Law of the Sea. The process of acceptance is slow and thorny and requires never ending campaigning by individuals, the civil society, and by NGOs that are particularly engaged in environmental matters. To them, the environment plays a central part in thinking; it has been lifted out of its marginal position precisely because of the important role it plays for our survival. The issues concern humanity as a whole, even if it is sometimes questioned '[w]hether or not such Doomsday predictions are valid,' (Holton, 1998, p.124). Relying on the market forces alone – this has been said above – does not suffice to solve a problem that is truly global and requires global politics.

The ecosystem is a vital part of human existence, but it cannot voice itself in terms politicians will understand. It requires the assistance of human mediators, people who recognize and formulate the problems in order to motivate the society to legislate in its favour. This, of course, means to introduce either new regulation, to maintain or to reinforce existing ones, a very unpopular measure in a period when rules have become unpopular. National parks and nature reserves have to be protected by the society (the state) with a legal base where the balance between different interests must be made. Is it more important, for example, to protect the Arctic National Wildlife Refuge in northern Alaska or to start exploiting the oil resources that lie below its surface? In the interests of the ecosystem, preservation can be the only answer to this question, and it has been granted more than forty years ago, in 1960, when President Eisenhower declared the area a wildlife refuge (Sierra Club, n.d.). Twenty years later, President Carter reaffirmed this decision. The issue of short-term profit against long-term environmental protection has been decided in favour of the latter. The plans to end the protection of this fragile ecosystem in favour of the exploitation of the oil reserves will eventually lead to its destruction, damage that will continue for a long time after the oil reserves have been exhausted. We might criticize the company managers and politicians who advocate the sacrifice of a national park in favour of the economy, of using domestic oil resources rather than purchase them on the global market, but criticism is a negative reaction. It would be much better to persuade the actors in question through pointing out, for example, that oil can be substituted by other means of combustion, that it should be used sparingly as a raw material for all kinds of manufactured products that could serve mankind more that just going up in fumes (or finishing in the oceans following a tanker accident), that it is a precious resource, etc. The demarginalization of the environment has to take place in the human mind.

PART III
RESPONSES TO MARGINALIZATION

Chapter 8

Policy Responses to Marginality

When it comes to dealing with marginality from a political perspective, its negative aspect usually prevails, and the positive side, i.e. the potential contained in it (see p.52), is rarely taken into account. Policy makers usually represent mainstream ideas and firmly believe that every element of the society (the social system) must find its place within this range. They want to impose their own values on the entire society. Flexibility is not in their particular interest as it upsets their carefully prepared plans. In certain systems, innovators thus face difficulties because their ideas may seem too far away from ordinary thinking, and the risk of failure is estimated to be too high. Inventors often lead a solitary life, unless they operate in a clearly defined economic environment, for example in the R&D section of a large firm – but true inventions (product and process innovations) cannot be 'bought' like goods in a shop, they can only emerge in surroundings where there is freedom of thought, time and material constraints.

In the same way, marginal regions cannot obtain a simple remedy, a prepared medicine to emerge from what may look a hopeless situation. This is particularly true when hopelessness is defined from the outside and not felt by the population within the marginal region. The case of the Amish presented briefly in Chapter 3 (p.75 f.) is but one example among many. Such groups are dominated by centripetal ideas, although they do not necessarily close their eyes to what is going on around them. Their region may come close to or even be an *isolat* or an *angle mort* (Chapter 2, Figure 2.3). According to the prevailing centrifugal economic thinking, however, trade and exchange are an absolute necessity for every society, and they are seen as the main instruments to overcome underdevelopment and marginality. The idea that trade is the solution to all problems of disparity occasionally takes grotesque forms when goods are transported across a country or even an entire continent, either to take advantage of minimal production cost differences or simply for the sake of a label (pigs from western Europe to Parma in Italy to be sold as Parma ham). If this philosophy would be strictly pursued, marginal regions would not exist at all, but all regions would be included in the centre-periphery model. No region would be excluded and disparities under such circumstances considerably reduced. The profits resulting from such kinds of trade are, however, fallacious because they are based on the assumption that the ecosystem is free of charge – pollution and resource depletion have no cost.

However, disparities are inherent in the human society; this point has been made explicit in Chapter 1 with reference to Rousseau (1987). Also innate in human society is the idea that disparities have to be reduced or even eliminated. The Islam, for example, has the obligation of alms to the poor, an obligation imposed by God (Quran, 9,60) who sees everything (Quran 2,271). Alms are a gesture to help the poor and deprived, the prisoners, those who are working for God, those who travel. Although alms do not eliminate the difference between the wealthy and the deprived, they help to lessen them to some extent and alleviate the situation of the poor. The wish to help is present in everybody, but the understanding of what help really means and to which end it is practiced varies between individuals, groups and societies.

This chapter looks at top-down attempts to reduce marginality. The first section is devoted to a brief discussion of the development question, a topic that is considered from a global perspective. the second section presents regional policy on a national scale, emphasizing in particular the case of Switzerland. A third section examines the efficiency of top-down measures. The chapter concludes with a few general remarks on regional policy and the political system.

The Development issue

There is no need to engage into an extensive development debate in this place; the literature on the topic is legion, and the perspectives vary according to the ideological and political position and the personal engagement of the individual authors, as a glance at a number of texts easily demonstrates (Smith, 1984; Cole, 1987; Adams, 1990; Hettne, 1990; Sen, 1999). In the present context, it is the widening gap between the rich and the poor around the world, between the North and the South at every scale (according to the metaphor used in Chapter 1 on pp.34 f.) that serves as our *leitmotif*. Despite many efforts, little has been achieved since development became an international issue in the 1960s. The reduction of the (economic) inequalities (we dare no longer speak of their elimination) through development has been declared a public issue on the national as well as on the global level, but whereas the per capita gross domestic product for the poor countries stagnates, that for the rich ones has been growing at an increasing rate since 1800 (Kreutzmann, 2003, p.3). Development measures looked very promising at the outset, but one has learnt to be very cautious about their outcome. Very much depends on the understanding and interpretation of the fundamental term of 'development'. In Chapter 2 (p.39), we have defined it as a process that leads from an unsatisfactory to a satisfactory situation, satisfaction being something perceived that cannot be objectively defined. The actual goals and the type of development are often not identified by the people concerned but by outside interests.

Aid to development can be motivated in many ways. It can be forced upon someone without his or her wish (the classic top-down approach), it can be asked for out of a feeling of distress, or it can even be cultivated within a depressed population devoid of external assistance (the bottom-up way). These three possibilities mirror different power-relationships. In the first case, help may be selfish, stimulated by some (overt or covert) hegemonic objective of a geopolitical nature (the carrot and stick mentality): 'The power to do good goes almost always with the possibility to do the opposite, ...' (Sen, 1999, p.xiii). In the second case, it would be our duty to aid someone who asks out of sheer despair, whereas in the third case, a population draws on its endogenous potential. Self-help does not mean that a society behaves as if it were a closed system; on the contrary, by watching the world around it, it mobilizes internal forces in order to escape from a depressed situation.

Development policies by governments take many forms; they have in common that they are usually top-down oriented, i.e. guided by a state's ideas about what development should be like and what could be undertaken to arrive at a specific goal (a normative or positivistic approach; Gustafsson, 1989, p.137). Governments do not act in complete independence but are themselves entangled in a myriad of connections both nationally and internationally, hence the definition of development is not necessarily founded on a domestic view but on a multiple perspective, influenced by internal and external political and economic actors and their specific goals. Without entering into more specific arguments, our position here is that a development process cannot be forced on a population and a region from above without even looking at their concrete needs and wants. Such assistance smells of hegemonic desire and is not motivated by humanitarian considerations. Humanitarian motives will usually serve as camouflage for the real intentions. During the Cold War, the geopolitical component of aid for development has been all too obvious, and interestingly enough, all countries (even those not directly allied to a particular camp) fell into the same (psychological) trap, viz. to help those whose ideologies were closest to their own. Real development assistance must, however, be guided by the principle of selflessness.

Since the 1960s, development in favour of the countries of the South has been strongly influenced by the initiatives taken by the United Nations. Its development agency, the UNDP, is monitoring the process and adapting the development goals according to the changing philosophies. The development goals of the 1960s and 1970s were very much under the impact of quantitative thinking, whereas from the 1980s onwards, qualitative goals have become more important. The millennium goals (see Appendix 4) therefore propagate improvements in the quality of life by formulating the quantitative aspects in a discrete way. Among the five millennium goals, the reduction or even elimination of hunger and access to clean water are certainly those that are the most urgent to be reached; once everybody (especially children) has enough to eat and drink (both quantitatively and qualitatively), there will be positive feedbacks on (infant) mortality and life expectancy. The educational goals, on the other hand, must be viewed with care. Literacy is certainly necessary. It is also necessary, however, to support and promote the local knowledge that is adapted to the natural and social environment of people in the South (see pp.162-6 on the Roma).

A recent development initiative has been started on that continent we called the most marginalized part of the world, in Africa. The New Partnership for Africa's Development (NEPAD; see Chapter 4, pp.91-3.) unites the countries of that continent and invites them to cooperate in order to ensure Africa a place in the global society. It is an achievement in itself to bring many diverse regimes and philosophies together, and the stakes are correspondingly high. To put all the goals with the respective measures into operation will require an almost superhuman effort and a massive investment that will be difficult to ensure. On the other hand, to do nothing at all would boost the downward spiral and result in total deprivation of the entire continent. As long as Africa is only appreciated for cash crops and its rich mineral resources, it will remain an exploited periphery and risks to be entirely marginalized once the resources are exhausted or substituted. The NEPAD initiative thus wants to halt this process and develop the continent's endogenous potential. However, it is not free from the idea of catching up with the rest of the world: the developed countries with their industries are still the model to be imitated. Modernization and the neoliberal paradigm dominate the programme – we admit that it is very difficult to evade it, given its global diffusion. What still remains somewhat obscure is the specificity of

Africa – but again, we have to admit that Africa is a continent, not a country or a small region with a clear-cut culture and tradition. The NEPAD is a formidable challenge and requires nothing less than abandoning conventional (egoistic) ways of thinking and substitute them with a new (altruistic) philosophy – something not even Europe has managed throughout its history. But no development is possible without Utopia.

The NEPAD is a continental initiative on the highest political level that has to build on the individual countries' experience and processes. It is not possible to develop Africa as such; the NEPAD can formulate guiding principles, propose specific actions and outline measures, but the member states alone are competent for their implementation. This can be interpreted as a provocation of the ruling classes: they have to review their power claims and take the cultural diversity into account as well. An important contribution to the development process could be furnished by the decentralization of authority away from an almighty and all-competent central government to lower spatial units (regions, provinces, local governments). Decentralization 'implies the transfer of powers of national government or its agents to the representatives of local territorial collectivities, whereby the latter are not directly responsible either to the national government or to its agent.' (Basta, 1999, p.30). It is a process that takes place on the basis of specific laws, and it can be reversed or modified at any time; there may thus be no long-term guarantee as to its functioning. Decentralized units do not have legislative power but are merely administrative in character (ibid.). A more radical solution on the constitutional level would be a federal system where the members of the federal state 'dispose of original autonomy' (ibid.). From the temporal and legal points of view, this is a much better solution, but it robs the central state of a considerable part of its authority and may render central policy rather complicated (see the examples from Switzerland quoted in the Prologue). A federal state cannot, however, be installed from scratch but requires time and culturally vested habits to evolve. Although it may be an ideal instrument to promote bottom-up or generative (Gustafsson, 1989, p.137) development policies, there is no guarantee that disparities can be eliminated within a short time. The federal state is a political construction, operating on the basis of consensus within and among its members. Interests as to development goals may conflict for the simple reason that there are too many actors involved. The advantage of decentralization – whether on the legal or on the constitutional level – is the ease with which cultural diversity, local and regional specificities can be taken into account.

Development processes could thus be managed on a bottom-up basis according to a society's needs.

However valuable a decentred regime might be, it still faces the obstacle left by the European colonization. It is easy to criticize the NEPAD initiative from the perspective of the North, but the critics should always bear in mind what the Europeans have done to Africa. The colonial powers have exploited the riches of the continent and left it with the cumbersome legacy of the nation-state, with its bureaucracy, its taxes, national boundaries and the request for a national identity. On a continent where the societies were traditionally organized on a local or regional level, and where family and clan ties determined allegiance, not membership in an anonymous political institution, this represents a particular challenge. Traditional bonds still exist, but they are superseded by the imported concept of an anonymous state. To find ways out of this dilemma requires a new beginning, a new definition of the state, and within this new political system also a distinct role for the various levels. Power relations must be clearly defined, and security must comprise material, psychological and moral sides to enable a society to function adequately. Finally, the collective assets 'ensure material, moral and spiritual security' (Sawadogo, 2002, p.26) and allow the construction of feelings of public responsibility, sharing and solidarity. These qualities are imperative to guarantee long-term development towards a better future. Above all, this process requires more time than our epoch is prepared to give the Africans. Development processes are slow, they have to be understood by the political actors as well as by the masses, and they have to overcome the mental inertia of everybody – Europeans ought to know that. We are used to see a quick result of every effort we undertake. Innovation is the key word, but for the sake of innovation we overlook the need for consolidation. The NEPAD initiative should therefore be encouraged, but the necessary mental conditions must also be developed in the North.

A final remark: development is a normative term and can be interpreted in different ways. Everybody, irrespective of his or her socio-economic status, education or place of residence, requires constant development, albeit not in the same domain. What has been said above is based on our northern view of development, a view that has been condescending for decades. However, the deficit that is to be filled through development is not confined to food and water shortage, lack of transport, low gross domestic products etc. Development must go beyond pure material goods. 'What the West has in abundance is lacking in the tribal society, and vice versa. Among the many things the tribal society could transfer to the modern world – because they exist in abundance – are mind and emotions' (Somé, 2000, p.105; transl. WL). Here lies a deficit the North has to overcome. While our conventional aid to development operates at the global scale, humane development has to take place at the local level and in the families.

Regional policies

Idealistic ideas also prevail when it comes to reduce disparities inside a country. Regional imbalances are usually identified on the economic level (unequal salaries or price levels, insufficient access to services, etc.), and they are often reduced to the urban – rural continuum (see Figure 2.7) with rural areas at a disadvantage. This is often true, but cities also have their problem areas, as has been discussed in the section on urban poverty (Chapter 6). The difference between rural and urban areas is a difference in scale with practical consequences: it means that not the same type of public actors is responsible for respective policies. Disparities in cities fall into the field of competence of urban authorities, whereas rural deprivation often covers large areas where one or several regional administrations have to cooperate, or where in the national interest a central government has to take action. To resort to the central state may often be the only way out when a region is lacking the internal dynamics and the funds needed to free itself from marginality.

Demarginalization in this context will be attempted through regional economic development or regional policy. The aim of this process is to balance the relations between centre and periphery and marginal regions, to eliminate undesired disparities, and to create an equilibrium situation against potential conflicts. Such policies can emerge from the inhabitants of the region concerned or be proposed by a central administration. The population may complain about the economic and social situation it lives in and formulates demands to the government (an input into the political system), or the government itself becomes aware of regional disparities, regards them as unwanted situations, and decides on political measures in order to reduce or eliminate them. Bottom-up and top-down motives are possible; the political organization of a country plays an important role in the way regional policy is discussed and implemented.

Efficient regional development policy must take the population into account and not only be conceived of as a top-down strategy. Any solutions must be anchored in the region and offer long-term prospects, i.e. a certain guarantee that jobs may last, inequalities be reduced, and the population can see a future. According to circumstances, this may take a long time –

the political mobilization of the inhabitants, the perception of the region by potential investors, the general economic situation, the situation of the public finances, oncoming elections, etc. are elements in this process. The time-perspective to be adopted therefore reaches far beyond the term-of-office time-span of a government or a parliamentarian. Quick action is usually required, but the long-term positive outcomes require time to manifest themselves.

Backgrounds to regional policy

Various processes, such as technological change or political events, can motivate regional policies as much as the political awareness of disparities. After the division of Germany in 1945, the German Federal Republic (FRG) declared a belt of 40 kilometres along the boundary with the German Democratic Republic (GDR) and Czechoslovakia an assisted area (*Zonenrandgebiet*; Ante, 1991, p.69). The new border cut through traditional economic and social networks and upset former centre-hinterland relationships. In both countries, new structures and relationships had to be built. For the GDR, the border zone was an area of restricted access, and no investments were directed there except for military purposes. The FRG, on the other hand, wanted to keep the entire zone alive (with the idea of re-unification as the ulterior motive); rapid and efficient action was thus required. As emergency measures, governmental subsidies were granted from the 1950s onwards, and in 1965, special planning laws and programmes were implemented in order to reduce excessive disparities (ibid., p.72 f.). While economic and social objectives dominated, this policy was also a political sign, both inward and outward directed: the *Zonenrandgebiet* was not to become an economic and political vacuum. As the outward border of NATO, it was of particular strategic importance, although there was no excessive military presence (Kim, 1990, p.141). An important task was the reorganization of the transport infrastructure that had been disrupted by the drawing of the boundary. The promotion of industrial development (new firms or branch plants of firms situated outside the region) and tourism (Ante, 1991, pp.71 f.; Kim, 1990, pp.146 ff.) became goals within the special aid programmes. However, the *Zonenrandgebiet* was not the only region with problems, in fact 'the landscape of the German-German border has become only one peripheral area among others' (Ante, 1991, p.73). The results of the various policy measures were not entirely positive: 'Despite current financial aid, the economic situation of the border region compared with the rest of the Federal Republic has not improved, and in some parts it has become worse' (ibid.). Besides, the policy of promoting the *Zonenrandgebiet* met with considerable opposition both in Germany and in the European Community (Kim, 1990, pp.152 f.). Needless to say that after reunification in 1990, the situation has changed radically, and regional policy in Germany reoriented itself according to the new reality, i.e. the structural weakness of the former German Democratic Republic.

Technological change has often led to the marginalization of places and regions that were once prosperous. When the railway was introduced, conventional road transport (mules, oxcarts, etc) disappeared. Valleys and passes in the Alps, that once had carried transit traffic, were deserted, and the populations who had made their living of it were forced to find other occupations. Tourism to some extent provided a way out of distress. The railway age brought new prosperity to other regions, but in the course of time, the change of technology worked against some of them.

> The village of Göschenen at the northern entrance to the Gotthard railway tunnel (1000 metres a.s.l.) in Switzerland, for example, was once an important station for the transport across the Gotthard pass. After the construction of the Gotthard railway tunnel (1872-80) and the opening of the railway line in 1882, it became a water-filling station for the steam-engines; the halt required for this operation permitted the travellers to have a meal at the station restaurant. When electric trains were introduced, the halt was reduced to a few minutes, and the station restaurant lost its former importance. Intercity trains nowadays no longer stop, and around 2015, the high-speed trains will use the new Gotthard tunnel that crosses the Alps between 470 (north) and 300 (south) metres a.s.l. The authorities will have to develop a survival strategy for the future.

Similar stories can be told of many villages and valleys that once served transit transport across the Alps and elsewhere and lost this function because of the construction of roads (see the example of Ernen in Leimgruber, 2003a, p.159). In such cases, individual local initiatives can often halt the marginalization process (Leimgruber and Imhof, 1998). When, finally, motorways competed successfully with the railways and the conventional roads, new forms of marginality threatened. The railways, partly deregulated (or re-regulated) fight for survival; to close down lines that are of limited regional importance signifies a reduction to the accessibility of centres and threatens jobs. The motorways help to reduce traffic on ordinary roads, but they also drain potential customers from places (towns, villages, local pubs) that many travellers used to pass.

Old industrial regions

What can be said for the local scale also holds good for entire regions. The old industrial regions around Europe – the Ruhr area in Germany, Wallonia in Belgium, Lorraine in France, Northern England, South Wales, etc. – have experienced the rise of coal and iron ore mining following the industrial revolution. They have all been plunged into a deep recession when coastal locations for the steel industry were preferred to inland sites (due to the import of ores from overseas with a considerably higher iron content than the domestic raw material), when coal mining declined, and when overseas steel-producers started to compete on the European market. Regional development policies had to be developed in order to avoid excessive disparities.

The Ruhr area in Germany experienced a slowdown in coalmining from the early 1960s onwards, and the slump was aggravated in the mid-1970s in the iron and steel industry. Unemployment and emigration were the consequence. Structural programmes in order to alleviate the situation focused on the tertiary sector, mainly by promoting education (five new universities were founded in that period), innovation and new technologies (Schrader, 1993, p.137 f.). As in most old industrial region, the infrastructure was outdated and required modernizing (ibid., p.138). The positive aspects of this crisis were the coordination of the sectoral policies, the growth of regional policy in the Bundesland Nordrhein-Westfalen, and participation of the private sector in these efforts. Policy measures from above and cooperation from below combined to ease the situation of a once prosperous region.

The Saar Bundesland on the western border of Germany faced similar problems. The economy of this region was also entirely oriented towards coal and steel, and it entered into crisis towards the end of the 1950s (Dörrenbächer et al., 1988, p.210). The state-owned Saarberg corporation controlled the entire mining sectors. From the 1960s onwards, it started to react to the problems that arose out of the slump in coal sales, first in a defensive way (adaptation), afterwards by consolidating the status quo (second half of the 1960s). In the 1970s, the corporation reorganized and focused on energy and energy processing and started to develop new technologies of energy production from coal. However, the uncertainty of the European and global energy market lessened the hopes that had been put in the new energy technologies (Dörrenbächer et al., 1988, pp.210-4). Between 1957 and 1986, the number of coalmines dropped from 18 to six, and the production from 15.8 to 10.3 million tons (ibid., p.212). In the 1980s it was the turn of the steel industry to be restructured: within three years, 1984-7, many plants were closed, and unemployment became a major problem. In 1988, it was 11.9 per cent as against 9.7 per cent national (FRG) average (Giersch, 1989, p.259). The quantitative reduction of steel plants was accompanied by investment in new technologies and an improvement in quality: new types of steel were developed that 'were not known ten years ago, which is a sign of the innovative strength of the company' (Dörrenbächer et al., 1988, p.219). Innovation in all respects was thus the magic word to support the industry. In addition, the central as well as the regional government engaged in economic programmes to the benefit of manufacturing industry. In particular the production of investment goods (automobile industry, machines, electro-technical industry) and of consumer goods and food products (Giersch, 1989, p.262) helped the Saar Bundesland to overcome the critical situation following the downsizing of the traditional pillars of its economy.

Britain went through a similar process since the 1960s, but on a different scale and volume from Germany. The less prosperous regions of the North West, Yorkshire and Humberside, the North, Wales, and Scotland, experienced net-emigration between 0.1 per cent (Wales) and 6.2 per cent (Scotland) in the period 1961-71, and their population increased by 2.4 per cent as against a national average of 5.2 per cent in the same period (data from Table 1.2 in Manners et al., 1980, p.8). The oil crises of the 1970s (1973-4 and 1979-80) that threw the entire European economy into turmoil, further aggravated the situation. De-industrialization characterized most of Britain, in particular north and west of a line from the Humber to the Severn estuary, whereas 'manufacturing industry in the South East and its three adjacent regions of East Anglia, the East Midlands and the South West recorded slower rates of decline' (Keeble, 1984, p.43). As a consequence, large parts of the traditional industrial regions had become assisted areas by the early 1980s. The issue of regional policy and assisted areas has since been a continuing debate, both as a question of domestic affairs and within EU regional policy. To revitalize the regions has become a prime political issue, as the recent White Paper (DTLR, 2002) and the ensuing report (DTI, 2003) demonstrate. While, however, early regional policies aimed at assisting poor areas in order to reduce the gap between them and the rest of the country, the new philosophy addresses the country as a whole, 'every nation and region' (ibid., p.1). To promote every part of the country and activate regional potentials can only be achieved by transferring some sort of responsibilities from the central government to the regions. The devolution and decentralization process had started in the 1990s and led to the constitution of regional assemblies in Scotland, Wales and Northern Ireland in 1999. Further such assemblies are planned for the English regions. While they are introduced in a classic top-down manner, they will nevertheless have to be confirmed regionally by a referendum (DTLR, 2002, p.59). These regional institutions shall then cooperate with both national and local ones to develop the endogenous potential and 'tackle the particular weaknesses of each areas' (DTI, 2003, p.3). In addition, they can also cooperate with the European Union (DTLR, 2002, p.60) and thus be able to obtain easier access to structural funds (ibid., p.39).

At its very beginning, the European Economic Community declared that 'reducing the differences existing between the various regions and the backwardness of the less favoured regions' (EU, 2002) was to be one of its most important targets. The founding fathers of the EEC were conscious of the fact that 'not all Europeans have the same advantages and chances of success' (ibid.), something valid at every spatial scale. The three objectives that were subsequently identified focus on helping regions that have lagged behind others (Objective 1), that are suffering from excessive dependence on one economic sector with structural problems (Objective 2), and on promoting training and employment (Objective 3; ibid.). Such a policy is in itself an enormous task and requires considerable time to be put into operation and show lasting results. It cannot be carried through unless corresponding national institutions are created (an act of re-regulation!) that are responsible for the implementation in the field. For the European Union it has been a formidable challenge from the very beginning, and it has become an even greater task since. After 46 years of existence, the adoption of new members, and following several internal transformations, we observe that 'overall disparities between regions within the EU15 remained virtually unchanged between 1995 and 2000' (EU, 2003, p.4). The ups and downs of the economy do not help to alleviate this situation, and the admission of ten new member states in 2004 will increase the disparities precisely at a time when little help from an economic boom is to be expected. It will, of course, be impossible to abandon the regional policy programme, but the expectations it has provoked may not be fulfilled in the short to medium term.

Regional policy in Switzerland

Regional policy is a national task, and it is the national institutions (Parliament and Government) that set the general framework. This is also the case in a decentralized country like the Swiss Confederation. An important question in regional development is the degree of autonomy granted to the regions, whether they are free to decide how and where to invest the money, which are their specific goals, and which is the time frame. The political system plays a key role in the shaping of regional policies. In the case of Switzerland, the cantons are the principal actors in regional development, but there are nevertheless a number of general principles to be followed. The measures have to take account of political, economic and ecological domains (Figure 8.1). The two goals are 1) to eliminate unwanted disparities, and 2) to preserve the highly desirable regional diversity.

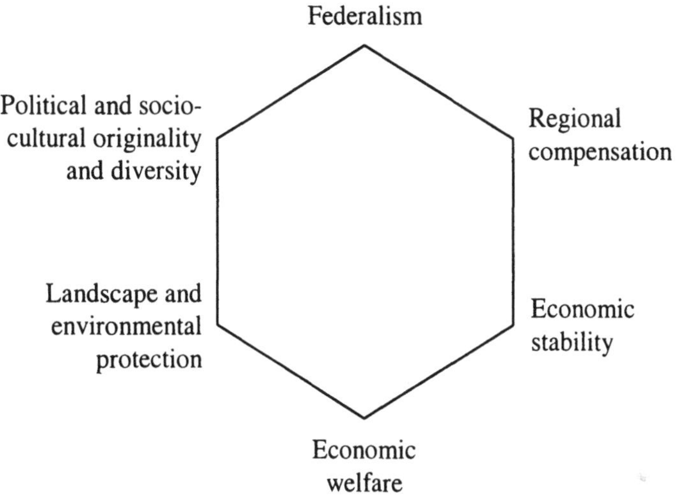

Figure 8.1 Goals of regional policy in Switzerland

Source: Brugger and Frey, 1985, p.47

This 'regional policy hexagon' cannot be taken as a general model but is restricted to the Swiss case because of the political background. The domains 'federalism' and 'political and sociocultural originality and diversity' do not automatically apply to other countries. The model can be considered an attempt at a balanced way of state assistance that has to take multiple interests into account.

In Switzerland, we can identify two aspects of inequalities, one based on the political organization of the country, the other on the unequal natural conditions for economic activities and social life. All instruments developed since the late 1940s have pursued the same goal, i.e. to offset unfavourable regional conditions, irrespective of their 'political' or 'natural' origin. It may, however, be useful to discriminate between these two as the instruments and their implementation are quite different.

The nature of disparities based on the *political organization* has been illustrated in Chapter 4. Although Switzerland is one of the richest countries in the world, there exist considerable differences between the individual cantons, and within the cantons also between municipalities. The uneven distribution of the population, the varying infrastructural challenges, and the different economic potentials are elements in this patchwork of disparities, to say nothing of the cultural diversity. A number of instruments have been developed since the 1950s with the aim to level out the disparities to some extent, but they depend on both the political will and the general economic situation. The most important is the law on financial equalization of 1959, which allowed the Confederation to link public funding to the financial strength of the individual canton (a principle also applied within most cantons with respect to the municipalities). This concerned subsidies and reimbursements as well as the cantonal quota of the direct federal tax (a sort of refund to the cantons that are collecting this tax), but also the contribution of the cantons to the social security system (Deiss, 1998, p.178; Gaudard, 1997, p.215). A more or less just distribution is achieved through the calculation of the financial strength of the individual cantons, an exercise repeated every two years and based on gross domestic income, fiscality, tax income and both surface and population in mountain regions. Weak cantons thus obtain payments from the Confederation as well as from the rich cantons. The classification into weak and strong may at times give rise to criticism, but this system is a sign of solidarity that must not be underrated from the perspective of internal cohesion. A revised law on financial equalization will be enforced in 2006 and include financial assistance also according to geographical and topographical criteria to offset disadvantages because of altitude (mountain regions), slope (difficulties for agriculture and forestry) or settlement density (excessive cost because of dispersed settlements). These criteria had been absent from the old law (Bottinelli, 2003, pp.24 ff.), and they mirror the role mountains play in Swiss thinking.

Naturally induced disparities are due to the mountainous nature of the country (about 75 per cent of the surface). Policy measures in favour of mountain regions are thus almost a matter of course. Mountains play an important role in the identity of the Swiss population, mainly for 'the fact that the original nucleus of Switzerland is composed of three Alpine States' (Chiffelle, 1983, p.28). The principal activity of the mountain population had always been farming, supplemented in certain areas by transit transport (before the arrival of the railway) and tourism (from the late 19th century onwards). At the outset, mountain policy was basically agricultural policy, and the first measures in favour of upland regions were laid down in the revised law on agriculture of 1929 (Leimgruber, 1985, p.100). During the period 1944-7, the Swiss government commissioned a detailed survey of the situation of Swiss agriculture, and in 1949 it issued a decree on assistance to farmers in mountain regions. By defining the standard limit of mountain regions (ibid., p.102), it paved the way for systematic support to farmers in upland regions. It was inspired by cultivation practices and based on five criteria (ibid., pp.102 f.):

- length of the vegetation season;
- distance of the individual farms to the nearest village;
- accessibility (distance to the nearest railway station);
- topography; and
- soils.

It soon became evident that this delimitation was not adapted to the differentiated nature of the Swiss mountain regions. A new decree of 1958 used animal husbandry as a starting point and arrived at a subdivision of the mountain region into three zones, to which a fourth was added in 1980. The list of criteria was considerably enlarged and included height, slope, climatic factors, percentage of cows of total livestock, milk sale, distance and freight rates to markets, etc. – 15 altogether (ibid., p.104). A major difficulty was to define a just delimitation of the mountain regions; this is a typical problem of the threshold values chosen. If, for example, a slope angle of 18 degrees is the lower limit, this does not mean that a slope of 17 degrees is easier to work. The sharp division between mountain and non-mountain regions was eventually reduced by the subsequent definition of intermediary zones: pre-alpine hill-zone, adjacent cattle-breeding zone, intermediary zone and extended intermediary zone (Leimgruber, 1986, p.49f.).

Both programmes were classic top-down policy measures, based on the perception of the needs and wants of mountain farmers by politicians mostly far away from the reality. They were a form of structural policy, directed exclusively at the farming population. The existence of other economic activities was ignored. The objective was to improve the economic situation via an increase of the farmers' income and by strengthening their competitiveness versus farmers in lowland regions. Social policy combined with income policy to reduce or stop the exodus from mountain regions.

It was only in the 1970s that a more radical change in thinking occurred. A purely top-down sectoral approach was no longer acceptable.

Economic and political circumstances called for an integrated policy that would provide for regional needs, going beyond an income guarantee for a small segment of the mountain population. The assistance provided under traditional mountain policy conflicted with the practice of subsidies for agriculture in general – the principle of an all-round distribution of money was increasingly criticized. Besides, little was done to improve the general living conditions of the mountain population: the rural exodus had continued despite the money poured into the mountain regions. New ways were sought, and one option was to provide a better infrastructure equipment through investment aid. The Law on Investment in Mountain regions (LIM) was implemented in 1974, and it became the cornerstone of regional policy in the country for the decades to come.

The aim of the LIM was to promote bottom-up development within a top-down regulation. The Confederation enacted the general law, but the proposals as to development measures had to be defined on a regional basis. Contrary to the programmes presented above, the LIM required that municipalities organized themselves in regions (usually within the same canton), the only spatial unit entitled to benefit from the programme. The idea behind the LIM was that good infrastructural equipment (schools, health service, sports and leisure facilities, etc.) increased the attractivity of a region and would thus not only prevent emigration but also attract investors. An investment in infrastructure would thus generate benefits for an entire region. The LIM-regions had to define a programme of infrastructural investments for 15 years that would help to improve the living conditions in mountain areas. Once these programmes were approved, credits could be obtained at favourable conditions. A major endogenous effort was required; the Confederation played a subsidiary role.

The LIM showed the strength of the federalist system: the best way of organizing regional development was to stick to the traditional spatial units, the municipalities, where there is a long tradition of self-government, and to base the regions on this experience. The individual LIM-regions remained within cantonal boundaries, because that facilitated the collaboration with cantonal authorities. Traditional structures were thus retained. The LIM is a good example of an integrated strategy.

The LIM met with considerable success: until 1983, 1,228 municipalities had joined together in 54 regions, roughly two thirds of the country in surface and a quarter in inhabitants (Bottinelli, 2003, p.30). The LIM fund, set aside by the Confederation, originally contained 500 million Swiss francs, but the sum was subsequently raised to 1.5 billion (ibid., p.30 f.). The success of the programme can be measured by the decision of the government to continue the programme. In the new LIM, enforced in 1998, the direct assistance is limited to basic infrastructures (water supply, sewage, electricity supply, compulsory schools) that do not bring direct economic advantages; in this way the regions can direct their own investments towards economically viable infrastructures (vocational training, tourist infrastructure). In particular, the cantons have also to contribute to the policy measures. The experiences of 20 years have left their traces.

Mountain regions, however, were not the only areas that faced difficulties. Further regional policy measures became urgent in the late 1970s as a consequence of the oil crises. Certain regions suffered from a recession in the secondary sector, particularly in manufacturing. The watch-making region of the Jura was the most outstanding example, but other regions had similar problems, albeit not on the same scale. Promoting enterprises and innovations thus became a key task for a focused regional policy in regions that threatened to lose out as a consequence of an unstable economy both in Switzerland and Europe.

Regional policy has so far been guided by the traditional pattern of cantons and directed towards regions with specific problems. Mountain regions were particularly favoured, as has been demonstrated. Since the 1990s, however, major social transformations have occurred that will eventually influence regional policy (Expertenkommission ... 2003). Such transformations are in particular:

- the growing role of conurbations in spatial development processes, where functional interrelationships override the local and cantonal outlook;
- new scales on the national and international level;
- the general economic slowdown; and
- the increasing regional disparities between conurbations and the periphery or marginal regions.

Regional policy will have to be reoriented along these lines. In particular, agricultural policy requires a profound revision. Farming practices, agricultural education, research and technology have to focus on the principle of sustainable development and on the new spatial dimensions. To this effect, the independent expert panel (Expertenkommission ..., 2003) has listed a number of proposals whose main points can be summed up in a simplified way (Table 8.1). Whether they can be incorporated into a new philosophy of regional policy that is politically acceptable remains to be seen. The political process may not be able to keep up with the rapid changes of the challenges.

Table 8.1 Ideas for a new orientation for Swiss regional policy

	Traditional regional policy	New regional policy
Target	Reduce regional disparities	Improve competitiveness of regions
Orientation	Promotion of infrastructure	Promotion of innovation, development of know-how networks
Spatial frame	Specified regions (LIM)	Whole country
Direction of stimuli	Top-down and bottom-up	Bottom-up (entrepreneurs)

Source: Expertengruppe … (2003), p.13 (abridged)

Regional policy is a permanent task, oriented towards medium- to long-term goals; but in a period when everything seems to change at an ever-increasing pace, it may become very difficult to measure its efficiency. Its orientation is predominantly economic and demonstrates the prevalence of economic thinking (Chapter 4), and it is likely to remain so, as Table 8.1 demonstrates. From this perspective, the claim to sustainability does not sound very credible, as the concept of sustainability is related to a shift on the value continuum from secular to sacred, from technocentric to ecocentric (see p.71 f.). It would also mean that the speed of a process was to be adapted to the rhythm of all components in the system. Since it is impossible to accelerate the ecosystem, the social system is likely to have to slow down. Such a deceleration is something decision-makers fear most as it looks like a step backwards. Innovation is the magic word, but constant innovation also produces intellectual waste (Reheis, 1996, p.118). The knowledge acquired will be outdated within no time, and time becomes a trap for the society: if the consumer no longer recognizes an innovation as such, the sum of efforts to create it has been useless (ibid., p.221). The constraint to innovate at an increasing pace will ultimately have negative consequences for the individual, the society, and the environment on all scales. To be profitable, innovations have to be accepted by the consumers who have to comprehend them and integrate them into their thinking (Gronemeyer, 2000, p.122). The human mind works at a pace that is radically different from industrial production and is controlled by the memory and past experiences.

Regional policy cannot simply cement past achievements but must face the challenges that arrive every day from around the world. However, it must not be an instrument of economic promotion but emphasize the interrelationships that are shown in the human ecology triangle (Figure 2.6) and – in the case of Switzerland – still consider the goals outlined in Figure 8.1. The critical question is, which among the many interests in this field should be privileged (Figure 8.2). Politicians have been elected to ensure the well-being of the entire society, hence they have to strike the balance between the various groups – not an easy task given the centrifugal interests.

From the wide range of interest groups it is obvious that a true consensus is difficult to find, as they are all situated close to the margin of the hexagon, i.e. in rather extreme positions. The only groups that are situated closest towards the centre are the trade unions and the right wing parties – two extremes in the political world. The figure shows that even the alternative factions follow diverging goals – precisely those movements that can be regarded as defending the environment and at the same time demanding new orientations and changes of attitude in the society.

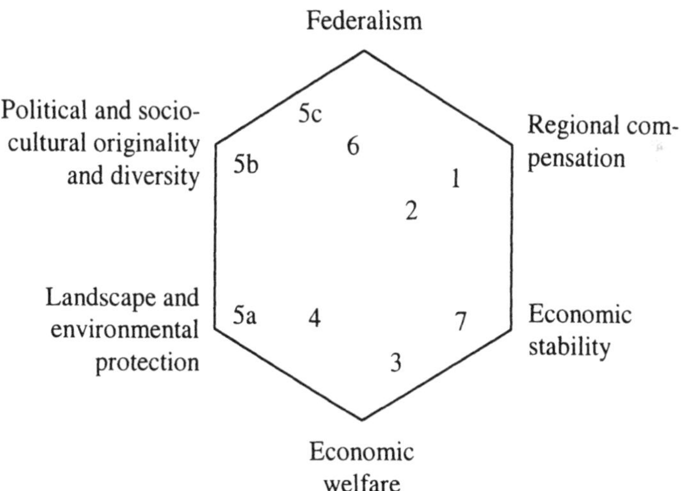

Figure 8.2 Regional policy goals and the spectrum of interest groups

1 Trade associations and enterprises in problem regions; 2 Trade unions; 3 National chambers of commerce (except sectoral associations); 4 Taxpayers and consumers in centres in the Swiss Plateau; 5 Alternative movements (5a Green Party; 5b "disembarkers", satisficers; 5c Democratic grassroot movements, citizen initiatives); 6 Right wing parties based in peripheral regions; 7 Left wing parties based in centres

Source: Brugger and Frey, 1985, p.62

Figure 8.2 is not an illustration of party programmes – they change with the terms of office – but of groups in the population that have diverging interests concerning the future evolution of the country. In the Swiss political reality, where a final decision is always a compromise, a new regional policy will eventually have to be defined, but all interest groups will have to review their claims.

How efficient are top-down policy measures?

Regional policy alone cannot bring about miracles. Its efficiency is limited because a variety of interests, perceptions and time frames collide and complicate its implementation. There is a tendency to perpetuate dependencies despite the existence of policy measures. However, it would be wrong to point out poor results only and overlook their real effect. Regional policy is conducted with a view to satisfy the needs of a population in a certain area. The inhabitants are at the same time regional and local politicians, the electorate, taxpayers, workforce and entrepreneurs, consumers, pupils, members of sports clubs, etc., a diversity that makes it difficult to satisfy all claims. However, on the regional scale it may be easier to marry diverging interests. The people are conscious of their multiple social roles, the regional economy is usually anchored in the population, and survival is more important than profit. The statement of the owner of a medium firm in Switzerland is a good example for this attitude. When he constructed an extension to his factory, he placed most of the orders with regional firms, even if the costs were somewhat higher than with external companies. His reasoning is interesting because it mirrors a philosophy that cannot be found with large firms, in particular with TNCs: 'We may have paid half a million francs more [than with external firms], but look at the satisfaction to the regional craftsmen, the quality of the work that was executed on time, and the publicity this created for us – the money has already come back. This is my understanding of sustainability.' (Gschwend, 1999, p.3/2; transl. WL). He has a very pragmatic interpretation of sustainability, and his statement demonstrates that he is anchored in his region and feels responsible not only for his workforce but the population as such.

To what extent did regional policy fulfil the expectations? We shall attempt to look at the outcomes in Britain and Switzerland and see how far goals like maintaining a population in the region have been reached. It must nevertheless be stated at the outset that the best regional policy cannot succeed if external agents disrupt it with decisions based on global conditions.

In the case of *Britain*, it has become evident that designating Assisted Areas since the 1960s has not resulted in a general improvement of the economic situation. This is neither the fault of the regions nor of regional policy; it is rather the perception of Britain, i.e. the subjective image people have of their country that may override rational arguments about regional development in depressed areas. A survey among school leavers in various British cities has demonstrated that Southern England ranks very prominent among the young generation. For most of the persons interviewed, 'the Southern Plateau of Desire is very prominent, and extending from it the East Anglian and Welsh Border Prongs define the now well-known Midland Mental Cirque. In the southeast the Metropolitan Sinkhole can be easily distinguished, while the highest gradient by far is over the Bristol Channel between England and Wales.' (Gould and White, 1974, pp.81-3). London was not the preferred goal, but the rural or less urbanized regions in the south figured high in the school leavers' minds. Migrating from northern regions to the south was seen as a viable alternative to staying in one's own region, in particular as unemployment in the southeast was lowest in the entire country at the time of Gould and White's survey (1967) and still is today. Although local peaks of preference existed, the overall tendency pointed to a potential migration towards the southeast of the young generation.

Through the devolution process, Scotland, Wales and Northern Ireland have received autonomy and regional assemblies. The British government has understood that decentralizing decisions may bring more success to regional development than national top-down programmes. It has applied the principle of subsidiarity, according to which decisions have to be taken on the political level that is most competent, and further devolution measures are now being planned, this time within England. The White Paper (DTLR, 2002) prepares the ground for such an important change. Regional policy will no longer focus on assistance to poor regions but encourage 'every nation and region of the UK to perform to its full economic potential.' (ibid., p.1). This sounds very much like one of the goals of a new regional policy in Switzerland (Table 8.1). National, regional and local levels must cooperate in order to stimulate the economy at every scale. The 'five drivers of productivity' (skills, investment, innovation, enterprise, and competition) are the focus of reforms at the microeconomic level (ibid., p.3). Although it is not said explicitly, this turnaround expresses dissatisfaction with what has been achieved (or rather not achieved) under the old system.

Such a radical change owes a lot to reflections made at the international level, to ideas about regional development formulated within the European Union. As a member state, Britain on the one hand benefits from the Structural Funds; on the other she can introduce her own ideas about EU regional policy (Chapter 4 in DTLR, 2002). The insight that 'modern regional policy must be locally led' (ibid., p.iii) has grown in countries that were strongly centralized and had to recognized that their pure top-down regime was no longer apt to the challenges of the present. Redistributing power is, of course, no easy task; it does not please the centralists, and it is a serious test for those who claim it. The devolution process that started in Britain in 1999 and is about to continue is in itself a top-down measure: the regions receive some autonomy at the discretion of the central government. The first step, Regional Development Agencies (1999), was only a partial measure, as decision-making was not decentralized and the agencies were primarily 'accountable to Ministers and Parliament', although they should also be 'responsive to regional views and that they give an account of themselves to those with an interest in their work' (Office of the Deputy Prime Minister, 2001). The road forward towards more regional responsibility seems to be the only way to improve efficiency and kindle the interest in regional potentials.

Regional policy in *Switzerland* is an ongoing process, as has been shown above. The steady examination of the efficiency of the various programmes has proved to be valuable in adapting them to changing circumstances. As the political process in Switzerland is relatively slow (due to the system of direct democracy), changes do not happen overnight, and it is possible to evaluate the results. The early sectoral programmes were too unilateral and did not take the complex regional realities into account. The measures in favour of mountain regions were in fact part of agricultural policy, destined to reduce the disadvantage farmers with difficult productions conditions faced as opposed to farmers in lowland areas. Financial contributions were paid according to the zone where a farm was located.

One of the goals of the policy was to maintain mountain agriculture and thus slow down emigration. It is hardly possible to measure success or failure by looking at census figures. The evolution of the population was in fact very uneven, periods of decline changed with times of growth. In general, however, the tendency was towards a population decline in the more remote areas, whereas regions with good accessibility could maintain or even slightly increase the number of inhabitants. Little can be said about the role of migration in this process, as reliable data are missing for the period before 1970. Most movements would in fact take place within municipalities or valleys, and the major centre of a valley might receive the migrants first. Besides, by looking at regional figures, we must always bear in mind that even in non-tourist areas there is a minor presence of tourism (for family holidays, for example) that offers some income. The exploitation of hydropower throughout the Alps also offers a few jobs, and so do small industries, often based on local resources (building, woodworking). The true impacts of the agriculturally oriented regional policy measures are therefore difficult to assess.

In the case of the Law on Investment in Mountain regions (LIM), the situation is different. To be recognized as such, every LIM-region had to set up a secretariat and draw up a development plan for 15 years. The planning and development process was carefully monitored, and at the end, a report on the achievements had to be compiled, together with proposals for future activities for the next 15-year period. Despite the careful organization, it is still difficult to discover the true effect of the LIM programme on regional development as opposed to external influences. The aim of this kind of regional policy was not to 'irrigate' the regions with money but to furnish an incentive to internal actions, to mobilize the endogenous potential present in every region. The evaluation of the programme is made difficult by the fact that some LIM-regions are far too large and heterogeneous to allow for a balanced development. At times certain parts of the LIM-region benefit from good accessibility whereas others are dead ends. Internal migrations may occur that do not show up in the overall balance. One has to look at internal subregions to find an answer that can be satisfactory to some extent. The case of the Sense region in the Canton of Fribourg permits to illustrate this point. The following account is based on the report elaborated in 1995 (Region Sense, 1995).

The Sense LIM-region (which comprises the district with the same name and parts of municipalities of the neighbouring district of Gruyères) lies halfway between the cantonal capital Fribourg and the national capital Bern. It runs from the Prealps (c. 2000 m a.s.l.) in the south to the hilly lowlands in the north (c. 500 m). From the topography and the accessibility to Fribourg, it can be subdivided into three subregions, Upper, Middle and Lower Sense Region (Figure 8.3). The motorway A 12 from Bern to Vevey on Lake Geneva runs through the northernmost Lower Region, whereas the Upper Region has no major transit road. This fact alone suffices to suggest differences inside this LIM-region (Table 8.2). Concerning employment, they can be summed up as follows: in the Upper Region, the tertiary sector dominates (thanks to the presence of tourism), whereas in the Lower Region, the secondary sector plays a major role. Motorway access is an important locational factor in this case. In between lies the predominantly agricultural Middle Region. Looking at the dynamics over the past few years, however, one recognizes that the primary sector has been diminishing in general, but least in the Middle Region (Figure 8.4).

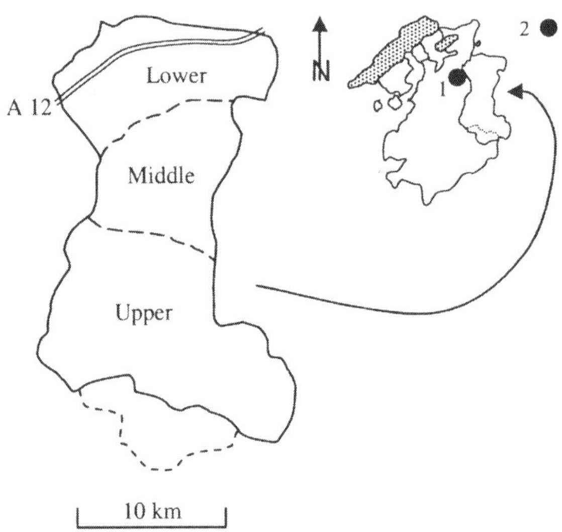

Figure 8.3 The Sense LIM-region and district

1 Fribourg, 2 Berne

Table 8.2 Employment in the Sense district and its three subregions, 1991 (in per cent)

	Primary	Secondary	Tertiary
Upper Region	27.7	24.2	48.1
Middle Region	32.2	27.6	40.2
Lower Region	14.3	42.5	43.2
Sense district	*21.0*	*35.6*	*43.4*

Source: Region Sense, 1995

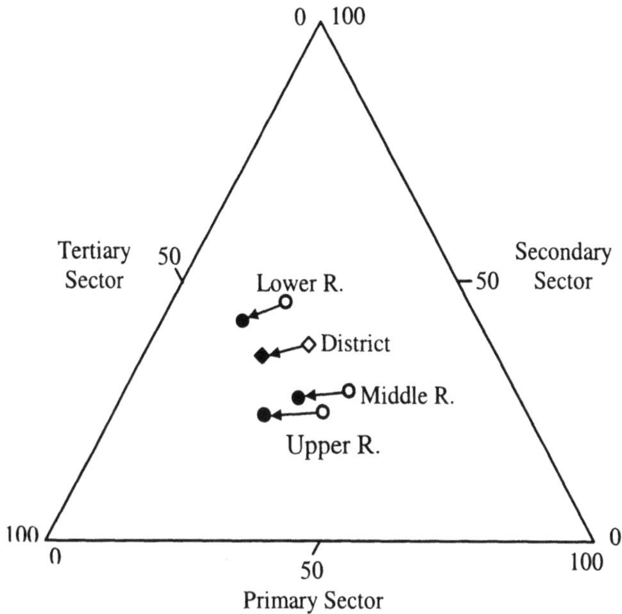

Figure 8.4 Sense district: sectoral shifts in employment (1985 [circles] to 1991 [dots] in per cent)

Source: Region Sense, 1995

Infrastructure is a very wide field, and the projects undertaken therefore cover a large spectrum, ranging from culture and education to sports, leisure and tourism, the health sector, water supply and sewage, and traffic. In the Sense LIM-region, the 171 realizations included the extension and renovation of a number of old people's homes, the foundation of a sheltered workshop for handicapped persons, the construction of multifunctional buildings for municipal administrations and the general public, the renovation of a chair lift, etc. The total sum invested amounted to 274.5 million Swiss francs, 19.1 per cent of which (52.4 million) were granted under the LIM scheme. This money was either interest-free or benefited from special interest rates – in this way, the region saved 31 million francs in interests.

The situation between Bern to the east and Fribourg to the west and a beautiful landscape make the district an attractive residential area for inhabitants working in either of the two towns. The region still offers relatively low prices for industrial land, it is within easy reach of the two centres by both rail and motorway, and the Canton's economic policy is favourable to the implantation of new manufacturing and service industries. The population has therefore increased considerably in the past fifty years (from 24,892 in 1950 to 38,398 in 2000), a trend that can be seen throughout the region, including the Upper Region which is most distant from the two centres and has a relative disadvantage from the economic point of view. Rurality is an increasing advantage to attract new inhabitants. The economic weakness in this latter region appears in the GDP per inhabitant that is clearly below its two counterparts (Table 8.3). The motorway and the main railway line from Bern to Lausanne and Geneva that runs roughly parallel to it privilege the Lower Region. However, the potential of the Upper Region can be seen in the sharp increase of the per capita GDP from 1970-93.

Table 8.3 Per capita GDI in the Sense district, its subregions, and the Canton of Fribourg (Index: Switzerland = 100)

	Upper Region	Middle Region	Lower Region	Sense district	Canton of Fribourg
1970	48	66	74	66	82
1980	56	63	81	67	81
1993	65	71	91	77	91
1970-93	+ 379 %	+ 281 %	+ 311 %	+ 315 %	+ 253 %
1980-93	+ 116 %	+ 112 %	+ 117 %	+ 116 %	+ 109 %

Source: Region Sense, 1995, p.37

Such figures demonstrate that in this particular case the regional policy via the LIM shows very encouraging results. We draw this conclusion despite the particular geographic location of the entire area close to the two major urban centres of Bern and Fribourg that makes it difficult to discriminate between free-market effects and regional policy. In 1964, the Fribourg government introduced a policy of economic promotion that propelled the canton into modernity. The regional policy at that time was only in favour of mountain farmers (see above). The LIM, that became effective in the Sense region in 1980, may well have been an additional incentive and have triggered off a virtuous cycle. Every improvement of the infrastructure increases the attractivity of a region for newcomers, whether they look for residences or want to create jobs. Industries in printing and publishing, in metalworking and machinery have emigrated from Bern into the Sense district, and the regional building sector has profited from both the arrival of new firms and the increasing demand for dwellings (Region Sense, 1995, p.45 f.).

While this sounds all very positive, one must not overlook certain negative sides of this growth process. Land for industrial purposes is becoming a rare resource, the landscape suffers from the increasing encroachment of roads and houses, and people do not necessarily like to be the outlier of Bern for industries that produce little added value. In order to balance gains and losses and to avoid long-term damage, regional policy must cooperate with cantonal and regional planning.

Conclusion

If we believe in the interplay between public and private actors, we must allow the state to show a concern for a balanced society. It is in its interest to avoid extreme disparities, and regional policy is an adequate instrument. The private sector, for its part, must agree to this form of regulation, even if it may run contrary to its objectives of free markets and optimum profits. After all, the political system is a sort of mediator or buffer between the population (the consumers, the workforce, and the electorate) and the economy (the production system and the source of state income; Figure 8.5). Whereas the state wants to defend the interests of the people, the economy opts for profit and free competition. Each side strives for some sort of legitimacy, and both are inextricably linked to the state.

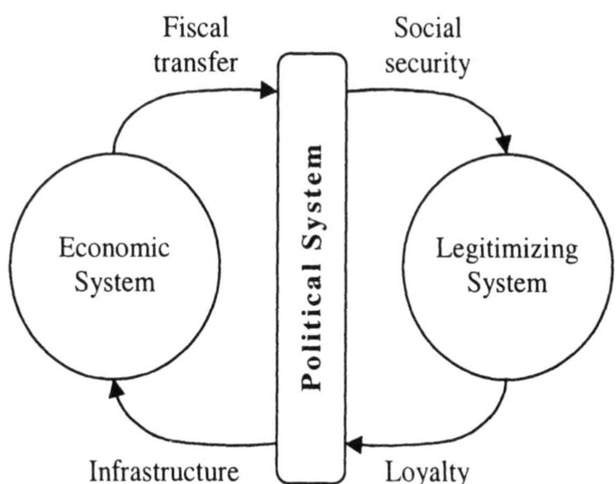

Figure 8.5 The political system as buffer in the social system

Source: Offe, 1973, in Grauhan and Linder, 1974, p.72

Deregulation in the proper sense of the word is not possible – this becomes obvious from the role the political system plays in the society. Every society requires a minimum of rules, and all actors have to comply with them. However, such regulations are not immovable but have to be adapted to changing circumstances. The advent of development and regional policies is an example where re-regulation was undertaken. They became a necessity when people realized that the disparities began to pose social problems on the global and on the national scale. The international community created development programmes that ought to improve life for populations in distress around the world, and regional policy measures were being taken on the national as well as on the international level (cf. the European Union) with the same fundamental goal. The choice of tools and the degree of enforcement vary in either case.

Two questions may be asked in this place, questions Piveteau (1971, pp.29-30) has formulated at the conclusion of a paper on regional economic disparities in Switzerland in the 1960s: Is it necessary to combat economic disparities at any cost, and which measures should be applied? There are no final answers. Disparities are almost unavoidable, and while the rich regions are favoured and tend to become richer, the efforts to improve the situation of the poor regions may be futile. A comparison of the 1960 situation with the present confirms this statement. The instruments applied 'are of a recent nature, limited in extent, and do not necessarily reach the goal which itself may be too vaguely defined.' (ibid., p.30). Twelve years later, Hanser (1983) looked at the question of regional policy again, and came to similar conclusions: '(t)he policy objective can hardly be achieved with the strategies in use today' (p.370). Something therefore is wrong with the idea of a top-down policy to reduce disparities and marginality.

The most important step forward has been when most agents involved have recognized the limits of the top-down policy in favour of a more flexible approach. True demarginalization (i.e. levelling out of disparities and escape from a hopeless situation) can only be achieved if the people concerned can participate in the designation of the goals to be reached and instruments to be applied. This is not to say that bottom-up strategies can bring about the solution of the problems; it is the combination of the two that can ensure broad acceptance of the policy and active contributions of the population concerned. It is to these 'grassroots' aspects that the last chapter is devoted.

Chapter 9

The Answers from Below

What can the people themselves do to escape the problem of (unwanted) marginality? Are there ways to change a desperate situation from the inside or is the helping hand of some external agent indispensable? How can the people make their voices heard? Can the state (the government) offer this way out?

In a period when anti-globalization protests dominate the headlines, the answers to these questions remain open. It is true that public protests serve the purpose to be heard, but do they really move the ideas of those few who decide on political actions, on the fate of the world? Demonstrations are spectacular events for the spectators at the television at home, they are a nightmare for the shop-owners along the route, and they offer the authorities an opportunity to show muscles. But how impressed are the leaders of the world's economy? And how much do the heads of state and governments see and understand, and how far can and are they prepared to act? How independent are they from the other mighty actor, the economy (Figure 8.5)?

It is not our purpose to enter into a debate of the globalized protests against globalization that have increased since Seattle (Smith, 2000); we consider them a sign of the powerlessness many people seem to feel. Since the 1990s, a global demonstration culture has emerged. Its efficiency has been facilitated by the use of the Internet and of mobile phones in the phase of organization, and its impact has been fuelled by real time transmissions via television. Protests on almost any problem can be viewed around the world – on TV, if the programme editor decides to include them in the news-bulletin, or on the Internet, of the organizers lack the support of the media. This demonstration culture has evolved parallel to the increasing awareness of the cleavage between rich and poor. The dictum of the two-class world society is not a whim of frustrated hotheads but a reality that can be proved statistically (Kreutzmann, 2003, p.3). The concern expressed by demonstrators is genuine and shows a heightened interest in politics of the young generation. Their claim to be heard is legitimate; they are protesting in favour of their own future, and in order to achieve results, they have to oppose the way our world is managed at present. However, apart from protestations, constructive actions are necessary, actions that may not be very spectacular and not make the headlines but that are efficient in the field. It is often the young generation to initiate and back bottom-up initiatives that may sometimes be crowned with success.

The helplessness manifested by many people is a consequence of the real power situation in society that can be summed up in a modified version of Offe's model (Figure 9.1). It reflects the general asymmetry of the power relations, although in some countries the situation may be more balanced than in others. This model may look exaggerated to some readers but it mirrors a widespread reality. It demonstrates that the political system has become more and more involved with the economic system, abandoning its true role as representative of the people and as a 'buffer' between legitimizing and economic systems and allying itself with the money-world.

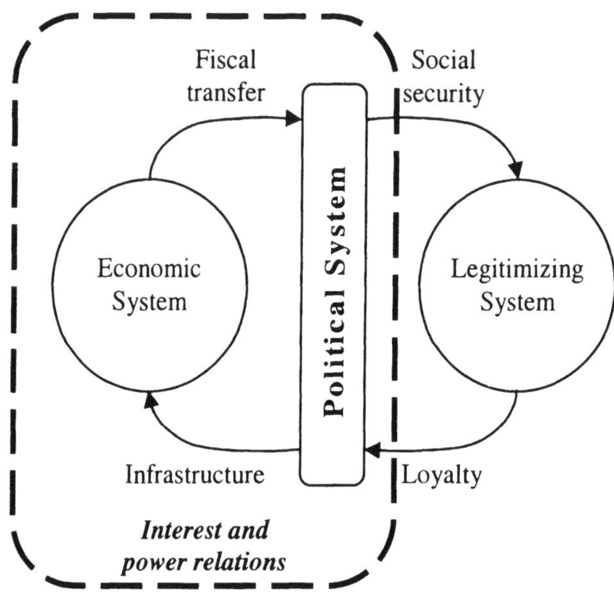

Figure 9.1 Interest and power relations between political, economic and legitimizing system

Source: Based on Figure 8.5

From this process we can understand the emergence of all kinds of grassroots movements around the globe, movements that are usually organized on a local to regional level and are basically citizens' initiatives in favour of particular issues. Many remain confined to the scale they were intended for, some may find their postulates fulfilled and will disappear, but some of these movements may gather momentum and develop into vast national, continental or even global networks. The new information and communication technology has played a major role in creating and expanding such networks. A number of them have become large organizations that play an increasing role in opinion building around the world and seek to (and sometimes manage to) influence political actions. These NGOs (Non Governmental Organizations) have grown to be important global players in questions of Human Rights and environmental protection. As such they merit to be discussed at some length.

Non-governmental Organizations (NGOs)

NGOs can be interpreted as the response of the legitimizing system to the disequilibria created by the combination of the political and the economic systems in the search for power and the pursuit of their particular interests. The ordinary citizen feels no longer represented by the political institutions but rather at the mercy of the seemingly all-powerful international corporations. His preoccupations are not the same as those of company managers and politicians on the national and international level. Citizens generally have an everyday and local to regional perspective, and their interests are essentially to ensure survival at decent conditions of living (which include a job with a just salary, adequate food and drinking water, a decent dwelling, good health and environmental conditions). Politicians and company managers, in particularly CEOs and chairs of the boards, although they are also citizens, must cultivate a strategic outlook, going beyond the daily chores and sorrows to long-term developments and encompassing an entire country, a continent or even the whole earth. Thanks to their leading position and their salaries they need not worry about their conditions of daily life. It is another story that they often work significantly longer hours, often on seven days a week, than an ordinary citizen who picks up his task in the morning and drops it when the siren goes, and this during five or six days only. As a consequence of their position and perspective, managers and politicians also have enhanced responsibilities: they must not just look after their personal advantage (i.e. salaries and bonuses) but be aware of the consequences their decisions will have on thousands and millions of people. But the latter should clearly dominate the former.

There exists a sort of vacuum between these two categories of people (citizens and leaders) that can only be filled by collective action of the people. Since individuals cannot really exercise pressure, they have to build up pressure groups of their own in order to be heard. Political parties can hardly play this role. Once they are part of the power complex, they are no longer open to the true problems of the electorate, and as long as they are small opposition parties, their voice will not be listened to. The NGOs thus occupy a particular place in the social system (Figure 9.2), and their role is to point to deficits in and violations of Human Rights in the widest sense of the term and to destructions of the environment, two domains that have no other defenders.

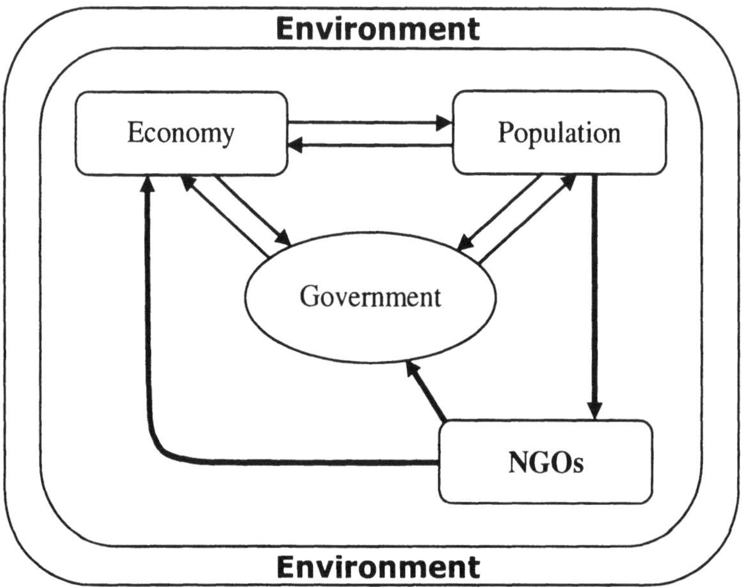

Figure 9.2 The position of NGOs in the social system

The history of the NGOs remains to be written; the phenomenon can look back on less than 150 of existence. For a long time, only one NGO was operating around the globe, the Red Cross, and its tasks were and still are essentially confined to the problems associated with armed conflicts. It was only after World War II that non-governmental bottom-up movements began to be founded around the globe, covering the two domains mentioned (Human Rights and care for the environment). Both fields have been and still are marginalized by many people, in particular by leading political powers. A brief presentation of a number of NGOs may help to understand their role.

The Red Cross

The International Movement of the Red Cross and the Red Crescent is probably the oldest NGO that was devoted to questions of Human Rights, long before they were universally recognized. It represents the type of bottom-up movement where an individual person (Henri Dunant) organized medical aid to wounded soldiers in the Battle of Solferino (1859) irrespective of the side they were fighting for. As an outsider from a neutral country (Switzerland), he emphasized the humanitarian problem, not the nationality of the victims. This philosophy prevails today; the ICRC is the only truly global organization that serves a humanitarian cause without discriminating between the 'good' and the 'bad'. This may not always please every party, as the repeated attacks on ICRC delegates (murder, kidnapping, extortion etc.) in recent years have proved, but it is an absolute necessity in our time.

The foundation of the International Committee for the Relief of the Wounded in 1863 (that was later to become the International Committee of the Red Cross, ICRC) and the subsequent international recognition in the Geneva Convention (1864) gave a voice to those victims of wars that would usually remain unattended. By creating an international institution, Dunant lifted these issues to the global level and created an organization that could receive international mandates on behalf of combatants, prisoners of war, refugees due to belligerent actions, and civil victims of wars. Although it is an NGO, it enjoys a good reputation with the political systems, and the tasks are conferred to the ICRC by the states that have adhered to the Geneva Conventions of 1949 and the Additional Protocols of 1977 (ICRC, 2002, p.6). As a truly international and neutral organization it is not bound by any ideological or other orientation. It operates according to the 'seven fundamental principles humanity, impartiality, neutrality, independence, voluntary service, unity and universality' (ibid., p.8), and its sole aim is 'to prevent and alleviate human suffering, without discrimination, and to protect human dignity' (ibid.). The Geneva Conventions and Protocols are part of the international law; as such they are only as efficient as the states that are party to this body of legislation permit. While the Red Cross is widely accepted, other agreements that pursue similar goals and complete its endeavours experience more difficulties. The Ottawa Convention on landmines (1997), for example, has not been signed by the entire international community – the producers of these lethal and long-time effective weapons have not committed themselves to a ban but continue to support this business.

The ICRC usually acts *ex-post* (following the outbreak of a conflict), but it also sees some opportunities for *ex-ante* activities. It groups its principal tasks into four categories (ICRC, 2002):

- respect for international humanitarian law (the law of armed conflict);
- protection in war (civilians, civilian internees, prisoners of war);

- assistance for conflict victims (providing vital support for survival, economic rehabilitation);
- preventive action (to diffuse the rules of war, influence governments to remind their soldiers of these rules, campaign against landmines).

To ease the multiple negative consequences of wars is a noble task, and the relief brought to the populations afflicted is genuine. However, the results would be even better if the prevention of wars was more successful. This, however, is not the fault of the ICRC.

Although the Red Cross and Crescent is primarily an organization with a global outlook, its actions are often local and regional in scope. The slogan 'think global, act local' perfectly mirrors the philosophy behind. It has, in addition, encouraged the creation of national associations that are motivated by the same ideas but can act more swiftly within state boundaries, based on the national laws. At this scale, the rights and obligations are clear and can be enforced. These national organizations act independently and can also intervene in cases of natural catastrophes. Besides they organize medical and paramedical education, blood donation campaigns and are active in information and documentation.

Amnesty International

Fighting for the respect for Human Rights is the core activity of Amnesty International, founded in London in 1961. The lawyer Peter Benenson had gathered experience with the defence of political prisoners and proposed to create an organization in favour of a better diffusion of the idea of Human Rights. Just as Henri Dunant with the Red Cross, it was an individual who set about the task to propagate the issue of human dignity globally. By monitoring countries and their policy, AI fills a vacuum in a field that is not covered by the ICRC. Its aim is 'to undertake research and action focussed on preventing and ending grave abuses of the rights to physical and mental integrity, freedom of conscience and expression, and freedom from discrimination, within the context of its work to promote all human rights' (AI, 2003c). With its perseverance, Amnesty International has gradually obtained international recognition. In the 1960s already it was given consultative status by the United Nations (1964), the Council of Europe (1965) and UNESCO (1969). In 1977 it received the Nobel Peace Prize, and in 1984 its battle against torture was to some extent rewarded by the UN Convention against torture. Unfortunately, this plight has not yet been eradicated, as the third campaign launched by AI in 2000 shows (AI, n.d.).

Doctors Without Borders (Médecins Sans Frontières)

Dissatisfied with the relative slowness of the Red Cross and its close relations to governments, ten French medical doctors founded an independent organization in 1971. Its principal goal is to offer medical assistance to endangered populations all over the world, if possible in cooperation with official health structures or ministries of health (MSF, 2000). Victims of armed conflicts, man-made and natural catastrophes are likewise entitled to receive relief from suffering – another vacuum that is filled in this way. Similarly, violations of Human Rights are denounced. The organization claims to be neutral and independent from governments, an attitude which gives it liberty in criticizing politicians and administrations, but makes it vulnerable to government actions, e.g. expulsion.

The charter of MSF comprises four criteria (MSF, 2002):

- assistance to populations in distress, to victims of natural or man-made disasters and of armed conflict, without discrimination and irrespective of race, religion, creed or political affiliation;
- neutrality and impartiality in the name of universal medical ethics;
- respect of the professional code of ethics, complete independence from all political, economic and religious powers;
- awareness of the risks and dangers of the missions.

WWF

The World Wide Fund for Nature (formerly World Wildlife Fund) originated in 1961. The year before, Sir Julian Huxley, the eminent British biologist, had discovered the great loss of wildlife occurring in Africa, essentially due to poaching. His public appeal through the press aroused considerable interest, and in 1961, an international organization was founded, backed by 'scientists and advertising and public relations experts' (WWF, 2003a). Very important were business and political leaders and in particular 'the support and guidance of HRH Prince Bernhard of The Netherlands and HRH The Duke of Edinburgh' (WWF, 2003b). The main task of the WWF is to give the ecosystem the voice among humans required to promote ecological behaviour. Information, lobbying and campaigning are activities undertaken, mostly in cooperation with governments. They concern conservation on an international and national scale, education, awareness, and conservation policy, as can be seen from the income and expenditure sheet (WWF, 2001, p.29). In its 40 years of existence, the WWF has become an unavoidable factor in environmental policy, active on all levels, from global to local.

Summing up

The choice of only four NGOs may look arbitrary and not do justice to the hundreds of such organizations that operate around the globe. The answer to this criticism is that this chapter does not seek to present an overall picture of NGOs but look at the principle of one specific type of bottom-up endeavours to counter unwanted marginality. To voice the concerns of those whose appeals are not heard is the chief motive of all NGOs. This includes the defence of the ecosystem, the question of human dignity, and help for those in distress. The globality of the causes for distress has led them to act internationally and be present in as many countries as possible. To condemn violations of the International Humanitarian Law and of Human Rights or to denounce the destruction of our environment (pollution, resource depletion) has become their daily bread. The efficiency of their activities cannot be measured on a daily, not even on an annual basis. Both political and ecosystems operate at slow rhythms, although it would be possible to make the former run faster. However, inertia is not only a physical law, it also holds good for the human world. Election years are unfavourable for grand decisions; a four-year term of office is too short for long-term thinking – the system blocks itself and does not really want to change. Change, after all, means to give up traditional habits and depart into the unknown, unforeseeable, something no politician is really willing to do.

A more speculative question arises: to what extent are NGOs really bottom-up movements? They are not tied to governments directly, they emerged independently of political pressure, but they are the creations of the elite, of persons with a lot of experience, persons who had perhaps already worked for or been in contact with the political world, and they are essentially based in the North. NGOs do not emerge as mass movements but in small circles (although they tend to receive the support of the masses), and they touch upon fundamental questions for humanity. They have become indispensable agents in national and international politics, touching sensitive questions and stirring the conscience of world leaders. They challenge the inertia of the political system, and their insistence will eventually lead to concrete results. Governments will increasingly have to watch over their image, something large companies have already learnt (see Chapter 1, p.26). A good image is a key element in economic development, vital to attract foreign investors and at the same time an assurance for domestic economic actors. As global players, the NGOs are also faced with the uncertainty about international law and the respect governments pay to it. This is probably a further point that fuels their actions. This role is not to the pleasure of everyone, as the attempts of the United States demonstrate to monitor the activities of NGOs, albeit 'without prejudice' (NGOWatch, n.d.). What for?

Finally, an interesting coincidence is worth mentioning: AI and WWF were founded in the same place (London) in the same year (1961), both arising out of the preoccupation of two individuals (a lawyer and a biologist) with some malfunctions in our society: malfunctions towards our fellow human beings and towards nature and the ecosystem. Together with the ICRC they demonstrate that a single person can move a lot if he or she is deeply convinced of the relevance of the issue. Henri Dunant, Julian Huxley and Peter Benenson stand for numerous other individuals who did the same.

Grassroots movements

NGOs and grassroots movements are in a way the same kind of organization – both are non-governmental – and pursue similar goals, yet we prefer to make a distinction between them. Whereas the NGOs originated from the concern with general (global) problems, grassroots movement arise out of local and regional issues. This is not to say that they will remain confined to this level; their ideas may travel round the world and stimulate other groups to pick them up. Contrary to NGOs, a grassroots movement will not create a general headquarter from where the activities are coordinated internationally. Their activity space is primary local and regional, at the utmost national.

Grassroots movements are the direct expression of the fears and sorrows of local/regional populations; they can be active where NGOs do not arrive, be it for financial reasons, lack of interest in the specific issues, insufficient accessibility, or lack of information. The number of grassroots movements around the world is unknown and is in fact irrelevant. They all have the same goal: to make life better for destitute groups of all kinds: ethnic minorities, women, children, consumers, etc. This goal encompasses both the care for the environment and the improvement of social relations. Such movements exist both in the North and in the South. There is no *limes* to this kind of initiatives as marginalized groups can be encountered everywhere. On the contrary, such movements can easily cooperate across the *limes* when their interests converge. The Max Havelaar foundation (see below) is an excellent example for such cooperation. At times, citizens' actions are temporary undertakings, limited to very specific issues, but in other cases, they focus on fundamental long-term goals that cannot be reached within a short time but require patience and perseverance and must accept setbacks as well. In this case, a grassroots movement must strive to become firmly anchored in people's minds to ensure long-term survival and success.

The history of real bottom-up movements is inextricably linked to the biographies of many individuals who have engaged themselves into the fight for a better world and have thus been able to build mass-organizations all over the planet, persons such Petra Kelly (Germany), Helen Caldicott (Australia), Astrid Einarsson (Sweden) and many others (Ekins, 1992, pp.43 ff.). The early issues were political (anti-nuclear, anti-war in general), but they often included ecological components. Public opinion has considerable effect on politics if it is prepared to defy a prevailing political view in a consistent and persevering manner. The peace and anti-nuclear movements of the 1970s and 1980s that developed almost simultaneously in Europe, Israel, Australia – indeed around the globe (ibid., p.40 ff.) – demonstrated, for example, that pressure from below has 'made it impossible for a conservative West German Chancellor to modernize tactical nuclear weapons, against the full weight of NATO pressure' (ibid., p.44). A mass protest in 1975, including the peaceful occupation of the site, prevented the construction of a nuclear power plant eight kilometres from Basle, at the periphery of an international conurbation of more than half a million inhabitants. The ferocity of the opposition was such that the construction permit was not granted.

The approach by grassroots movements is interesting. They do not simply oppose the 'adversary' but act in an autonomous way. Instead of demanding action from the 'other', measures will be taken to reassure this 'other'. The root cause of a problem may often lie within one's self, usually 'defined as clean and proper' (Sibley, 1991, p.33) and part of our general attitude towards ideas, material objects and peoples. To accept this means a gigantic step forward.

The scale of operation and the topics of grassroots movements are highly varied. Human Rights, development, and the environment figure prominently in Ekin's (1992) account, which includes also health, education, housing, and spiritual and cultural renewal. He describes two specific Human Rights issues: women's liberation, particularly in Muslim countries (pp.76 ff.) with reference to the religiously veiled patriarchy (which, however, is also present as a relic in certain Christian societies), and the defence of indigenous peoples (p.81 ff.). The former is chiefly concerned with Islamic countries (nation-states), the latter is a global issue, culturally and regionally focused and also challenging nation-states.

Other bottom-up movements can be found in the domain of development. They all emphasize the use of the endogenous potential, the integration of traditional knowledge with modern techniques, self-reliance, and they focus on the real needs of small communities rather than on large-scale projects and the world market. This kind of grassroots movement has diffused over the entire South, and fortunately, some of the northern donors have understood the message. They have switched from simple assistance to cooperation, demanding personal efforts from the beneficiaries of aid and assisting them in the realization of the projects. Encouraging local cycles (money, food, resources etc.) is a primary goal of this new approach. Examples are the farmers' organizations in the Senegal (Hammer, 1997, p.25 f.), the organizations Naam and Six 'S' in Burkina Faso (the latter also in neighbouring countries; ibid., pp.244-68; Ekins, 1992, pp.112 ff.), the Grameen Bank in Bangladesh (Ekins, 1992, p.122 ff.) and others. Embedded in their local or regional culture, they will contribute to achieve what the Sarvodaya Shramadana Movement in Sri Lanka has identified as the ten basic human needs (Ekins, 1992, p.101):

- a clean and beautiful environment;
- a clean and adequate supply of water;
- minimum clothing requirements;
- a balanced diet;
- a simple house to live in;
- basic Health care;
- simple communications facilities;
- minimum energy requirements;
- total education;
- cultural and spiritual needs.

Village and tribal cultures are the driving forces, and behind them lies the conviction that development has first and foremost to do with quality, not with quantity, with non-material as well as with material aspects. The bottom-up local/regional scale is essential to the development process, as 'development starts in the local territory' (Andersson, forthcoming). This is also one of the *leitmotifs* of consumer organizations worldwide that observe the market, the multitude of goods offered for consumption, and the potentially negative outcomes of some of them. They are primarily a northern phenomenon, destined to protect the individuals in the consumer societies, but through other agencies they reach out into the South. From a different perspective, they pursue the same goals, concerning the social and the environmental problems of our time.

It is not our intention to present an overall picture of grassroots movements; the book by Ekins (1992) is a very good reference. Just as a few NGOs were highlighted above, the basic ideas of a selected number of true bottom-up initiatives shall be presented in order to complete the picture. The choice is evidently subjective, and the presentation rudimentary.

Sarvodaya Shramadana (Sri Lanka)

The grassroots movement of Sarvodaya Shramadana was founded in 1958, ten years after independence, by the teacher A.T. Ariyaratne (Ekins, 1992, p.100). The objective of the organization, that was recognized as an approved charity by parliament (Lanka Jathika Sarvodaya Sharamadana Sangamaya) in 1972 (Sarvodaya, 2003), is contained in its name: 'Awakening of all by voluntarily sharing people's resources, especially their time, thoughts and efforts' (Ekins, 1992, p.100). This denomination stands for an ambitious programme that wants to include the entire hierarchy from the human personality to the family and village communities, the urban and national communities, and the world community (Sarvodaya, 2003). The programme revolves around the villages, attempting to bring the material and the non-material human needs into equilibrium. It is essentially driven by sacred values (see Chapter 3), but its open character prevents it from becoming entrenched in pig-headed conservationism. As a non-violent and non-partisan (ideology-free) organization, Sarvodaya follows in the footsteps of Mahatma Gandhi. It is a very idealistic movement, to some extent utopian, but – as has been said above – without utopia and idealism humanity will not advance. Setbacks are inevitable, and the Sarvodaya movement has notably been criticized that during the civil war it 'never succeeded in finding an adequate cultural base for its development programmes, which could withstand the combined hostility of the Sri Lankan government and the internal, especially ethnic, contradictions within Sri Lankan society.' (Ekins, 1992, p.111). This may be true, but such reproaches can be taken as a positive input into the ongoing process of awakening or awareness. The movement carries on despite the internal strife and the assault of the western life style. Its spiritual foundations help to defend 'community and environmental co-operation … against invasion from the outside' (Miller, 1998, p.262). As long as these roots can be maintained, the risk of their culture to be watered down by western superficiality is small. It would be wrong to give up and yield to the influences of modernization simply because external forces may slow progress down. Internal progress is also possible, and it may be useful for the movement to rethink the objectives and the approach chosen in order to fortify itself against this onslaught.

Naam – Six 'S' (Burkina Faso and West Africa)

Another teacher, Bernard Lédéa Ouedraogo, became the founder of two successful movements in Burkina Faso that eventually spread over large parts of West Africa. The two groupings, Naam and Six 'S' (*Se Servir de la*

Saison Sèche en Savane et au Sahel – 'Association for Self-Help during the Dry Season in the Savannahs and the Sahel'; Ekins, 1992, p.114) are separate and yet interconnected. The former are true bottom-up village groups, the latter is 'a federation of peasant organizations like (and including) NAAM' (ibid., p.115). The Naam groups are self-help groups, essentially producer organizations (Hammer, 1997, p.245), whereas Six 'S' plays the role of a support organization and serves as a link to external donors (ibid.).

The Naam groups are successors to the traditional Kombi-Naam, truly egalitarian age groups of the village youths (boys and girls), associations whose duration was usually limited to the rainy season when they took an active part in the work on the fields. It was on this traditional basis that Ouedraogo built when he started to stimulate school leavers of the Rural Education Centres (*Centres d'Education Rurale*) to organize groups of young farmers (1967). His aim was to enlarge the scope of these groups' activities from pure work in the fields during the rainy season to long-term engagements throughout the year in farming, horticulture, and infrastructure projects. At the same the groups became more formalized and lost their age-class orientation (ibid, p.249). The new Naam groups (recognized under this name in 1978; ibid., p.250) are a means to combat the rural exodus and to introduce new ways of thinking while maintaining useful elements of the traditional society. The young generation frequents schools and introduces new ideas, but it is the elders who will have to decide about their implementation. The members of a Naam group thus also have an educational function, and they must convince the traditional decision-makers of the feasibility of their projects and ideas. The groups are part of the village society, and by inviting the elders to their meetings, they usually manage to obtain their formal support for their ideas. In this way, the social hierarchy is respected and innovative activities receive the consent of the traditional decision-makers (Félix, 1992, p.17).

The activities of the Naam-groups are based on the principle of improving traditional techniques (Hammer (1997, p.252). Vegetable farming has a long tradition among the Mossi, and the introduction of new species is nothing fundamentally new. Similarly, erecting stone rows to prevent soil erosion has been practised for centuries, but the young members of the community have learnt to improve the technique of anchoring the stones in order to increase their stability. Land use planning and fighting desertification are the dominant domains among the Naam activities, and they receive priority with the support granted by Six 'S' (ibid.).

The Six 'S' was founded in 1976, primarily as an organization to provide the Naam groups with financial support in order to improve their activities. The philosophy behind this organization was to halt the rural exodus and to find means to occupy the rural population during the dry season when there was no work on the fields (Hammer, 1997, p.253). Although Six 'S' is a grassroots movement, it has no projects of its own but supports and promotes existing farmers' associations with the explicit goal of allowing them to realize their own ideas (ibid., p.256). 'For this reason, Six 'S' primarily offers services such as training, logistics support, counselling and funds that are not destined to a specific purpose, while material support such as the transportation of material for the construction of a village dam ranks second.' (ibid.). A further field of action is the promotion of village banks to mobilize the savings potential that does exist in these rural areas. Auto financing is an important goal that is to be achieved within the concept of self-determined development (Hammer, 1997, p.259).

The Max Havelaar foundation

Consumer associations are bottom-up organizations that defend the right of the citizens to obtain quality goods at reasonable prices and under sound conditions. This kind of defence against the abuse of power (by the producers or the wholesalers and distributors) is imperative in an epoch when egoistic monetary profit ranks higher than altruistic general satisfaction. Every country in the North has its own association because the general framework of consumption varies from one country to another. Increasingly, consumer associations have also turned their attention to the way the goods are produced, and the working conditions in the countries of the South have aroused their particular attention. The various aid agencies are offering goods produced by local groups in Nepal, the Sahara etc., but these actions are very 'spotty' in that they concern specific crafts or village groups but not large populations, and the products (handmade paper, silver jewellery, etc.) are not necessarily destined to everyday consumption.

More in-depth action was therefore required. When in 1986, Mexican coffee producers demanded fair prices for their product rather than outside help (Kici, n.d.), enterprising persons in the Dutch development organisation *Solidaridad* created the Max Havelaar foundation in the same year. Its goal was to promote a fair price to these coffee farmers and encourage them in this way to join together and found cooperatives. This goal has since been widened to encompass fair trade and marketing a variety of quality products from tropical countries to consumers in Europe. Fair trade means to offer market access to small producers and pay them prices that are not dictated by the world market but reflect the willingness of the consumers to ensure a decent income to small farmers and farmer cooperatives who would otherwise be shut off from the world market and might prefer to grow narcotics (Max Havelaar, n.d.). The foundation met with considerable success, although the prices are slightly higher than for 'ordinary' goods. In the beginning, the Max Havelaar products could only be bought in special shops (Third-World-Shops, World Shops etc.), but when supermarket chains began to adopt these products, they helped to diffuse the idea – and at the same time they improved their image. Thus both local economies and the ecosystem are promoted: emphasis is laid on high quality products, if possible organically produced. The range of products has widened since the first coffee under this label was sold in 1988, and currently comprises pineapple, mangos, bananas and orange juice, coffee, tea and cocoa or chocolate, rice, honey, sugar, and flowers.

The Max Havelaar foundation is a good example of transnational cooperation of bottom-up movements, something that modern development assistance has adopted as its philosophy throughout the North. In this field, such private initiatives are often more efficient and closer to the people than the top-down measures by individual countries or international organizations such as the World Bank or the IMF.

Long-term efficiency of bottom-up movements (?)

Contrary to state-run policy programmes that are anchored in the legislation and usually operate on a medium- to long-term base, bottom-up movements have to fight for survival, particularly in their first years of existence. They lack the institutional backing and often depend on the enthusiasm, the engagement and the conviction of a few individuals who had the initial idea and give their time and energy to carry it through. Once they are established in the mind of the general public, their cause will usually receive increasing support. To be recognized by the government as an official charity may on the one hand signify certain restrictions to their activities, but it can also ensure more support and, for example, exempt such an organization from taxes.

We have to discriminate between the globally active NGOs and the locally and regionally anchored grassroots movements. A clear distinction is often not possible and may not even be necessary. Many NGOs started from very modest beginnings as a sort of grassroots movements and subsequently expanded their range of action over the whole world. The Sarvodaya movement includes villages over the entire country and has an office in the United States, and the Max Havelaar foundation has spread over Europe and is based on national organizations.

There are many other examples for bottom-up initiatives that have been crowned with success, demonstrating that personal conviction and a good example are qualities that have not yet been forgotten. Organic farming is such an example. The early organic farmers were viewed as queer: to give up rationalized farming for maximum profit and replace it by manual work for good quality products without residues of chemical fertilizers and pesticides was seen as contrary to the overall tendency of agriculture. The success story of organic farming in the 1990s (Leimgruber et al., 1997), however, has silenced most critics, some of whom have even been converted. Acceptance by official agricultural policy even in the European Union has increased, and the future promotion of farmers depends on their readiness to engage in environmentally friendly farming practices. Again, persistence and conviction were the qualities required on the road to success.

Organic products, local/regional cycles and quality labels have become elements in regional development policy. The Biosphere label in the Swiss Biosphere Reserve of the Entlebuch (see p.216), but also the LaNaTour project in the Swiss canton of Valais (Landscape-Nature-Tourism; Leimgruber and Imhof, 1998, p.392 f.) emphasize the environmentally compatible character of their products: short transport routes, small scale production, often organic, worked by local craftsmen and small enterprises rather than by large outside industries ... these are elements of a positive image that can be marketed not only with the local consumers but also with tourists and in the gastronomy sector. Parallel to this goes the rediscovery of local specialities that permit to rediscover the diversity in the food sector. This is an important innovation, and it should also be propagated in countries of the South.

The discovery of the stakeholder by the business community represents a sort of a revolution in economic thinking. The stakeholders are all those affected directly or indirectly by a company's activities: 'shareholders, customers, employees, suppliers, the local community and the natural environment' (Ekins, 1992, p.128). A firm is in fact committed to all of them; its activities should therefore try to strike a balance between them, with as little negative effects as possible. To adopt a stakeholder attitude signifies 'development of new forms of corporate governance where these multiple interests are represented in organizational decision making.' (Foote Whyte and Kochan, 2002). It obviously also means to depart from the short-term profit oriented thinking to 'a longer and broader view of corporate objectives' (ibid.) where the survival of the firm may be more important than the dividend in one particular year or the shareholder value. Such a shift in thinking requires considerable courage. The entrepreneur quoted in Chapter 8 (p.251) is such an example. His resolutely regional focus, despite a European or even global market, is an indicator of the confidence he puts into his location, a location that is not just a physical place but a social and natural environment.

Other company managers pursue similar goals – indeed it can be said that gradually a shift in thinking is occurring. They are supported in their efforts by the ISO certificates that permit a company to improve its image. Together with environmental reporting, this can be very important for the future of a firm. The president and CEO of a large Swiss firm takes this long-term view: 'By usefully linking ecology and economics, we can increase the value of our company in the long term' (Georg Fischer, 2000) However, setbacks are inevitable, and yet they must not serve as pretexts to stop all endeavours towards ecologically and socially friendly behaviour. Usually they last for a short time only and can be overcome. They require confidence in medium- to long-term perspectives and a sense for all stakeholders.

The power of the marginalized

This chapter has attempted to show that marginalized groups are not without prospects but have means and ways to make themselves heard. Whether this is via a global organization or as a locally or regionally organized movement is not important. Every individual has the potential to launch an initiative and to create a group that can pursue a specific goal. Sometimes external inputs (ideas, material help) may trigger off such an initiative, sometimes it grows out of a population. The examples in the two sections on NGOs and grassroots movements have demonstrated that there is not simply a gradient between the political and economic power on the one side and the 'ordinary people' on the other, but that interaction and cooperation between centre and periphery/margin are possible. The results may at times look insignificant, but the forces of change cannot be halted. Change requires time, and every process has its own rhythm, its own time scale. Progress may thus at times meet with obstacles, but it will eventually overcome them and continue.

The marginal are not powerless, as long as they believe in their own strength and potential. Perseverance and patience are virtues that are required if a group or a region will try and find its way out of a situation of distress, and the time frame must not be set too short. Similarly, expectations must not be directed towards goals that are impossible to achieve but proceed step by step. Respect for Human Rights and for the environment can only be achieved if violations and pollutions are constantly denounced, if politicians are constantly reminded of their responsibility. It is a never ending task that requires strength and determination – and the hope, that insight will come about. Here lies the true strength of the marginalized.

Conclusion

What is the moral of this book? How strongly has the author been influenced by a preconceived interpretation of the terms used in the subtitle? Is an objective assessment possible at all? This conclusion tries to look back at the preceding chapters and establish a sort of balance sheet.

Approach to marginality

The subtitle of the book refers to globalization and deregulation, two concepts discussed in Chapter 2. The subsequent chapters on the four topics of the economy, politics, social and cultural life, and the environment have tried to take them into account. It may not always have succeeded well, but we have been faithful at least to the title and chosen our examples from different spatial scales.

We have deliberately chosen a very broad outlook at the theme of marginality. The sectoral approach has permitted us to demonstrate that marginality is not only an economic phenomenon (related to differences in wealth, income, or job opportunities) but covers the entire spectrum of human life. Maybe it has not always become quite clear how strongly the various domains are interconnected. On the other hand it may have become apparent that marginality is an inevitable part of every system and that, as systems are dynamic, marginal regions participate in the processes all systems pass through. There is no final state of marginality, but processes of marginalization and demarginalization follow each other.

It is with reference to these statements that the value of Reynaud's (1981) approach to the Centre-Periphery model can be understood (Figure 2.1, Table 2.1). His descriptive model goes beyond the simple dependency interpretation and sees centres and peripheries in a dynamic interrelationship. Besides, centres and peripheries are not fixed categories but are diversified. This is particularly important for the peripheries where dependency is only one possibility among several. Just as the peripheries are dynamic, also marginal regions (exemplified by Reynaud as *isolat* and *angle mort*) need not be final states. There is always the possibility of a change, and this possibility concerns all components of a system, including the centres. This statement is contrary to the view of development theory that uses the centre-periphery model 'as a means of continuing and maintaining the differences' (Miller, 1998, p.259). Such an understanding considers development a static phenomenon, not a constant process that is contained in the word itself (see p.39) This static perspective has therefore been vehemently criticized, for example by Colin Leys (ibid.).

The geography of difference

Studying marginal regions and marginality calls for a differentiated and dynamic approach that takes the actors themselves into account. The inhabitants of a region outsiders would call marginal are not dejected and miserable people but do have an identity of their own, a 'sense of the local environment' and a 'sense of cultural validity and solidarity' (Miller, 1998, p.262). They are different from others, and it is a quality of outsiders to accept this difference. In this context, the notion of 'geography of difference' has been suggested (Miller, 1998, p.262; Andersson, forthcoming). What does this term mean? After all, geography is first and foremost about differences, 'the uneven distribution of phenomena over the earth surface' (Forsberg, 1998, p.19), but difference is not limited to structures but includes processes. The author continues by pointing out that 'one would find different kinds of development when looking at different places in different parts of the country' (ibid.) – we would say 'of the world'. Change (development) is not the prerogative of a particular type of region (e.g. the centre) but can be found in all regions. We can even imagine that wealthy regions have more difficulties with change than poor ones where change usually is equivalent to improvement. With the notion of 'geography of difference', the two authors mentioned above want to emphasize that 'the value of being different' is recognized (Miller, 1998, p.262). What is new in this context is that difference is a positive quality, something that has entered the philosophy of Swiss regional planning in the 1980s when the goal was 'to eliminate undesired (economic and social) disparities by maintaining the desirable (cultural) diversity' (see p.243).

The differences in question are to be found with the people, the inhabitants of a region, and they are naturally diverse. It is the perception the population of a given region has of its own living space and of neighbouring regions, the comparisons they make and the conclusions they draw out of them that will lead to actions or to passivity. There will always be regions 'which will not be regarded as very attractive places to live in and with hardly any opportunities to become more attractive' (Forsberg, 1998, p.28). Attractivity, of course, is not an objective quality but is defined by every individual anew, hence the second part of this sentence sounds to me a bit too categorical and too pessimistic. The shifting perceptions of mountain regions (Figure 7.1) shows that attractivity is not really measurable.

The cultural and historical background of a society plays an important role, too. Derounian (1994) paints a vivid picture of the way idyllic images of the English countryside (through children's tales, tourism leaflets and general advertisements) shape and mislead public opinion by painting the rural areas in rose colours. Problems like low pay, high cost of living, limited choice of jobs, etc. (ibid., p.205) render rural areas marginal, but they are largely missing from general knowledge or awareness, they are not part of the general expectations of what the countryside should be like. Immigration into rural areas thus does not contribute to solve these problems; rather does it accentuate them. The demand for rural housing puts pressure on the prices, and to purchase a house lies beyond the means of local populations. Most immigrants tend to be elderly, well-off middle class persons, i.e. usually people without children; they will thus not contribute to a demographic revival. Very often they maintain their urban life style with mobility and consumer habits, thereby not increasing the demand for 'jobs, public transport, schools or village shops', things the rural population needs most (Derounian, 1994, p.206). The Swedish case, described by Forsberg (1998), shows a different picture. The countryside has been subject to emigration and immigration in turn: rural exodus, followed by counter-urbanization, and again rural depopulation (p.19). During the second phase, the Swedish researchers found three different age-groups among the immigrants: 'a) family establishers, b) children, and c) newly retired people' (ibid., p.22). The immigrants came from all kinds of occupation (except, of course, farmers), thus increasing the social differentiation. From her observations, Forsberg concludes that a variety of factors help to distinguish the Swedish case from England, for example. The rural tradition in Sweden (self-employed farmers) differs markedly from England (feudal farming tradition), industrialization started in the countryside and took place in a very decentralized way, access to the land is free, irrespective of landownership, and the presence of vast open spaces prevents excessive pressure on land prices (Forsberg, 1998, p.24 f.). Immigrants into rural areas (the case study concerned central and eastern Sweden, around Stockholm) were thus not intruders who aggravated a marginal situation but on the contrary people who contributed to keep the countryside alive. The only question is to what extent the phenomenon has a medium- to long-term future.

Perspectives on marginality and marginal regions

The book may have created the impression that we look at marginality from an outsider position, 'top-down', so to speak. Marginality is indeed mainly depicted as a deplorable state of things that has to be avoided or repaired. For most of the text this perspective has doubtlessly prevailed, although a timid approach has been made in Chapter 2 (on p.52) to point to some positive aspects. This unilateral perspective is in fact true for the bulk of the work of the IGU Commission. Even in the volume edited by Jussila, Leimgruber and Majoral (1998), marginality is viewed from the top-down perspective. The perceptions mirror a certain type of power relationships that have been summed up by Cullen and Pretes (1998, p.191) as 'Defining others as marginal enables one to have power over them.' Inventors and innovators are very often marginal people, 'crazy', as one might call them, and they usually do accept their position that runs against established opinions. This has been true for organic farmers, who in the beginning were looked upon as 'queer', or of those farmers that start building ecological retention ponds (see Chapter 7, p.218), whose less ecologically minded colleagues would not understand this measure. But in the long run, the usefulness of these strange innovations has been and will be recognized and imitated: organic farming is receiving high political priority, and the positive side of the retention pools will become obvious to the sceptics as well (Stähelin, 2003). Change of perception and attitude inevitably occurs. Andersson (forthcoming) makes this essential point when he says, 'the perspective of the evaluator is of vital importance.'

Which, then, is the bottom-up perspective of marginality? Do marginal regions (more precisely: their inhabitants) have the same idea of marginal as the outsiders who define it? This may indeed be the case where entire populations were deprived of their basic rights, as has happened to many indigenous societies. The Australian government became aware of the many problems the Aborigines faced in the white society (drugs, suicides) at about the same time when they 'were themselves demanding a greater role in decision-making.' (Rugendyke, 1998, p259). Top-down awareness met with bottom-up requests – this may be an ideal combination to build up integrated development policies. Marginality has also been felt in rural and urban Zimbabwe as a consequence of ESAP (Economic Structural Adjustment Policy) of 1991 (Potts and Mutambirwa, 1999). There has been no organized reaction, but the authors refer to the increase of urban agriculture that can be seen as a sort of spontaneous answer to the rise of food prices (ibid., p.202). A bottom-up view of marginality may also be detected with the inhabitants of the Bondolfi area in Zimbabwe who discovered the potential of beekeeping and honey production in their area in order to improve living conditions in their region (Nel, 2002, p.164 f.) – although the Catholic mission doubtlessly played a role in the early stage of the development of high-quality honey.

Answers to the questions above can therefore be found at least in part from the policy responses, the non-governmental initiatives undertaken on various scales, and from the numerous grassroots movements. A look at such bottom-up movements that operate outside formal institutions (albeit within the legal framework conceded by the individual countries) shows that populations often want to maintain certain specificities without renouncing 'progress'. To be entirely detached from the world may look tempting for some, but to leave the status of marginalization seems to prevail. However, the questions concern not the countries of the South alone.

The violent demonstrations during WTO conferences, the Davos Global Forum, and the G-8 summit meetings are testimonies to a growing uneasiness about globalization and deregulation, processes which are intertwined and mirror the predominance of the neoliberal dogma. Free trade and free markets are considered the remedies to inequality in this world, but no consideration is given to the fact that people are not born equal but are endowed with different gifts, and that a minimum of social regulations are essential. Disparities are a natural phenomenon that cannot be eradicated through free markets. The feeling of being excluded, marginalized, has become very strong, in particular among the young generation who are increasingly becoming politically mobilized. Anti-globalization (better: anti-neoliberal) protests have become a sort of ritual that has to be celebrated somewhere close to the location of the meetings of the decision-makers, and they inevitably attract not only peaceful protesters but also small violent groups whose pleasure it is to break shop windows and fight with the police. More impressive have been the worldwide peace demonstrations before the 2003 Iraq war that showed a much more genuine political mobilization of the young, out of the feeling that the entire international community was being marginalized (indeed made a fool of) by the only remaining superpower of the world.

Epilogue

It is very difficult to conclude such a book at a time when marginalization is a recurring theme and is becoming an increasingly serious problem on our planet. We do not contest the value of marginality as such – someone who temporarily plunges into distress may find the strength to get out of it again. For a region and its inhabitants, to become marginal may also be a chance to develop new energy and overcome the dissatisfaction with the current situation. What, however, must be countered is the exploitation of a marginal situation by the powerful, and this on all levels and with respect to every topic we have discussed in this book. The remedy – a change of attitude and perception – may look far from reality, but events have proved that this change is possible, indeed that it does occur every now and then. In the end, those who marginalize others risk to become marginalized themselves – something that can be found throughout history. Attitudes and perceptions are cyclical as are all other phenomena; it is only the time frame that may pose problems to us. Particular efforts are therefore needed to protect human dignity and the environment in order not to draw out the imbalances created in the ecosystem through human activities. This is the challenge of our time.

Appendices

Appendix 1 The OECD

Member country	Inhabitants 2000 ('000)	Per capita GDP US$ 2000	Membership in trade blocs
Australia	19,157	26,100	APEC
Austria	8,110	26,000	EU
Belgium	10,251	26,300	EU
Canada	30,750	28,100	APEC, NAFTA
Czech Republic	10,272	14,000	CEFTA, EU candidate
Denmark	5,340	28,300	EU
Finland	5,181	24,900	EU
France	58,892	23,200	EU
Germany	82,205	24,900	EU
Greece	10,543	16,000	EU
Hungary	10,024	12,200	CEFTA, EU candidate
Iceland	281	27,500	EFTA
Ireland	33,787	28,500	EU
Italy	57,189	24,500	EU
Japan	126,926	25,600	APEC
Korea	47,245	17,700	APEC
Luxemburg	439	45,000	EU
Mexico	97,379	9,000	APEC, NAFTA
Netherlands	15,926	27,500	EU
New Zealand	3,831	19,400	APEC
Norway	4,491	29,400	EFTA
Poland	38,646	9,300	CEFTA, EU candidate
Portugal	10,008	17,600	EU
Slovak Republic	5,401	11,400	CEFTA, EU candidate
Spain	39,466	19,300	EU
Sweden	8,872	24,400	EU
Switzerland	7,184	30,100	EFTA
Turkey	67,461	6,800	ECO
United Kingdom	59,766	23,900	EU
United States	275,372	36,000	APEC, NAFTA

Appendix 2 Trade blocs

Acronym	Name	Year	Members
CARICOM	Caribbean Community and Common Market	1973	Barbados, Guayana, Jamaica, Trinidad & Tobago
MERCOSUR	Mercato Común del Cono Sur	1991	Argentina, Bresil, Chile, Paraguay, Uruguay
NAFTA	North American Free Trade Area	1992	Canada, Mexico, USA
SICA	Sistema de Integración Centroamericana	1993	Costa Rica, El Salvador, Guatemala, Honduras, Nicaragua, Panama
EU	European Union	1957	Austria, Belgium, Denmark, Eire (Republic of Ireland), Finland, France, Germany, Greece, Italy, Luxemburg, Netherlands, Portugal, Spain, Sweden, United Kingdom
EFTA	European Free Trade Association	1960	Iceland, Liechtenstein, Norway, Switzerland
CEFTA	Central European Free Trade Agreement	1992	Czech Republic, Hungary, Poland, Slovakia
ECOWAS	Economic Community of West African States	1976	Benin, Burkina Faso, Cape Verde, Gambia, Ghana, Guinea, Guinea-Bissau, Ivory Coast, Liberia, Mali, Mauretania, Niger, Senegal, Sierra Leone, Togo
PTA	Eastern and Southern African Preferential Trade Area	1983	Angola, Burundi, Djibuti, Ethiopia, Kenya, Comoro Is., Lesotho, Malawi, Mauritius, Mozambique, Namibia, Rwanda, Somalia, Sudan, Swaziland, Tanzania, Uganda, Zambia, Zimbabwe
ECO	Economic Co-operation Organization	1985	Afghanistan, Azerbeijan, Iran, Kazachstan, Kirgistan, Pakistan, Tajikistan, Turkey, Turkmenistan, Uzbekistan
APEC	Asia Pacific Economic Forum	1989	Australia, Brunei-Darussalam, Canada, Chile, China (People's Republic), China Taipeh, Korea (Republic), Hong Kong, Malaysia, Mexico, New Zealand, Papua New Guinea, Philippines, Singapore, Thailand, USA
AFTA	ASEAN Free Trade Area	1992	Brunei, Cambodia, Indonesia, Laos, Malaysia, Myanmar, Philippines, Singapore, Thailand, Vietnam

Appendix 3 Human Development Index in Africa, 1975–2000

Rank 2000	Country	1975	1990	1995	2000
High human development					
47	Seychelles	n.d.	n.d.	n.d.	0.811
Medium human development					
64	Libyan Arab Jamahiriya	n.d.	n.d.	n.d.	0.773
67	Mauritius	0.630	0.723	0.746	0.772
97	Tunesia	0.514	0.646	0.682	0.722
100	Cape Verde	n.d.	0.626	0.678	0.715
106	Algeria	0.501	0.639	0.663	0.697
107	South Africa	0.649	0.714	0.724	0.695
111	Equatorial Guinea	n.d.	0.553	0.582	0.679
115	Egypt	0.435	0.574	0.605	0.642
117	Gabon	n.d.	n.d.	n.d.	0.637
119	São Tomé and Principe	n.d.	n.d.	n.d.	0.632
122	Namibia	n.d.	n.d.	0.629	0.610
123	Morocco	0.429	0.540	0.569	0.602
125	Swaziland	0.512	0.615	0.620	0.577
126	Botswana	0.494	0.653	0.620	0.572
128	Zimbabwe	0.547	0.597	0.563	0.551
129	Ghana	0.438	0.506	0.525	0.548
132	Lesotho	0.478	0.574	0.572	0.535
134	Kenya	0.443	0.533	0.523	0.513
135	Cameroon	0.410	0.513	0.499	0.512
136	Congo	0.417	0.510	0.511	0.512
137	Comoros	n.d.	0.502	0.506	0.511
Low human development					
139	Sudan	0.346	0.419	0.462	0.499
141	Togo	0.394	0.465	0.476	0.493
147	Madagascar	0.399	0.434	0.441	0.469
148	Nigeria	0.328	0.425	0.448	0.462
149	Djibouti	n.d.	n.d.	n.d.	0.445
150	Uganda	n.d.	0.388	0.404	0.444
151	Tanzania, U.Rep. of	n.d.	0.422	0.427	0.440
152	Mauritania	0.337	0.390	0.418	0.438
153	Zambia	0.449	0.468	0.432	0.433
154	Senegal	0.313	0.380	0.400	0.431
155	Congo, Dem.Rep. of	n.d.	n.d.	n.d.	0.431
156	Côte d'Ivoire	0.369	0.415	0.416	0.428
157	Eritrea	n.d.	n.d.	0.408	0.421
158	Benin	0.288	0.358	0.388	0.420
159	Guinea	n.d.	n.d.	n.d.	0.414

160	Gambia	0.272	n.d.	0.375	0.405
161	Angola	n.d.	n.d.	n.d.	0.403
162	Rwanda	0.336	0.346	0.335	0.403
163	Malawi	0.316	0.362	0.403	0.400
164	Mali	0.252	0.312	0.346	0.386
165	Central African Republic	0.333	0.372	0.369	0.375
166	Chad	0.256	0.322	0.335	0.365
167	Guinea-Bissau	0.248	0.304	0.331	0.349
168	Ethiopia	n.d.	0.297	0.308	0.327
169	Burkina Faso	0.232	0.290	0.300	0.325
170	Mozambique	n.d.	0.310	0.313	0.322
171	Burundi	0.280	0.344	0.316	0.313
172	Niger	0.234	0.256	0.262	0.277
173	Sierra Leone	n.d.	n.d.	n.d.	0.275

Source: UNDP, 2002b, pp. 149-156

Note

No HDI is available for Liberia and Somalia. The two countries are, however, in an appalling situation, as can be seen from infant mortality rates (157 and 133 per Thousand respectively) and undernourished persons (42 and 75 per cent respectively), for example (UNDP, 2002b, p.251)

Appendix 4 Progress of the effort to attain the UN Millennium Development Goals

Rank 2000	Country	Goal 1	Goa 2a	Goal 2b	Goal 3a	Goal 3b	Goal 4	Goal 5
47	Seychelles	--	--	OT	--	--	OT	--
64	Libyan Arab Jamahiriya	--	--	--	--	--	OT	FB
67	Mauritius	OT	OT	OT	A	A	OT	A
97	Tunesia	--	A	OT	OT	OT	OT	--
100	Cape Verde	--	--	--	OT	A	OT	--
106	Algeria	OT	OT	OT	OT	OT	SB	OT
107	South Africa	--	OT	--	OT	A	SB	--
111	Equatorial Guinea	--	--	--	--	--	OT	--
115	Egypt	OZ	OT	--	OT	OT	OT	OT
117	Gabon	OT	--	--	--	--	FB	--
119	São Tomé and Principe	--	--	--	--	--	FB	--
122	Namibia	FB	OT	--	A	A	FB	L
123	Morocco	OT	OT	FB	OT	OT	OT	OT
125	Swaziland	FB	OT	FB	OT	OT	SB	--
126	Botswana	SB	SB	OT	A	A	SB	--
128	Zimbabwe	FB	--	--	OT	FB	SB	OT
129	Ghana	A	--	--	--	--	L	OT
132	Lesotho	L	SB	--	A	A	FB	OT
134	Kenya	FB	--	--	A	OT	SB	L
135	Cameroon	OT	--	--	--	--	SB	OT
136	Congo	FB	--	--	OT	FB	FB	--
137	Comoros	--	--	--	--	OT	OT	A
139	Sudan	OT	--	--	OT	OT	FB	OT
141	Togo	OT	OT	--	FB	FB	FB	FB
147	Madagascar	SB	SB	--	OT	A	FB	FB
148	Nigeria	A	--	--	--	--	FB	L
149	Djibouti	--	FB	SB	FB	OT	FB	A
150	Uganda	FB	--	--	OT	FB	L	FB
151	Tanzania, U. Rep. of	SB	FB	FB	OT	OT	FB	FB
152	Mauritania	OT	--	SB	OT	FB	FB	FB
153	Zambia	FB	SB	--	OT	--	SB	OT
154	Senegal	FB	OT	OT	OT	FB	FB	OT
155	Congo, Dem. Rep. of	SB	--	--	--	--	FB	--
156	Côte d'Ivoire	OT	FB	FB	FB	FB	SB	OT
157	Eritrea	--	FB	--	--	--	OT	--
158	Benin	OT	OT	--	FB	FB	FB	--
159	Guinea	OT	FB	--	OT	FB	OT	FB
160	Gambia	OT	--	--	OT	OT	FB	--
161	Angola	OT	--	--	--	--	SB	--

162	Rwanda	SB	--	--	--	--	SB	--
163	Malawi	OT	--	--	OT	OT	L	L
164	Mali	FB	FB	OT	OT	SB	FB	OT
165	Central African Republic	FB	--	--	--	--	FB	FB
166	Chad	OT	FB	FB	FB	FB	FB	--
167	Guinea-Bissau	--	--	--	--	--	FB	--
168	Ethiopia	--	FB	--	SB	SB	FB	FB
169	Burkina Faso	OT	FB	--	FB	--	FB	--
170	Mozambique	OT	SB	--	FB	FB	FB	--
171	Burundi	SB	--	--	FB	--	FB	--
172	Niger	FB	FB	OT	FB	OT	FB	FB
173	Sierra Leone	L	--	--	--	--	FB	--
Other countries								
--	Liberia	SB	--	--	--	--	FB	--
--	Somalia	SB	--	--	--	--	FB	--

Explanations:

Millennium Development Goals

Goal 1 Halve the proportion of people suffering from hunger

Goal 2 Ensure that all children can complete primary education
2a: net primary enrolment ration
2b: children reaching grade 5

Goal 3 Eliminate gender disparity in all levels of education
3a: female gross primary enrolment ratio as % of male ratio
3b: female gross secondary enrolment rate as % of male ratio

Goal 4 Reduce under-five and infant mortality rates by two-thirds

Goal 5 Halve the proportion of people without access to improved water sources

Abbreviations

A Achieved (target already achieved)

FB Far behind (less than 70 per cent of required progress rate achieved)

L Lagging (70-89 per cent of required progress rate achieved)

OT On track (90-100 per cent of required progress rate achieved)

SB Slipping back (level of achievement achieved in 2000 is at least five per cent below that of 1990)

-- no information

Source: UNDP 2002b, pp.46-49; 259

Appendix 5 The cantons of Switzerland

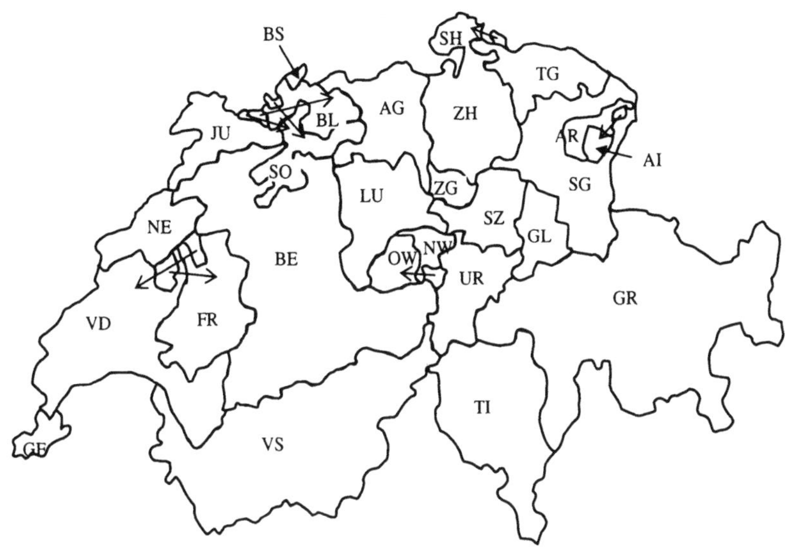

Abbreviations:

AG Aargau, AI Inner Appenzell, AR Outer Appenzell, BE Bern, BL Basel, Countryside (Baselland), BS Basel City, FR Fribourg, GE Geneva, GL Glarus, GR Grisons, JU Jura, LU Lucern, NE Neuchâtel, NW Nidwalden, OW Obwalden, SG St. Gallen, SH Schaffhausen, SO Solothurn, SZ Schwyz, TG Thurgau, TI Ticino, UR Uri, VD Vaud, VS Valais/Wallis, ZG Zug, ZH Zürich

Bibliography

Abt T. (1988), *Progress without loss of soul. Towards a wholistic approach to modernisation planning*, Chiron Publications, Wilcott IL

Adamo F. (2001), Geopolitica e geoeconomica. Dal colonialismo al globalismo, *Bollettino della Società Geografica Italiana*, Ser. XII, vol. VI, pp.589-610

Adams W.M. (1990), *Green development. Environment and sustainability in the Third World*, Routledge, London

Afangbedji M.Y. (2000), *Structures agricoles des regions de montagne et motivation de la reconversion des exploitations en agriculture biologique. Exemple du district de la Gruyère*, Unpublished diploma thesis, Geography Institute, University of Fribourg/CH

AI (Amnesty International) (2003a), The Death penalty: list of abolitionist and retentionist countries (1 Januar 2003), Internet, www.amnesty.org (19.05.2003)

AI (2003b), Facts and figures on the death penalty, Internet, www.amnesty.org (19.05.2003)

AI (2003c), Fact and Figures: The work of Amnesty International, Internet, www.amnesty.org (01.06.2003)

AI (n.d.), Amnesty International timeline, Internet, www.amnesty.org (01.06.2003)

Amat J.-P. (1987), Guerre et milieux naturels: les forêts meurtries de l'est de la France, 70 ans après Verdun, *L'Espace Géographique* 3, pp.217-233

Amat J.-P. (1988), La forêt et la guerre, un exemple de sylvofaciès sur les champs de bataille de la Grande Guerre en Argonne, *Bull. Assoc. Géogr. Franç.* 3, pp.191-201

Andersen A. (1987), Heimatschutz: Die bürgerliche Naturschutzbewegung, in Brüggemeier and Rommelspacher (eds.), pp.143-157

Anderson E. (1994), The code of the streets, *Atlantic Monthly*, May, pp.81-94

Andersson L. (1992), Further discussion of an illustration of the characteristics of local development in marginal areas, *Occasional Papers in Geography and Planning* 4, Appalachian State University, Boone N.C., pp.10-23

Andersson L. (forthcoming), PIMA 1989-1999. From 'Development in marginal areas' to 'Marginal rural areas in the new millennium', Paper presented during the Stockholm Conference of the IGU Commission on Evolving issues of geographical marginality in the early 21st century world, 2001

Andersson L. and Blom T. (eds. 1998), *Sustainability and development. On the future of small society in a dynamic economy*, Regional Science Research Unit, University of Karlstad, Sweden

Andreoli M. (1992), An analysis of different kinds of marginal systems in a developed country: the case of Italy, *Occasional Papers in Geography and Planning*, 4, Department of Geography and Planning, Appalachian State University, Boone N.C., pp.24-44

Andreoli M., Brunori G.L., Campus F. and Tellarini V. (1989), I sistemi agricoli in aree marginali, aspetti socio-economici, in Zanchi C. (ed.), *Sistemi agricoli marginali. Mugello – Alta Romagna – Garfagnana – Alto Reggiano*, CNR, Rome, pp.281-475

Annaheim H. (1950), Die Raumgliederung des Hinterlandes von Basel, *Wirtschaft und Verwaltung* 3, pp.85-122

Annaheim H. (1952), *Basel und seine Nachbarlandschaften. Eine geographische Heimatkunde*, Lehrmittelverlag des Kantons Basel-Stadt, Basel

Ante U. (1991), Some developing and current problems of the eastern border landscape of the Federal Republic of Germany: the Bavarian example, in Rumley and Minghi (eds.), pp.63-85

Aubert C. (1996), Chinese agriculture, the limits of growth, in *Itinéraires, Notes et Travaux*, No. 46, pp.3-34, Institut Universitaire d'Études du Développement (IUED), Geneva

Bähr J. and Mertins G. (2000), Marginalviertel in Grossstädten der Dritten Welt, *Geographische Rundschau* 52/7-8, pp.19-26

Barberis C. (2002), Introduction: landscape, taste and smell, in Montanari A. (ed.), pp.7-9

Basavarajappa K.G., Beaujot R.P. and Samuel T.J. (1993), *Impact of migration in the receiving countries. Canada*, CICRED, Paris, and IOM, Geneva

Basta L.R. (1999), Decentralization – key issues, major trends and future developments, *SDC Publications on Development* 2, pp. 28-43, Swiss Agency for Development and Cooperation, Berne

Baudot J. (2000), Betterment of the human condition and the spirit of time, *SDC Publications on Development* 3, pp.62-71, Swiss Agency for Development and Cooperation, Berne

Bauman Z. (1998), *Globalization. The human consequences*, Polity Press, Cambridge

BBC News (1998), Brent Spar gets chop, Internet, http://news.bbc.co.uk/1/hi/world/europe/221508.stm (10.03.2003)

Beattie J. (1964), *Other cultures. Aims, methods and achievements in social anthropology*, Routledge, London

Beck U. (1992), *Risk society: towards a new sociology of modernity*, Sage, London

Beck U. (2000), *What is Globalization*? Polity Press, Cambridge

Bendixson T. (1977), *Instead of cars*, Penguin Books, Harmondsworth

Berié E., Jansen S., Kobert H. and Rudloff F. (eds. 2002), *Der Fischer Weltalmanach 2003*, Fischer Taschenbuch Verlag, Frankfurt/M. (CD-Rom edition)

Berthe J.-P. (2002): Las Casas (B. de), *Encyclopédie Universalis*, DVD version 8, Paris

Berthe J.-P., Lengellé M. & Nicolet C. (2002): Esclavage, *Encyclopédie Universalis*, DVD version 8, Paris

Bidima J.-G. (1995), *La philosophie négro-africaine*, Presses Universitaires de France, Paris

Birindelli A.M. and Bonifazi C. (eds. 1993), *Impact of migration in receiving countries. Italy*, CICRED, Paris, IOM, Geneva

BISN, Border Information & Solutions Network (1995), *Information & telecommunications assessment of the Texas/Tamaulipas border region*, Mimeo

Bitterli U. (ed. 1980), *Die Entdeckung und Eroberung der Welt. Dokumente und Berichte. Band I, Amerika, Afrika*, Beck, München

Bitterli U. (ed. 1981), *Die Entdeckung und Eroberung der Welt. Dokumente und Berichte. Band II, Asien, Australien, Pazifik*, Beck, München

Blake G.H. (1998), Globalisation and the paradox of enduring national boundaries, in Lee B.-T. and Bahrin T.S. (eds.), *Vanishing borders: the new international order of the 21st century*, Ashgate, Aldershot, pp.247-256

Blom T. (1998), Microperipherality in metropolitan areas, in Andersson and Blom (eds.), pp.164-175

Blom T. (2000), Microperiphery in megacities, in Jussila H., Majoral R. and Delgado-Cravidão F. (eds.), pp.178-193

Blum W. (2002), Rogue state. A guide to the world's only superpower, New, updated edition, Zed Books, London

Blume H. (1978), *USA. Eine geographische Landeskunde. Bd. I: Der Grossraum in strukturellem Wandel*, Wissenschaftliche Buchgesellschaft, Darmstadt

Boesler K.-A. (1983), *Politische Geographie*, Teubner, Stuttgart

Bottinelli L. (2003), *L'evoluzione del rapporto centro-periferia in ticino: implicazioni per la politica regionale*, Unpublished diploma thesis, Department of Geosciences, Geography Unit, University of Fribourg/CH

Brodbeck K.-H. (1998), *Die fragwürdigen Grundlagen der Ökonomie. Eine philosophische Kritik der modernen Wirtschaftswissenschaften*, Wissenschaftliche Buchgesellschaft, Darmstadt

Brücher, Wolfgang (1989): Saar-Lor-Lux: Grenzregion, Peripherie oder Mitte der Europäischen Gemeinschaft? *Geographische Rundschau* 41/10, pp.526-529

Brücher W. (1992), *Zentralismus und Raum. Das Beispiel Frankreich*, Teubner, Stuttgart

Brüggemeier F.-J. and Rommelspacher T. (eds. 1987), *Besiegte Natur. Geschichte der Umwelt im 19. & 20. Jahrhundert*, C.H. Beck, München

Brugger E.A. and Frey R.L. (1985), *Regionalpolitik der Schweiz. Ziele, Probleme, Erfahrungen, Reformen*, Paul Haupt, Bern

Brugger E.A., Furrer G., Messerli B. and Messerli P. (eds. 1984), *The transformation of Swiss mountain regions. Problems of development between self-reliance and dependency in an economic and ecological perspective*, Bern, Paul Haupt

Bufon M. (1994), Local aspects of transborder cooperation. A case study in the Italo-Slovene border landscape, in Gallusser W.A., Bürgin M. and Leimgruber W. (eds.), pp.19-29

Bull C., Daniel P. and Hopkinson M. (1984), *The geography of rural resources*, Oliver & Boyd, Edinburgh

Buttimer A. (2001), Geography for the Third Millennium: inventory and prospect, in Palacio-Prieto J.L. and Sánchez-Salazar M.T. (eds.), *Geografía para el Tercer Milenio – Geography for the Third Millennium*, Instituto di Geografía, Universidad Nacional Autónoma de México, México, pp. 9-16

Bylund E. (1989), Regional policy and regional research in Sweden, in Gustafsson G. (ed.), pp.29-36

Camartin I. (1982), Die Beziehungen zwischen den schweizerischen Sprachregionen, in Schläpfer R. (ed.), *Die viersprachige Schweiz*, pp.301-351, Benziger, Zürich

Cameron S. (2002), *Year in review 1999: World affairs. Cuba*, Encyclopedia Britannica, DVD-edition

Campus F., Cannata G., Marini M., Schiavoni N. and Tellarini V. (1987), I sistemi agricoli in aree marginali, in Consiglio Nazionale di Ricerche (ed.), *I sistemi agricoli marginali*, Rapporto intermedio, CNR, Roma, pp.229-295

Capra F. (1982), *The turning point: science, society, and the rising culture*, Simon and Schuster, New York

Cencini C. (1999), Il paesaggio come patrimonio: i valori naturali, *Bollettino della Società Geografica Italiana* Ser. XII, vol. IV, pp.279-294

Chang-Yi C., Sue-Ching S. & Yin-Yuh L. (eds., 1994), *Marginality and Development Issues in Marginal Regions*. Proceedings of the Study Group on Development Issues in Marginal Regions, Taiwan 1 - 7 August 1993, National Taiwan University, Taipei

Chaussade J. (1997), *Les ressources de la mer*, Flammarion, Paris

Chazel F. (1988), Normes et valeurs sociales, in *Encyclopédia Universalis*, vol. 13, Paris, pp.124-127

Chiffelle F. (1983), Swiss agricultural policy in mountainous areas, in Koutaniemi L., (ed.), *Northern and mountain villages under the pressure of change*, Geographical Society of Northern Finland, Oulu, pp.27-31

Christaller W. (1968), *Die zentralen Orte in Süddeutschland*, Wissenschaftliche Buchgesellschaft, Darmstadt (reprint of the original edition, 1933)

Clark G., Groenendijk J. and Thissen F. (1984), *The changing countryside*, Geo Books, Norwich

Coelho, P. (1997), *The Alchemist*, Thorsons, London

Cohen S.E. (1993), *The politics of planting. Israeli-Palestinian competition for control of the land in the Jerusalem periphery*, Geography Research Paper No. 236, University of Chicago Press

Cohen E., Delvert J., Godement F., Mengin F., Rabut I. and Sigwalt P. (2002), Taïwan, *Encyclopédia Universalis*, DVD-Rom V. 8, Paris

Cole J. (1987), *Development and underdevelopment. A profile of the Third World*, Methuen, London

Coleman A. (1985), *Utopia on trial. Vision and reality in planned housing*, Hilary Shipman, London

Cordey P.-A. (2000), *L'évolution de la question amériendienne dans le contexte canado-québecois*, Unpublished diploma thesis, Geography Institute, University of Fribourg/CH

Courthiade M. (2001), Rom, *Encyclopédia Universalis*, DVD-Rom V. 7, Paris

Coy M. and Pöhler M. (2002), *Condomínios fechados* und die Fragmentierung der brasilianischen Stadt. Typen – Akteure – Folgewirkungen, *Geographica Helvetica* 57/4, pp.264-277

Cullen B.T. and Pretes M. (1998), Perceptions of marginality in the United States and Canada, in Jussila H., Leimgruber W. and Majoral R. (eds.), pp.183-194

Cumbers A. and Farrington J. (2000), Keeping privatization on track: the active state, the unwilling investor and the case of rail freight in the UK, *Area* 32/2, pp.157-167

Cunha A. (1988), Systèmes et territoire: valeurs, concepts et indicateurs pour un autre développement, *L'Espace Géographique* No. 3, pp.181-198

Dahlgren B. (1997), On development in large and small places, *Occasional Publications* 1, pp.83-91, Department of Geography and Regional Planning, Indiana University of Pennsylvania

D'Aponte T. (1999), I territori del paesaggio, *Bollettino della Società Geografica Italiana* Ser. XII, vol. IV, pp.253-267

Darby H.C. (1976), *A new historical geography of England*, Cambridge University Press, Cambridge

Deiss J. (1998), *Politique économique et sociale de la Suisse*, Éditions Fragnières, Fribourg/CH

Demeter D. and Pancera M. (2000), *I frontalieri nel 1999*, 1° Annuario, Ufficio di Statistica, Bellinzona

Derounian J.G. (1994), Rural society in fact and fiction: the extent of deprivation in England's countryside, in Wiberg U. (ed.), pp.197-208

DeWoskin K.J. (2002), *China: The economy*, Encyclopedia Britannica, DVD-edition

Diamond D. (1991), Managing urban change. The case of the British inner city, in Bennett R. and Estall R. (eds.), *Global change and challenge. Geography for the 1990s*, Routledge, London, pp.217-241

Dicken P.(1998), *Global shift. Transforming the world economy.* 3rd edition, Paul Chapman, London

Dickens, C. (1961), *Hard Times*, Collins Classics 483, Collins, London,

Dieren W. van (1995), *Taking nature into account. A report to the Club of Rome*, Copernicus-Springer, New York

Dilsaver L.M. (1992), Conflict in Yosemite Valley, in Janelle D.G. (ed.), pp.133-136

Dörrenbächer P., Bierbrauer F. and Brücher W. (1988), The external and internal influences on coal mining and steel industry in the Saarland/FRG, *Zeitschrift für Wirtschaftsgeographie* 32,3, pp.209-221

Douglas M. (1996), *Natural symbols. Explorations in cosmology*, Routledge (new edition), London

Downs R.M. and Stea D. (1973), *Image and environment. Cognitive mapping and spatial behavior*, Aldine, Chicago

DTI, Department of Trade and Industry (2003), *A modern regional policy for the United Kingdom*, HMSO, Norwich

DTLR, Department of Transport, Local Government and the Regions (2002), *Your region, your choice. Revitalizing the English regions,* White Paper, DTLR, Regional Policy Unit, London

EALTA, East Arnhem Land Tourist Association (2002), Nhulunbuy – Grove Pensinsula, Internet, www.ealta.org/nhulumbuy.html (05.01.2003)

Efionayi D. and Piguet E., (2000), Entre restrictions administratives et forces du marché: l'intégration économique des requérants d'asile et des réfugiés, in Centlivres P. and Girod I. (eds.), *Les défis migratoires*, pp. 120-133, Seismo, Zürich

Ekins P. (1992), *A new world order. Grassroots movements for global change*, Routledge, London

Else R. (1984), Return to Consett, A Coast to Coast Special, BBC Northern TV

EU, European Union (2002), Regional policy – Inforegio, Internet http://europa.eu.int/comm/regional_policy/intro/ (29.04.2003)

EU, European Union (2003), *Second progress report on economic and social coherence (January 2003). Unity, solidarity, diversity for Europe, its people and its territory*, EU Regional Policy Unit, Brussels

Ewald, K.C. (1978), *Der Landschaftwandel. Zur Veränderung schweizerischer Kulturlandschaften im 20. Jahrhundert*, Berichte EAFV Nr. 191, Birmensdorf

Expertenkommission Überprüfung und Neukonzeption der Regionalpolitik (2003), *Neue Regionalpolitik. Zusammenfassung Schlussbericht*, BHP – Brugger und Partner, Zürich

Featherstone M. (ed. 1990), *Global culture. Nationalism, globalization and modernity*, Sage, London

Félix J. (1992), *Le mouvement associatif au Burkina Faso. La Fédération des unions des groupements Naam. Groupements du Yatenga et de la Comoé.* Unpublished diploma thesis, Geography Institute, University of Fribourg/CH

Felten, A. von (1998), *Essai d'évaluation de l'impact de l'homme sur le territoire suisse dans la première moitié des annés 80. Carte d'hémérobie*, Unpublished diploma thesis, Geography Institute, University of Fribourg/CH

Fernandes J.L. and Carvalho Tomás P. (2001), Conservation, development and the environment: a conflictual relationship or a different view for new geographies? Paper, presented to the IGU Commission on Evolving Issues of Geographical Marginality in the Early 21st Century World, Stockholm, 25-29 July

Fischer G. (1983), *Räumliche Disparitäten in der Schweiz*, Paul Haupt, Bern and Stuttgart

Fischer R. (1980), *Das Selbstbild von biologisch wirtschaftenden Bauern*, Ph.D. Thesis, Federal Polytechnic, Zürich

Fischer G., Rutishauser P. and Baumeler J. (1983), *Räumliche Einkommensdisparitäten in der Schweiz: das persönlich verfügbare Einkommen nach Regionen, 1970 und 1980*, Arbeitsberichte, Nationales Forschungsprogramm,Regionalprobleme in der Schweiz', vol. 40, Bern

Fischer K. and Parnreiter C. (2002), Transformation und neue Formen der Segregation in den Städten Lateinamerikas, *Geographica Helvetica* 57/4, pp.245-252

Florini A.M. (ed. 2000), *The third force. The rise of transnational civil society*, Japan Center for International Exchange, Tokyo, and Carnegie Endowment for International Peace, Washington D.C.

Florini A.M. and Simmons P.J. (2000), What the world needs now? In Florini (ed.), pp.1-15

Fluder R., Nolde M., Priester T. and Wagner A. (eds. 1999), *Armut verstehen – Armut bekämpfen. Armutsberichterstattung aus der Sicht der Statistik*, Office Fédéral de la Statistique, Neuchâtel

Foord J., Robinson F. and Sadler D. (1985), *The quiet revolution. Social & economic change at Teesside 1965-1985*, A special report for BBC North East, mimeo

Foote Whyte W. and Kochan T.A. (2002), *Industrial relations*, Encyclopedia Britannica, DVD-edition

Forsberg G. (1998), The differentiation of the countrysides, in Andersson and Blom (eds.), pp.19-29

FOS, Federal Office of Statistics (2002), *Statistical Yearbook of Switzerland*, Neue Zürcher Zeitung, Zürich

Fox M. and Sheldrake R. (1996), *The physics of angels. Exploring the realm where science and spirit meet*, Harper, San Francisco

Frankel J. (2002), *War*, Encyclopedia Britannica, DVD-edition

Freeman D. and Pankhurst A. (2001), *Living on the edge: marginalized minorities of craftworkers, and hunters in southern Ethiopia*, Department of Sociology and Social Administration, College of Social Sciences, Addis Ababa University

Frey M. and Mammey U. (1996), *Impact of migration in the receiving countries. Germany*, CICRED, Paris, and IOM, Geneva

Friedman J. (1990): Being in the world: globalization and localization, in Featherstone M. (ed.), pp.311-328

Friedmann J. (1966), *Regional Development policy: a case study of Venezuela*, MIT Press, Cambridge/MA. and London

Friedmann J. (1973), *Urbanization, planning and national development*, Sage, Beverley Hills/CA.

Furlani de Civit M.E, Pedone C. and Dario Soria N. (eds. 1995), *Development issues in marginal regions II: policies and strategies*, Universidad Nacional de Cuyo, Facultad de Letras, Mendoza, Argentina

Gaboriau P. and Gouguet J.-P. (2001), Pauvreté et exclusion, *Encyclopédie Universalis*, DVD version 7, Paris

Gade O. (1989), Regional development in a southern state: the North Carolina experience, 1970-1988, in Gustafsson G. (ed.), pp.37-46

Galante E. and Sala C. (1987), Introduzione, in Consiglio Nazionale di Ricerche (ed.), *I sistemi agricoli marginali*, Rapporto intermedio, pp.9-31, CNR, Roma

Gallusser W.A., Bürgin M. and Leimgruber W. (eds. 1994), *Political boundaries and coexistence*, Peter Lang, Berne

Galtung J. (1979), *The true worlds*, Free Press, New York

Galtung J. (1993), *Eurotopia. Die Zukunft eines Kontinents*, Promedia, Wien

Galtung J. (1998), *Die andere Globalisierung: Perspektiven für eine zivilisierte Weltgesellschaft im 21. Jahrhundert*, Agenda Verlag, Münster

Garine I. de (1991), Les changements dans la politique et les habitudes alimentaires en Afrique. Aspects des sciences humaines, sociales et naturelles, in Garine I.de (ed.), *Les changements des habitudes et des politiques alimentaires en Afrique : aspects des sciences humaines, naturelles et sociales*, pp.15-53, Publisud, Paris

Gaudard, G. (1997), *Théorie de l'espace économique et structure économique régionale*, Centre de Recherches en Économie de l'Espace de l'Université de Fribourg, Fribourg/CH

Gent W. van (2000), *Der Geruch des Grauens. Die humanitären Kriege in Kurdistan und im Kosovo*, Rotpunktverlag, Zürich

Georg Fischer (2000), Environmental Report 2000, Internet, www.georgfischer.com (03.06.2003)

George S. (2003), Globalizing Rights? In Gibney M.J. (ed.), *Globalizing rights. The Oxford Amnesty Lectures 1999*, Oxford University Press, Oxford

Ghosh J. and Pyrce V.J. (1999), Canadian immigration policy: responses to changing trends, *Geography* 84,3, pp.233-240

Giddens A. (1984), *The constitution of society. Outline of the theory of structuration*, Polity Press, Cambridge

Giersch V. (1989), Saarwirtschaft im Wandel: Vom Montanstandort zu einer modernen Industrieregion, in Soyez D., Brücher W., Fliedner D., Löffler E., Quasten H. and Wagner J.M. (eds.), *Das Saarland, Band 1: Beharrung und Wandel in einem peripheren Grenzraum*, Arbeiten aus dem Geographischen Institut des Saarlandes Bd. 36, pp.257-268

Gigon N. (1991), Der französisch-schweizerische Jura. Grenze und Städtesystem, *Geographische Rundschau* 43/9, pp.514-519

Gloor D., Hohermuth S., Meier H. and Meier H.-P. (1996), *Fünf Idiome – eine Schriftsprache? Die Frage einer gemeinsamen Schriftsprache im Urteil der romanischen Bevölkerung*, Bündner Monatsblatt/Desertina, Chur

Gmünder M., Grillon N. and Bucher K. (2000), Gated communities. Ein Vergleich privatisierter Wohnsiedlungen in Südkalifornien, *Geographica Helvetica* 55/3, pp.193-203

Gosar A. and Klemencic V. (1994), Current problems of border regions along the Slovene-Croatian border, in Gallusser W.A., Bürgin M. and Leimgruber W. (eds.), pp.30-42

Gottlieb R.S. (ed. 1996), *This sacred earth. Religion, nature, environment*, Routledge, New York

Gottmann J. (1980, Confronting centre and periphery, in Gottmann J. (ed.), *Centre and periphery. Spatial variation in politics*, Sage, Beverly Hills and London, pp.11-25

Gould P. (1990), *Fire in the rain. The democratic consequences of Chernobyl*, Polity Press, Cambridge

Gould P. and White R. (1974), *Mental maps*, Penguin, Harmondsworth

Gourlay D. (ed., 1998), *Marginal rural areas in the new millennium. New issues? New opportunities*? The PIMA98 International Conference at the Scottish Agricultural College, Aberdeen, Scotland, June 26th-29th 1998

Gourlay D. (1998), Regional development aid in an expanding European Union, in Gourlay D. (ed, pp.226-242

Gramm M. (1979), *Das belgisch-niederländisch-deutsche Dreiländereck*, Paul List, München

Grandgirard, V. and Schaller, I. (1995), Espace et paysage, deux concepts-clé de l'approche géographique, *UKPIK, Cahiers de l'Institut de Géographie de Fribourg*, 10, pp.25-37

Granö J.G. (1922), Eesti maastiküksused [Landscape units of Estonia; in Estonian], *Loodus*, 2(4-5), pp.105-123, 194-214, 257-281

Grauhan R.R. and Linder W. (1974), *Politik der Verstädterung*, Athenäum Fischer, Frankfurt/M.

Grimal P. and Lévy M. (2002), Jardins (Art des), *Encyclopédie Universalis*, DVD version 8, Paris

Groenewold J. and von Praet S. (1997), Falling through the net. Outcast and marginalized populations, in MSF, Médecins Sans Frontières (ed.), pp.58-78

Gronemeyer M. (2000), *Immer wieder neu oder ewig das Gleiche. Innovationsfieber und Wiederholungswahn*, Wissenschaftliche Buchgesellschaft, Darmstadt

Grosjean G. (1984), Visual and aesthetic changes in landscape, in Brugger E.A., Furrer G., Messerli B. and Messerli P. (eds.), pp.71-99

Gschwend H., Mugglin M. and Müller S. (1999), *Wirtschaft und Ethik*, Doppelpunkt SR DRS 1, 7.3./14.3./21.3./28.3.1999, Manuscript of a Swiss Radio programme

Guerrier Y., Alexander N., Chase J. and O'Brien M. (eds. 1995), *Values and the environment. A social science perspective*, John Wiley, Chichester

Guichonnet P. and Raffestin C. (1974), *Géographie des frontières*, PUF, Paris

Gupta V. (2000), Privatisation of health, Internet, www.bicusa.org/ptoc/htm/gupta_privatisation.htm (29.12.2002)

Gustafsson G. (1989, ed.), *Development in marginal areas*, Research Report 89:3, University of Karlstad

Gustafsson G. (1989), An environmental ethics strategy for the development of marginal areas. A general development strategy? in Gustafsson G. (ed.), pp.135-146

Haggett P. (1972), *Geography, a modern synthesis*, Harper & Row, New York

Häkkilä M. (1974), Certain regional features in the growing stock of the Finnish forests, *Fennia* 135, Helsinki

Hall D. (2000), World Bank – politburo of water privatisation, Internet, www.bicusa.org/ptoc/htm/psiru_water.htm (29.12.2002)

Hammer T. (1997), *Aufbruch im Sahel. Fallstudien zur nachhaltigen ländlichen Entwicklung*, LIT, Hamburg

Hammer T. (2001), Biosphärenreservate und regionale (Natur-)Parke – neue Konzepte für die nachhaltige Regional- und Kulturlandschaftsentwicklung? *Gaia* 10/4, pp.279-285

Hammer T. (2002a), Das Biosphärenreservat-Konzept als Instrument nachhaltiger Regionalentwicklung? Beispiel Entlebuch, Schweiz, in Mose I. and Weixlbaumer N. (eds.), *Naturschutz: Grossschutzgebiete und Regionalentwicklung*, pp.111-135, Academia Verlag, Sankt Augustin

Hammer T. (2002b), Naturschutz als Überlebensstrategie? Zum Wandel der Bedeutung von Natur, Raum und Entwicklung im westafrikanischen Sahel, in Erdmann K.-H. and Bork H.-R. (eds.), *Naturschutz – neue Ansätze, Konzepte und Strategien*, pp.189-205, Bundesamt für Naturschutz, Bonn-Bad Godesberg

Hanser C. (1983), Regional policy in Switzerland: an analysis of ist effectiveness and an outline of a strategic reorientation, *Geoforum* 14/4, pp.363-373

Harrison C.M. and Burgess J (1994), Social constructions of nature: a case study of conflicts over the development of Rainham Marshes, *Transactions of the Institute of British Geographers* NS 19/3, pp.291-310

Harrison Church R.J. (1981), West African boundaries and the cultural landscape, *Regio Basiliensis* XXII/2+3, pp.258-267

Hartfiel G. and Hillmann K.-H. (1982), *Wörterbuch der Soziologie*, Kröner, Stuttgart

Hellie R. (2002), *Slavery*, Encyclopedia Britannica, DVD-edition

Hemmer H.-R. and Wilhelm R. (2000), *Fighting poverty in developing countries. Principles for economic policies*, Schriften zur internationalen Entwicklungs- und Umweltforschung, Zentrum für internationale Entwicklungs- und Umweltforschung, Justus-Liebig-Universität Giessen, vol. 1, Peter Lang, Frankfurt/M.

Hettne B. (1990), *Development theory and the three worlds*, Longman, Harlow

Hillmann, K.-H. (1989), *Wertwandel. Zur Frage soziokultureller Voraussetzungen alternativer Lebensformen*, Wissenschaftliche Buchgesellschaft, Darmstadt

Hinkelammert F.J. (1998), Globalisierung und Ausschluss aus lateinamerikanischer Sicht, in Fornet-Betancourt R. (ed.), *Armut im Spannungsfeld zwischen Globalisierung und dem Recht auf eigene Kultur.* Dokumentation des VI. Internationalen Seminars des philosophischen Dialogprogramms, pp.92-104, IKO-Verlag für Interkulturelle Kommunikation, Frankfurt/M.

Holdar S. (1994), Donbass: on the border of Ukraine and Russia, in Gallusser W.A., Bürgin M. and Leimgruber W. (eds.), pp.43-51

Holton R.J. (1998), *Globalization and the nation-state*, Macmillan, Basingstoke

Hormann H. (1997), *Die Oekologischen Auswirkungen des Variantenskifahrens und der maschinellen Beschneiung im Testgebiet Mürren*, Unpublished diploma thesis, Geography Institute, University of Fribourg/CH

Horton R. (1969), Types of spirit possession in Kalahari region, in Beattie J. and Middleton J. (eds.), *Spirit mediumship and society in Africa*, Routledge and Kegan Paul, London

Hoskins W.G. (1969), *The making of the English landscape*, 8th impr., Hodder & Stoughton, London

House J.W. (1982), *Frontier on the Rio Grande. A political geography of development and social deprivation*, Clarendon Press, Oxford

Houtum, Van H. (1998), *The development of cross-border economic relations*, Center for Economic Research, Tilburg University, The Netherlands, Dissertation Series No. 40

Howitt R. (1992), Weipa: industrialization and indigenous rights in a remote Australian mining area, *Geography* 77/3, pp.223-235

Hughes J.D. (1966), from *American Indian ecology*, in Gottlieb R.S. (ed.), pp.131-146

Huigen P. (1984), Access in a remote rural area, in Clark, Groenendijk and Thissen (eds.), pp.87-97

Hultkrantz Å. (1960), *General ethnological concepts*, Rosenkilde and Bagger, Copenhagen

Huonker T. and Ludi R. (2000), *Roma, Sinti und Jenische. Schweizerische Zigeunerpolitik zur Zeit des Nationalsozialismus*. Beiheft zum Bericht 'Die Schweiz und die Flüchtlinge zur Zeit des Nationalsozialismus', ed. Unabhängige Expertenkommission Schweiz – Zweiter Weltkrieg, BBL/EDMZ, Bern, and Chronos Verlag, Zürich (2001)

ICRC (2002), *Discover the ICRC*, Geneva

IGU Newslewtter (2001), United Nations Secretary General asks geographers to work on climate change, envirionmental degradation, and sustainable development, Nr. 3-4, December

Ingold T. (1986), *The appropriation of nature. Essays on human ecology and social relations*, Manchester University Press, Manchester

Inoue N. (1997a), Globalization's challenge to indigenous culture, in Inoue N. (ed.), pp.7-19

Inoue N. (1997b), The information age and the globalization of religion, in Inoue N. (ed.), pp.80-96

Inoue N. (ed., 1997), *Globalization and indigenous culture*, Institute for Japanese Culture, Kokugakuin University, Tokyo

Jäger H. (1994), *Einführung in die Umweltgeschichte*, Wissenschaftliche Buchgesellschaft, Darmstadt

Janelle D.G. (ed. 1992), *Geographical snapshots of North America*, The Guildford Press, New York

Janoschka M. (2002), Die Flucht vor Gewalt? Stereotype und Motivationen beim Andrang auf *barrios privados* in Buenos Aires, *Geographica Helvetica* 57/4, pp.290-299

Jargowsky P.A. (1996), *Poverty and place. Ghettos, barrios, and the American city*, Russell Sage Foundation, New York

Jatar A.J. (2002), *Year in review 2001: World affairs. Cuba*, Encyclopedia Britannica, DVD-edition

Jones G. and Morris A. (eds. 1997), *Issues of environmental, economic and social stability in the development of marginal regions: practices and evaluation*, Glasgow, Department of Geography, University of Strathclyde, and Department of Geography, University of Glasgow

Joye D., Busset T. and Schuler M. (1992), Geographische und soziale Trennlinien der Schweiz, in Hugger P. (ed.), *Handbuch der schweizerischen Volkskultur*, 3 vols., vol. 2, pp.661-676, OZV Offizin, Zürich

Jussila H., Leimgruber W. and Majoral R. (eds. 1998), *Perceptions of marginality. Theoretical issues and regional perceptions of marginality in geographical space*, Ashgate, Aldershot

Jussila H., Majoral R. and Cullen B. (eds., 2002), *Sustainable development and geographical space. Issues of population, environment, globalization and education in marginal regions*, Ashgate, Aldershot

Jussila H., Majoral R. and Delgado-Cravidão F. (2000), *Environment and marginality in geographical space. Issues of land use, territorial marginalization and development in the new millennium*, Ashgate, Aldershot

Jussila H., Majoral R. and Delgado-Cravidão F. (eds. 2001), *Globalization and marginality in geographical space. Political, economic and social issues of development in the new millennium*, Ashgate, Aldershot

Jussila H., Majoral R. and Mutambirwa C.C. (eds. 1999), *Marginality in space – past, present and future. Theoretical and methodological aspects of cultural, social and economic parameters of marginal and critical regions*, Ashgate, Aldershot

Jussila H., Majoral R. and Cullen B. (eds. 2002), *Sustainable development and geographical space. Issues of population, environment, globalization and education in marginal regions*, Ashgate, Aldershot

Kamdem M.S. (1994), Some characteristics of Kousseri population: a border town in Cameroon, in Gallusser W.A., Bürgin M. and Leimgruber W. (eds.), pp.224-232

Kanitscheider S. (2002), *Condominios* und *fraccionamientos cerrados* in Mexiko-Stadt – Sozialräumliche Segregation am Beispiel abgesperrter Wohngebiete, *Geographica Helvetica* 57/4, pp.253-263

Kant E. (1926), *Tartu. Linn kui ümbrus ja organism Linna geograafiline vaatlus, ühtlasi lisang kultuurmaastiku morfoloogiale.* [Tartu, Etude d'un environnement et organisme urbain], K.-U. 'Postimehe' trukk, Tartu

Kant E. (1935), *Bevölkerung und Lebensraum Estlands. Ein anthropoökologischer Beitrag zur Kunde Baltoskandias*, Akadeemiline Kooperatiiv, Tartu,

Keeble D. (1976), *Industrial location and planning in the United Kingdom*, Methuen, London

Keeble D. (1984), Industrial location and regional development, in Short J.R. and Kirby A. (eds.), *The human geography of contemporary Britain*, pp.40-51, Macmillan, Basingstoke

Keller G. (19XX), *I neo-rurali*, Unpublished diploma thesis, Geography Institute, University of Fribourg/CH

Kent P. (1980), *Minerals from the marine environment*, Edward Arnold, London

KICI (n.d.), KICI steunt Max Havelaar!, Internet, www.kici.nl (06.06.2003)

Kim J.-H. (1990), *Die Auswirkung der Grenzziehung auf die Grenzgebiete, ein Vergleich zwischen Südkorea und der Bundesrepublik Deutschland*, Materialien zur Raumordung vol. XXXIX, Geographisches Institut der Ruhr-Universität, Bochum

Kirk W. (1980), The rural-urban continuum: perception and reality, in Enyedi G. and Mézaróz J. (eds.), *Development of settlement systems*, Studies in Geography in Hungary vol. 15, pp.11-19, Akadémiai Kiadó, Budapest

Klein N. (2001), It's not the trade; it's the trade-offs, Internet, www.globsalisation debate.be/2001 (08.11.2002)

Klemencic V. and Bufon M. (1991), Geographic problems of frontier regions: the case of the Italo-Yugoslav border landscape, in Rumley D. and Minghi J.V. (eds.), pp.86-103

Knowles R. and Farrington J. (1998), Why has the market not been created for Channel Tunnel regional passenger rail services? *Area* 30/4, pp.359-366

Kohler P. (2002), Geschlossene Wohnkomplexe in Quito – Naturraum und rechtliche Rahmenbedingungen als Einflussgrössen für Verbreitung und Typisierung, *Geographica Helvetica* 57/4, pp.278-289

König F. (2001), Nestlé: Profit statt Gesundheit. Noch fehlt der Gegenbeweis, Med in Switzerland 5, Internet, www.medicusmundi.ch/med/med015.htm (25.02.2003)

Koter M. (1994), Transborder 'Euroregions' round Polish border zones as an example of a new form of political coexistence, in Gallusser W.A., Bürgin M. and Leimgruber W. (eds.), pp.77-87

Krebs P. (1997): Il carbone di legna dall'età della pietra all'età del barbecue. Indagine bibliografica e indagine sul terreno alla ricerca della storia del carbone di legna tra cultura natura ed economia, con particolare riguardo al Cantone Ticino e alle sue montagne. Unpublished diploma thesis, Geography Institute, University of Fribourg/CH

Kreutzmann H. (2003), Theorie und Praxis in der Entwicklungsforschung. Einführung zum Themenheft, *Geographica Helvetica* 58,1, pp.2-10

Krippendorf J. (1984), The capital of tourism in danger. Reciprocal effects between landscape and tourism, in Brugger E.A., Furrer G., Messerli B. and Messerli P. (eds.), pp.427-450

Kurs O. (2002), Scientific heritage of Edgar Kant, 'From native and landscape research to urban and regional studies, Internet, www.geo.ut.ee/KANT/absthema1.html (20.12.2002)

La Liberté 09.09.1999: Innovation: la Suisse risque d'être marginalisée, Fribourg/CH

Lafarge (n.d.), Lafarge and WWF: the partnership, Internet, www.lafarge.com (11.04.2003)

Läubli M. (2003), Eine riesige Wunde in der irakischen Wüste, Tages-Anzeiger 11.04.2003, p.74, Zürich

Lea D.R. (1994), Christianity and Western attitudes towards the natural environment, *History of European Ideas* 18/4, pp. 513-524

Le Heron R. and Roche M. (1995), A 'fresh' place in food's space, *Area* 27/1, pp.23-33

Leimgruber W. (1981), Political boundaries as a factor in regional integration, examples from Basle and Ticino, *Regio Basiliensis*, pp.192-201

Leimgruber W. (1985), What is a mountain region? *Innsbrucker Geographische Studien* 13, pp.99-107

Leimgruber W. (1986), From plain to mountain: zoning in Switzerland, *Nordia* 20:1, pp.49-56

Leimgruber W. (1987), *Il confine e la gente. Interrelazioni spaziali, sociali e politiche fra la Lombardia e il Canton Ticino*, Lativa, Varese

Leimgruber W. (1990), Image et localisation: la promotion économique, *Rapports et Recherches, Institut de Géographie, Fribourg/CH*, vol. 2, pp.123-131

Leimgruber W. (1991), Boundary, values and identity: the Swiss-Italian border region, in Rumley D. and Minghi J.V. (eds.), pp.43-62

Leimgruber W. (1992a), *Impact of migration in the receiving countries. Switzerland,* CICRED Paris, and IOM, Geneva

Leimgruber W. (1992b), Man and the mountains – the mountains and man: the changing role of mountain areas, *Occasional Papers in Geography and Planning,* vol. 4, pp.121-135, Appalachian State University, Boone N.C.

Leimgruber W. (1994), Marginality and marginal regions: problems of definition, in Chang-Yi C., Sue-Ching S. & Yin-Yuh L. (eds.), pp.1-18

Leimgruber W. (1994/5), Local efforts and people participation in the development of marginal regions. A summary of 10 years' research at the Fribourg Department of Geography, *Boletín de Estudios Geograficos, Universidad Nacional de Cuyo, Facultad de Filosofía y Letras, Instituto de Geografia*, vol. 26/tomo 2, p. 391-408, Mendoza (Argentina)

Leimgruber W. (1995), L'espace des finances publiques. Les aspects fiscaux en géographie politique: le cas de la Suisse, *Geographica Helvetica* 50,1, pp.12-20

Leimgruber W. (1998a), From highlands and high-latitude zones to marginal regions, in Jussila H., Leimgruber W. and Majoral R. (eds.), pp.27-33

Leimgruber W. (1998b), Defying political boundaries: transborder tourism in a regional context, *Visions in Leisure and Business* 17/3, pp.8-29

Leimgruber W. (1999), Border effects and the cultural landscape: the changing impact of boundaries on regional development in Switzerland, in Knippenberg H. and Markusse J. (eds.): *Nationalising and denationalising European border regions, 1800-2000: views from geography and history*, GeoJournal Library 53, p.199-221, Kluwer, Dordrecht

Leimgruber W. (2000), Land use and abuse. On the ecological and spiritual marginalization of land, in Majoral R., Jussila H. and Delgado-Cravidão F. (eds.), pp.7-24

Leimgruber W. (2001), In harmony with nature – the importance of the productivist cycle for agriculture, in Pelc S. (ed.), *Developmental problems in marginal rural areas: local initiative versus national and international regulation*, pp.15-26, University of Ljubljana

Leimgruber W. (2003a), Heritage conservation awards: stimuli to life for remote settlements? in Leimgruber W., Majoral R. and Lee C.-W. (eds.), pp.149-163

Leimgruber W. (2003b), Marginality and diversity: do marginal regions ensure biological and cultural diversity? in Leimgruber W., Majoral R. and Lee C.-W. (eds.), pp.239-255

Leimgruber W. (forthcoming), Values, migration, and environment. An essay on driving forces behind human decisions and their consequences, in Krool, M. et al. (eds.), *Environmental change: implications for human migrations*, Papers presented to the Wengen Workshop on Global environmental change, 2001

Leimgruber W. (2004), The right to diversity. Human Rights from a Geographical Point of View, in Bohnet A. and Höher M. (eds.), *The role of minorities in the development process*, Schriften zur Internationalen Entwicklungs- und Umweltforschung, Zentrum für Internationale Entwicklungs- und Umweltforschung, University of Giessen, pp.21-36

Leimgruber W. and Hammer T. (2002), Biosphere reserves – sustainable development of marginal regions? in Jussila H. et al. (eds.), pp.129-144

Leimgruber W. and Imhof G. (1998), Remote alpine valleys and the problem of sustainability, in Andersson L. and Blom T. (eds.), pp.385-396

Leimgruber W., Dettli W. and Meusy G. (1997), Organic farming: a solution to marginality? in Jones G. and Morris A. (eds.), pp.160-169

Leimgruber W., Majoral R. and Lee C.-W. (eds. 2003), *Policies and strategies in marginal regions. Summary and evaluations*, Ashgate, Aldershot

Léman sans frontière (2003), Internet, www.leman-sans-frontiere.com/e/car tezoom.html (15.03.03)

Lévy E. (1995), *Nouvelle histoire de l'antiquité, vol. 2, La Grèce au Ve siècle : de Clisthène à Socrate*, Editions du Seuil, Paris

Leu R.E. (1999), Konzepte der Armutsmessung, in Fluder R., Nolde M., Priester T. and Wagner A. (eds.), pp.39-64

Leuthardt B. (1999), *An den Rändern Europas. Berichte von den Grenzen*, Rotpunktverlag, Zürich

Lezzi M. (2000), *Porträts von Schweizer EuroRegionen. Grenzüberschreitende Ansätze zu einem europäischen Föderalismus – Transboundary cooperation in Switzerland. A training for Europe*, Schriften der Regio 17, Helbing & Lichtenhahn, Basel

Livet P. (2001), Valeurs (philosophie), *Encyclopédie Universalis*, DVD version 7, Paris

Livingstone D.N. and Withers C.W.J. (1999), Introduction: On geography and Enlightenment, in Livingstone D.N. and Withers C.W.J. (eds.), pp.1-31

Livingstone D.N. and Withers C.W.J. (eds. 1999), *Geography and Enlightenment*, The University of Chicago Press, Chicago and London

Loís-Gonzáles R. (2002), Economic growth, ecological consequences, and depopulation in the rural areas of Galicia (Spain), in Jussila H., Majoral R. and Cullen B. (eds.), pp.37-48

Lurati O. (1982), Die sprachliche Situation der Südschweiz, in Schläpfer R. (ed.), *Die viersprachige Schweiz*, pp.211-252, Benziger, Zürich

Maček I. (2000), *War within. Everyday life in Sarajewo under siege*, Uppsala Studies in Cultural Anthropology 29, Uppsala

Maide J. (1931), *Eesti minemi raioonid* [The agricultural market regions in Estonia], Publicationes Seminarii Universitatis Tartuensis Oeconomico-Geographici, nr. 2

Majoral R.M., Font Garolera J. and Sánchez-Aguilera D. (1996), Regional development policies and incentives in marginal areas of Catalonia, in Furlani de Civit M.E., Pedone C. and Soria N.D (eds.), pp.27-48

Malinen P., Jussila H. and Häkkilä M. (eds. 1993), *Finland's national rural policy facing the challenge of European integration*, Research Report 114, Research Institute of Northern Finland, University of Oulu

Mandela N. (1995), *Long walk to freedom*, Abacus, London

Manners G., Keeble D., Rodgers B. and Warren, (1980), *Regional development in Britain*, 2nd ed, John Wiley, Chichester

Markovits C. (2000-2001), La partition de l'Inde – The partition of India, *Transeuropéennes*, vol. 19/20, pp.65-79

Massey D. (2002), Globalisation: what does it mean for geography? *Geography* 87/4, pp.293-296

Matsuo Y. (1998), Maintaining or reviving the cultural landscape, in Andersson L. and Blom T. (eds.), pp.217-229

Matznetter J. (1981), Der strukturelle Dualismus Afrikas in Spiegel seiner Staatsgrenzen, *Regio Basiliensis*, XXII/2+3, pp.258-267

Maude B. (1975), *The turning tide. Towards the post-surplus society*, Faber & Faber, London

Max Havelaar (n.d.), Mission of the Max Havelaar Foundation, Internet, www.maxhavelaar.ch (02.06.2003)

Meadows D., Meadows D., Zahn E. and Milling P. (1972), *The limits to growth*, Universe Books, New York

Meadows D.H., Meadows D.L. and Randers J. (1992), *Beyond the limits*, Chelsea Green Publishing, Post Mills VT.

Mehretu A., Pigozzi B.W. and Sommers L.M. (1996), Issues of urban marginality in the Greater Detroit area, in Jones G. and Morris A. (eds.), pp.112-121

Mehretu A., Pigozzi B.W. and Sommers L.M. (2003), Analysis of spatial marginality in Michigan: some empirical illustrations, in Leimgruber W., Majoral R. and Lee C.-W. (eds), pp.325-338

Mehretu A. and Sommers L.M. (1997), Towardws an analysis of microspatial marginality in the Detroit region, Michigan, *Occasional Publications* 1, p. 65-73, Department of Geography and Regional Planning, Indiana University of Pennsylvania

Mehretu A., Pigozzi B.W. and Sommers L.M. (2002), Spatial shifts in production and consumption: marginality patterns in the new international division of labour, in Jussila H., Majoral R. and Cullen B. (eds.), pp.195-208

Menon R. (2000-2001), La dynamique de la division – The dynamics of division, *Transeuropéennes*, vol. 19/20, pp.155-173

Miller G.T. (1996), *Living in the environment*, 9th ed., Wadsworth, Belmont

Miller V.P. Jr. (1992), Towards the further understanding of rural sustained development: the case of the Amish in Indiana County, Pennsylvania, in Ó Cinnéide M. and Grimes S. (eds.), *Planning and development of marginal areas*, pp.143-150, Centre for Development Studies, Galway

Miller V.P. Jr. (1997), Landscape as cultural identity: some preliminary comments on the usefulness of semiotics as a tool of development, *Occasional Publication* no. 1, pp.1-19, Department of Geography and Regional Planning, Indiana University of Pennsylvania, Indiana PA.

Miller V.P. Jr. (1998), Where shall we go from here? A position paper on future directions within PIMA, in Gourlay D. (ed.), pp.257-274

Moizo B. (1989), Où en sont les aborigines d'Australie? *Hérodote* no.52, pp.156-166

Montanari A. (ed.), *Food and environment. Geographies of taste*, Home of Geography Publication Series II, Società Geografica Italiana, Rome

Moseley M. (1979), *Accessibility: the rural challenge*, Methuen, London

Moseley M. and Packham J. (1984), Mobile services and the rural accessibility problem, in Clark, Groenendijk and Thissen (eds.), pp.79-86

MSF, Médecins Sans Frontières (1997), *World in crisis. The politics of survival at the end of the 20th century*, Routledge, London

MSF (2000), The MSF role in emergency medical aid, Internet, www.msf.org (01.06.2003)

MSF (2002), The charter of Médecins Sans Frontières, Internet, www.msf.org (01.06.2003)

Moutouh H. (2000), *Les tsiganes*, Flammarion, Paris

Mutambirwa C. and Mehretu A. (2003), Residential marginalization in the post-colonial city of Harare, in Leimgruber W., Majoral R and Lee C-W. (eds.), pp.215-227

Myers F.R. (1982), Always Ask: resource use and land ownership among Pintupi Aborigines of the Australian Western Desert, in Williams N and Hunn E.S. (eds.), pp.173-195

Nabalco (2000), Internet, www.nabalco.com.au (05.01.2003)

Nel E. (2002), Global change and community self-reliance strategies in Southern Africa, in Jussila H., Majoral R. and Cullen B. (eds.), pp.159-172

Nel E. and Hill T. (2001), De-industrialisation and local economic development alternatives in KwaZulu-Natal, South Africa, *Geography* 86/4, pp.356-359

Nelson J.G. and Butler R.W. (1974), Recreation and the environment, in Manners I.R. and Mikesell M.W. (eds.), *Perspectives on environment*, Association of American Geographers Publication No. 13, pp.290-310, Association of American Geographers, Washington D.C.

NEPAD, New Partnership for Africa's Development (2001), Information brochure (available on Internet, www.dfa.gov.za or www.africanrecovery.org)

Neubauer G. (1988), Regulierung und Deregulierung im Gesundheitswesen der Bundesrepublik Deutschland, in Thiemeyer T. (ed.), pp.9-49

Newman D. and Kliot N. (1999), Introduction: globalisation and the changing world map, *Geopolitics*, 4/1, p.1-16)

NGOWatch (n.d. [2003]), NGOWATCH.ORG. A project of the American Enterprise Institute and The Federalist Society, Internet, www.ngowatch.org (12.07.2003)

O'Brien M. and Guerrier Y. (1995), Values and the environment: an introduction, in Guerrier et al. (ed.), pp.xiii-xvii

Ó Cearbhail, D. (1998), Training for sustainable development in disadvantaged rural areas of rural Ireland, in Andersson L. and Blom T. (eds.), pp.341-350

O'Loughlin J. and van der Wusten H. (1990), Political geography of panregions, *The Geographical Review*, 80/1, pp.1-20

OECD (2000), *OECD in figures. Statistics on the member countries*, OECD, Paris

Offe C. (1973), Krisen und Krisenmanagement, in Jänicke M. (ed.), *Herrschaft und Krise*, UTB, Opladen

Office of the Deputy Prime Minister (2001), Regional Development Agencies. General Information, Internet, www.local-regions.odpm.gov.uk/rda/info/ (29.05.2003)

Ohshima G. (1986), Between Australia and New Guinea – ecological and cultural diversity in the Torres Strait with special reference to the use of marine resources, *Geographical Review of Japan*, vol. 59 (Ser. B), No. 2, pp.69-82

Orwell G. (1954), *Nineteen eighty-four*, Harmondsworth, Penguin

Orwell G. (1975), *The Road to Wigan Pier*, Harmondsworth, Penguin

Pankhurst A. (forthcoming): Social exclusion and cultural marginalization: minorities of craft workers and hunters in Ethiopia, Paper delivered to the Symposium 'The Role of minorities in the development process', University of Giessen, October 2002

Papademetriou D.G. and Hamilton K.A., (1995), *Managing uncertainty: regulating immigration flows in advanced industrial countries*, International Migration Policy Program 1, Carnegie Endowment for International Peace, Washington DC

Parker G. (1979), *The countries of Community Europe. A geographical survey of contemporary issues*, Macmillan, London

Parolini J. (2001), *La Cuba fidelista: un mito al tramonto o una solida realtà?* Unpublished diploma thesis, Institute of Geography, University of Fribourg/CH

Persson L.O. (1998), Clusters of marginal microregions, in Jussila H., Leimgruber W. and Majoral R. (eds.), pp.81-99

Persson L.O. (1999), State support and rural dynamics, in Jussila H., Majoral R. and Mutambirwa C. (eds.), pp.273-283

Persson L.O. and Österberg R. (1998), Deregulation in marginal areas: dynamics or decline? in Gourlay D. (ed.), pp.77-88

Persson L:O. and Wiberg U. (1995), *Microregional fragmentation. Contrasts between a welfare state and a market economy*, Physica, Heidelberg

Pfeiffer J. (1997), The plurilingual European tradition as a challenge to globalization, in Inoue N. (ed.), pp.187-202

Piaget J. and Inhelder B. (1948), *Die Entwicklung des räumlichen Denkens beim Kinde*, Gesammelte Werke vol. 6, Klett, Stuttgart

Pieroni O. and Andreoli M. (1989), *Agricoltura, marginalità e sistemi socio-economici locali. Guida all'utilizzo di un sistema di indicatori per l'analisi socio-economica degli scenari I.P.R.A*, CNR., Roma

Pilger J. (2002), *The new rulers of the world*, Verso, London

Pitte J.-R. (2002), Geography of taste between globalization and local roots, in Montanari A. (ed.), pp.11-28

Piveteau J.-L. (1971), Les trois dimensions des disparités économiques régionales et leurs relations avec les milieux naturels et humains, *Revue économique et sociale*, 29/1, pp.21-30

Piveteau J.-L. (1995), *Temps du territoire. Continuités et ruptures dans la relation de l'homme à l'espace*, Éditions Zoé, Genève

Poon J.P.H., Thompson E.R. and Kelly P.F. (2000), Myth of the triad? The geography of trade and investment 'blocs', *Trans Inst Br Geogr* NS 25, pp.427-444

Potts D. and Mutambirwa C.C. (1999), 'Basics are now a luxury': Perceptions of ESAP's impact on rural and urban areas in Zimbabwe, in Jussila H., Majoral R. and Mutambirwa C.C. (eds.), pp.179-210

Prebisch R. (1959), Commercial policy in the underdeveloped countries, *The American Economic Review, Papers and Proceedings* 49, pp.251-273

Ragaz C. (1988), *Geography and the conceptual world. The significance of place to Aboriginal Australians with reference to the historical Lakes Tribes of South Australia*, PhD Thesis, Zürich, Zentralstelle der Studentenschaft

Raffestin C. (1980), *Pour une géographie du pouvoir*, Librairies Techniques, Paris

Region Sense (1995), *Entwicklungskonzept 2010 der Region Sense*, Region Sense, Tafers

Reheis F. (1996), *Die Kreativität der Langsamkeit. Neuer Wohlstand durch Entschleunigung*, Wissenschaftliche Buchgesellschaft, Darmstadt

Reitel, François (1989): Die Entwicklung der Eisen- und Stahlindustrie im Saar-Lor-Lux-Raum, *Geographische Rundschau* 41/10, pp.555-565

Rekacewicz P. (2001), Des millions de réfugiés, un fardeau pour le Sud, *Le Monde Diplomatique*, Avril, pp.18 f.

Reynaud A. (1981), *Société, espace et justice. Inégalités régionales et justice socio-spatiale*, PUF, Paris

Reynolds H. (1982), *The other side of the frontier. Aboriginal resistance to the European invasion of Australia*, Penguin Books, Ringwood, Victoria

Reynolds H. (1987a), *Frontier. Aborigines, settlers and land*, Allen and Unwin, SydneyReynolds H. (1987b), *The law of the land*, Penguin Books Australia, Ringwood

Rishi P.W.R. (1996), *Roma. The Panjabi emigrants in Europe, Central and Middle Asia, the URSS and the Americas*. 2nd. ed, Publication Bureau Punjabi University, Patiala

Ritzer G. (2000), *The McDonaldization of society*, New century edition, Pine Forge Press, Thousand Oaks, California

Robinson G.M. (1988), *Agricultural change. Geographical studies of British agriculture*, North British Publishing, Edinburgh

Rostovtseff M. (1970), *Geschichte der Alten Welt: Rom*, Schünemann, Bremen

Rousseau J.J. (1987), *Discours sur l'origine et les fondements de l'inégalité parmi les hommes* [1755], ed. by F. Khodoss, Bordas, Paris

Rufin J.-C. (1991), *L'empire et les nouveaux barbares. Rupture Nord-Sud*, Hachette/Jean-Claude Lattès, Paris

Rugendyke B.A. (1996), *Arrunge Aboriginal Corporation Community Development Planning Process. Final Report*, Unpublished consultancy report, University of New England, Armidale

Rugendyke B.A. (1998), Community participation as empowerment? Planning for change in remote Aboriginal Australia, in Jussila H., Leimgruber W. and Majoral R. (eds.), pp.257-278

Rumley D. and Minghi J.V. (eds. 1991), *The geography of border landscapes*, Routledge, London

Rusanen J., Muilu T., Colpaert A. and Naukkarinen A. (2002), Local poverty in Finland in 1995, in Jussila H., Majoral R. and Cullen B. (eds.), pp.49-64

Ryu J.-H. (2000), *Reading the Korean cultural landscape*, Hollym, Elizabeth NJ and Seoul

Saar E. (2003), Setos – eine südestnische Sprache und Kultur an der estnisch-russischen Grenze. Ein Beispiel für Ethnogenese durch Grenzverschiebungen, in Brücher W. (ed.), *Grenzverschiebungen. Interdisziplinäre Beiträge zu einem zeitlosen Phänomen*, Annales Universitatis Saraviensis, Philosophische Fakultäten, vol. 21, pp.103-118, St. Ingbert, Röhrig Universitätsverlag

Sarvodaya (2003), Sarvodaya and its objective, Internet, www.sarvodaya.org (03.06.2003)

Savadogo R.A. (2002), 'Refounding' the African state, decentralization and civil society, *SDC Publications on Development* 5, pp. 18-28, Swiss Agency for Development and Cooperation, Berne

Save the Children (2001), Expenditure Cuts and User Charges in Education and Health Services in Transition Countries, Internet, www.bicusa.org/ptoc/htm/stc_expenditurecuts.htm (29.12.2002)

Schäfer S. (1996), *Kulturraum Oberrhein. Grenzüberschreitende Kulturarbeit in der deutsch-französisch-schweizerischen EuroRegion*, Schriften der Regio 15, Helbing & Lichtenhahn, Basel

Schaller I. (1993), *Grenzwahrnehmung und Verhalten an einer Binnengrenze – empirische Untersuchung an der Kantonsgrenze Luzern-Bern*, Unpublished diploma thesis, Department Geography, University of Fribourg/CH
Schaller I. (1998), *Vom Wissen-Handeln-Graben zum sozialen Wandel*, Rapports de Recherche vol. 9, Institut de Géographie, Fribourg/CH
Schätzl L. (1981), *Wirtschaftsgeographie 1: Theorie*, 2nd ed., Schöningh, Paderborn
Schmähl W. (1988), Übergang zu Staatsbürger-Grundrenten. Ein Beitrag zur Deregulierung in der Alterssicherung, in Thiemeyer T. (ed.), pp.83-138
Schmidt M. (1998), An integrated systemic approach to marginal regions: from definition to development policies, in Jussila H., Leimgruber W. and Majoral R. (eds.), p. 45-66
Scholz U. (2000), Wege aus der Armut im ländlichen Indonesien. Wirtschaftlicher und sozialer Wandel in einem javanischen Reisbauerndorf, *Geographische Rundschau*, 52/4, pp.13-20
Schrader M. (1993), Altindustrieregionen der EG, in Schätzl L. (ed.), *Wirtschaftsgeographie der Europäischen Gemeinschaft*, pp.111-166, Schöningh, Paderborn
Schwind M. (1972), *Allgemeine Staatengeographie*, de Gruyter, Berlin
Schwind M. (1981), Kulturlandschaftliche Entwicklungen zu Seiten der deutsch-deutschen Grenzen, *Regio Basiliensis* XXII/2+3, pp.152-165
Scott P. (2001), People who were not there but are now. Aboriginality in Tasmania, in Jussila H., Majoral R. and Delgado-Cravidão F. (eds.), pp.248-266
Seitz J.L. (1995), *Global issues. An introduction*, Blackwell, Cambridge, Mass. and London
Sen A. (1999), *Development as freedom*, Oxford University Press, Oxford
Shiva V. (1993): *Monocultures of the mind. Perspectives on biodiversity and biotechnology*, Zed Books, London and New York, and Third World Network, Penang
Sibille J. (2000), France: Le statut des langues régionales, in *Encyclopédia Universalis*, *Universalia 2000*, pp.193-196, Paris
Sibley D. (1991), The boundaries of the self, in Philo C. (ed.), *New words, new worlds: reconceptualising social and cultural geography*, pp.33-35, Department of Geography, Lampeter
Sierra Club (n.d.), Arctic National Wildlife Refuge, Internet, www.sierraclub.org/wildlands/arctic/ (03.05.2003)
Simmons I.G. (1989), *Changing the face of the Earth. Culture, environment, history*, Blackwell, Oxford
Simon H. (1957), *Models of man*, Wiley, New York
Siuruainen E. (1987), Social aspects of regional development in Finland, in Varjo and Tietze (eds.), pp.416-429
Smith A.D. (1990), Towards a global culture? in Featherstone M. (ed.), pp.171-191
Smith D. (1977), *Human geography. A welfare approach*, Arnold, London
Smith D. and Braein A. (2003), *The atlas of war and peace*, Earthscan, London
Smith N. (1984), *Uneven development. Nature, capital and the production of space*, Blackwell, Oxford

Smith N. (2000): Global Seattle, *Society and Space (Environment and Planning [D])*, 18/1, p.1-5

Smith S.J. (1989), *The politics of 'race' and residence. Citizenship, segregation and white supremacy in Britain*, Polity Press, Cambridge

Somé M.P. (2000), *Die Kraft des Rituals. Afrikanische Traditionen für die westliche Welt*, Hugendubel, München

Sommers L.M. and Mehretu A. (1994), Patterns of microgeograpic marginality, in Wiberg (ed.), pp.45-59

Sommers L.M, Mehretu A. and Pigozzi B. (1995), Factors in microspatial marginality in southeastern Michigan, in Furlani de Civit M.E., Pedone C. and Dario Soria N. (eds.), pp.249-259

South Commission (1990), *The challenge to the South. The report of the South Commission*, Oxford University Press, Oxford

Speiser B. (1995), *Europa am Oberrhein. Der grenzüberschreitende Regionalismus am Beispiel der oberrheinischen Kooperation*, Schriften der Regio 13, Helbing & Lichtenhahn, Basel

Stähelin P. (2003), *Umweltwahrnehmung von Landwirten. Ökologische Retentions weiher als Beitrag zur Sanierung der Luzerner Mittellandseen*, Unpublished diploma thesis, Department of Geosciences, Geography Unit, University of Fribourg/CH

Steiner D. (1988), Das Dreieck und der Kreis, in Steiner D, Jaeger C. and Walther P. (eds.), *Jenseits der mechanistischen Kosmologie – Neue Horizonte für die Geographie?* Berichte und Skripten, Nr. 36, pp.147-165, Geographisches Institut ETH, Zürich

Suter C. and Mathey M.-C. (2002), *Wirksamkeit und Umverteilungsaspekte staatlicher Sozialleistungen*, Federal Office of Statistics, Neuchâtel

Sutton P. and Rigsby B. (1982), People with 'Politicks': management of land and personnel on Australia's Cape York Peninsula, in Williams N.M. and Hunn E.S. (eds.), pp.155-171

Taaffe E.J., Morrill R.L. and Gould P. (1963), Transport expansion in underdeveloped countries: a comparative analysis, *Geographical Review* 53, pp.503-29

Taylor P.J. (1993), *Political geography. World-economy, nation-state and locality*, 3rf. ed., Longman, Harlow

Tecflux (1999), Tecflux-Expeditionen – wissenschaftliche und technische Großprojekte im Pazifik vor der Küste des US-Staates Oregon, Internet, www.geomar.de/pressemk/mai99_tecflux.html (21.11.2001)

Thiemeyer T. (ed., 1988), *Regulierung und Deregulierung im Bereich der Sozialpolitik*, Schriften des Vereins für Socialpolitik, vol. 177, Duncker & Humblot, Berlin

Thomas W.L. (1956), *Man's role in changing the face of the earth*, University of Chicago Press, Chicago

Thomas-Hope E.M. (1994), *Impact of migration in the receiving countries. The United Kingdom*, CICRED, Paris, and IOM, Geneva

Thornes J. (1995), Global environmental change and regional response: the European Mediterranean, *Trans Inst Br Geogr* NS 20, pp.357-367

Thünen J.H. von (1826), *Der isolirte Staat in Beziehung auf Landwirthschaft und Nationalöconomie*, G.B. Leopolds Universitäts-Buchhandlung, Rostock

Tiefenthaler H. (1990), Verkehr ohne Grenzen - der Belastbarkeit. Die Umweltbelastung des Alpentransits, in Meyer-Tasch P.C., Molt W. and Tiefenthaler H. (eds.), *Transit. Das Drama der Mobilität. Wege zu einer humanen Verkehrspolitik*, pp.149-161, Schweizer Verlagshaus, Zürich

Torricelli G.P., Thiede L. and Scaramellini G. (1997), *Atlante socioeconomico della Regione insubrica*, Casagrande, Bellinzona

Trap R. (2003), *EFQM – Excellence Model and Athena Foods Ltd., a local company in Ghana. Athena Foods Ltd. in its organizational environment. Impact, strategy and organizational behavior*, Unpublished diploma thesis, Department of Geosciences, Geography Unit, University of Fribourg/CH

Troughton M.J. (1978), Persistent problems of rural development in 'marginal areas' of Canada, in *Nordia* vol. 12, No. 1, pp.97-107

Trudgill P. (1991), Language maintenance and language shift: preservation versus extinction, *International Journal of Applied Linguistics* 1/1, pp.61-69

Trudgill P. (2002), *Sociolinguistic variation and change*, Edinburgh University Press, Edinburgh

Tsao F.-F. (1997), Preserving Taiwan's indigenous languages and cultures: a discussion in sociolinguistic perspective, in Inoue N. (ed.), pp.97-112

TSRA, Torres Strait Regional Authority (1998), Torres Strait Development Plan, Internet, www.tsra.gov.au (18.01.2003)

TSRA, Torres Strait Regional Authority (2001a), Native title, Internet, www.tsra.gov.au/Nativetitle.htm (25.07.2001)

TSRA, Torres Strait Regional Authority (2001b), A Torres Strait Territory government, TSRA News No. 40, October, Internet www.tsra.gov.au, (18.01.2003)

TSRA, Torres Strait Regional Authority (2001c), Torres Strait regional sea claim lodged, *TSRA News* No. 41, November-December, Internet, www.tsra.gov.au (18.01.2003)

TSRA, Torres Strait Regional Authority (2002a), Regional sea claim highlights Torres Strait islanders' strong cultural links to the sea, *TSRA News* No. 45, June, www.tsra.gov.au (18.01.2003)

TSRA, Torres Strait Regional Authority (2002b), BSF helps Peter Nai work his way, TSRA News No. 46, July, Internet, www.tsra.gov.au (18.01. 2003)

Tuan Y.-F. (1974), *Topophilia. A study of environmental perception, attitudes, and values*, N.J., Prentice Hall, Englewood Cliffs

UNDP, United Nations Development Programme (2002a), *Avoiding the dependency trap. The Roma in Central and Eastern Europe*, UNDP, Bratislava

UNDP, United Nations Development Programme (2002b), *Human development report 2002*, Oxford University Press, New York and Oxford

UNDP, United Nations Development Programme (2003a), UNDP bolsters NEPAD with nearly $2 million financial contribution, Press release 13 February, Internet, www.undp.org (17.02.2003)

UNDP, United Nations Development Programme (2003b), TICAD III preparatory meeting examines NEPAD and priority development issues for Africa, Press release 4 March, Internet, www.undp.org (01.05.03)

UNESCO (1996), *Réserves de biosphère: la stratégie de Séville et le cadre statutaire du réseau mondial*, UNESCO, Paris

UNESCO (2003a), Frequently asked questions on biosphere reserves, Internet, www.unesco.org/mab/nutshell.htm (31.03.2003)

UNESCO (2003b), World network of biosphere reserves, Internet, www.unesco.org/mab/ (10.03.2003)

UNHCR, United Nations High Commisariate for Refugees (2003), Basic facts, Internet, www.unhcr.ch/cgi-bin/ (06.01.2003)

Van den Bosch K. (1999), Comparative research on poverty: issues and problems, in Fluder R., Nolde M., Priester T. and Wagner A. (eds.), pp. 93-109

Varjo U. and Tietze W. (eds., 1987), *Norden. Man and environment*, Gebr. Borntraeger, Berlin

Verhofstadt G. (2001), Open letter 2001: The paradox of anti-globalisation. A message to the anti-globalisation protesters, September 2001, Internet, www.globsalisationdebate.be/2001 (08.11.2002)

Verhofstadt G. (2002), Open letter 2002 to the alternative globalists. "From Doha to Cancun: the hypocrisy behind Western compassion", Internet, www.globalisation debate.be/2002 (20.11.2002)

Voyé L. (1997), Religion in modern Europe: pertinence of the globalization theories? in Inoue N. (ed.), pp.155-186

Wachter D. (1995), *Schweiz, eine moderne Geographie*, Neue Zürcher Zeitung, Zürich

Walker A.M. (1994), EuroAirport Bale – Mulhouse – Freiburg: strengths and weaknesses of a bi-national airport, in Gallusser W.A., Bürgin M. and Leimgruber W. (eds.), pp.279-283

Walker A.M. (1995), *Chance Regio-Flughafen. Wechselwirkungen zwischen dem EuroAirport Basel-Mulhouse-Freiburg und der Regio, Analysen und Szenarien*, Schriften der Regio 14, Helbing & Lichtenhahn, Basel

Walter F. (1996), *Bedrohliche und bedrohte Natur. Umweltgeschichte der Schweiz seit 1800*, Chronos, Zürich

Wälty S. (2000), Kintamani – Kultur und Entwicklung in einem Berggebiet Balis, *Geographische Rundschau* 52/4, pp.28-31

Warnier J.-P. (1999), *La mondislisation de la culture*, La Découverte, Paris

Watkins K. (1997), Globalisation and liberalisation: implications for poverty, distribution and inequality, UNDP Occasional Paper 32, Internet, http://hdr.undp.org/ (05.03.2003)

WBCSD, World Business Council for Sustainable Development (2002), *The cement sustainability, our agenda for action*, Earthprint, Stevenage

Webb L.J. (1981), The irreplaceability of natural rainforests in Australia, *Geographical Society of New South Wales, Conference Papers*, No. 1, p. 47-55, North Ryde N.S.W.

Weber M. (1962), *The protestant ethic and the spirit of capitalism*, Routledge, London

Weber M. (1980), *Wirtschaft und Gesellschaft*, 5th ed, Mohr, Tübingen

Weiss, R. (1947), Die Brünig-Napf-Reuss-Linie als Kulturgrenze zwischen Ost- und West-Schweiz auf volkskundlichen Karten, *Geographica Helvetica* II, pp.153-175

Weller B., Serow W.J. and Sly D.F. (1994), *Impact of migration in the receiving countries. United States*, CICRED, Paris, and IOM, Geneva

Werlen B. (1998), Thesen zur handlungstheoretischen Neuorientierung sozialgeographischer Forschung, in Sedlacek P. and Werlen B. (eds.), *Jenaer geographische Manuskripte: Texte zur handlungstheoretischen Geographie*, Band 18/1998, Universität Jena

Weston B.H. (2002), *Human Rights*, Encyclopedia Britannica, DVD-edition

Wiberg U. (1994), Swedish marginal regions and public sector transformation, in Wiberg U. (ed.), pp.259-272

Wiberg U. (ed. 1994), *Marginal areas in developed countries*, CERUM, Umeå

Wilber K. (ed., 1982), *The holographic paradigm and other paradoxes*, Shambhala Publications, Boston

Wilbur C.M., Young E.P. and Lieberthal K.G. (2002), *China: History*, Encyclopedia Britannica, DVD-edition

Williams M. (1987), Sauer and 'Man's Role in Changing the Face of the Earth', *Geographical Review* 77,2, pp.218-231

Williams N. (1982), A boundary is to cross: observations on Yolngu boundaries and permission, in Williams N.M. and Hunn E.S. (eds.), pp.131-153

Williams N.M. and Hunn E.S. (1982, eds.), *Resource managers: North American and Australian hunter-gatherers*, Westview Press, Boulder, and Australian Institute of Aboriginal Studies, Canberra (1986)

Wilson G.A. (2001), From productivism to post-productivism ... and back again? Exploring the (un)changed natural and mental landscapes of European agriculture, *Trans Inst Br Geogr* NS 26, pp.77-102

Winckler E.A. (2002), *Taiwan: History*, Encyclopedia Britannica, DVD-edition

Withers C.W.J. (1999), Geography, Enlightenment, and the paradise question, in Livingstone D.N. and Withers C.W.J. (eds.), pp.67-92

Wolfe, R.I. (1962), Transportation and politics: the example of Canada, *AAAG* 52.2, pp.176-190

Wolpert J. (1964), The decision process in a spatial context, *AAAG*, 54, pp.537-558

Wolpert J. (1965), Behavioral aspects of the decision to migrate, *Papers and Proceedings of the Regional Science Association*, 15, pp.159-169

Wood T.F. (1987), Thinking in geography, *Geography* 72/4, pp.289-299

World Commission on Environment and Development, The (1987), *Our Common Future*, Oxford University Press, Oxford

WTO, World Trade Organization (2002), Trade statistics, Internet, www.wto.org/english/res_e/statis_e/ (15.03.2003)

WWF (2001), *Annual report 2001*, Gland (Switzerland)

WWF (2003a), A history of WWF: the Sixties, Internet, www.panda.org (01.06.03)

WWF (2003b), A History of WWF: WWF's 40th Anniversary 1961 – 2001, Internet, www.panda.org (01.06.2003)

Young R.L. (1988), Language maintenance and language shift in Taiwan, *Journal of Multilingual and Multicultural Development* 9/4, pp.323-337

Zerbi M.C. (1999), Il patrimonio paesaggistico: i valori della cultura, *Bollettino della Società Geografica Italiana* Ser. XII, vol. IV, pp.269-277

Index

Printed and bound by CPI Group (UK) Ltd, Croydon, CR0 4YY

22/10/2024

01777628-0011